T0138934

Optical
Communication Systems

Optical Communication Systems
Limits and Possibilities

edited by

Andrew Ellis
Mariia Sorokina

JENNY STANFORD
PUBLISHING

Published by

Jenny Stanford Publishing Pte. Ltd.
Level 34, Centennial Tower
3 Temasek Avenue
Singapore 039190

Email: editorial@jennystanford.com
Web: www.jennystanford.com

British Library Cataloguing-in-Publication Data
A catalogue record for this book is available from the British Library.

Optical Communication Systems: Limits and Possibilities

ISBN 978-981-4800-28-0 (Hardcover)
ISBN 978-0-429-02780-2 (eBook)

Contents

Preface

Optical fiber systems are the backbone of the global tele-communication networks. Optical fibers enable information transmission from dense ultrashort cables in data-centers to transoceanic distances around the globe, connecting billions of users and linking cities, countries and continents. It is hard to overstate the **impact** that fiber-optic systems have made on the economy, healthcare, public and government services, society, and almost every aspects of our lives.

The exponential surge in the global data traffic driven by the skyrocketing proliferation of different bandwidth-hungry online services brings about the escalating pressure on the speed (capacity) and quality (bit error rate) characteristics of information flows interconnecting individual network participants. Examples for these bandwidth-hungry online services include cloud computing, on-demand HD video streams, online business analytics and content sharing, sensor networks, machine-to-machine traffic arising from data-center applications, the Internet of Things, and various other broadband services. It is well recognized nowadays that rapidly increasing data rates in the core fiber communication systems are quickly approaching the limits of current transmission technologies, many of which were originally developed for communication over *linear channels* (e.g., radio). However, optical fiber channels are fundamentally different.

Nonlinear effects in optical fiber have a major effect on transmission speed in modern fiber-optic communication systems. Unlike wireless communications, where signal quality can be enhanced by increasing the optical power at the transmitter, in fiber-optics the increase of power leads to stronger nonlinear signal impairments. Consequently, there is a clear need for the development of radically different methods for signal processing.

This book gives an overview of the current research by experts in this field. The first chapter, by Hadrien Louchet, Nikolay Karelin, and André Richter from VPIphotonics, covers key requirements and challenges for accurate modeling of nonlinear effects for achieving high-capacity transmission. In Chapter 2, Mohammad Ahmad Zaki Al-Khateeb, Abdallah Ali, and Andrew Ellis from Aston University discuss theoretical models that can predict the maximum performance and discuss optical fiber nonlinear limits.

This is followed by Chapter 3 by Danish Rafique from Adva Optical Networking, who reviews different methods of fiber nonlinearity compensation. Chapter 4 is focused on the method of phase-conjugated twin waves and phase-conjugated coding, presented by Son Thai Le from Nokia Bell-Labs.

The information-theoretic treatment, focused on channel models and their limits for estimating transmission throughput of fiber-optic channels, is given in Chapter 5 by Mariia Sorokina and Metodi P. Yankov from Aston University and Technical University of Denmark, respectively. In Chapter 6, Ivan B. Djordjevic from the University of Arizona reviews coding algorithms for fiber-optic systems, one of the key enabling technologies to improve the spectral efficiency and extend the transmission distance. The nonlinear Fourier transform method is described in Chapter 7 by Jaroslaw E. Prilepsky, Stanislav A. Derevyanko, and Sergei K. Turitsyn from Aston University and Ben-Gurion University of the Negev, who discuss challenges and advantages of its application on communication systems.

Finally, spatial multiplexing is considered here (i) from the technology perspective in Chapter 8 by Yongmin Jung, Qiongyue Kang, Shaif-ul Alam, and David J. Richardson from the University of Southampton, who discuss the prospects of scaling as well as potential energy and cost savings of this technology and (ii) from the modeling perspective by Filipe Ferreira, Christian Costa, Sygletos Stylianos, and Andrew Ellis from Aston University, who overview different modeling approximation and their limits.

To conclude, the book presents a broad overview of the contemporary research on communications limits of fiber-optic systems, discussing different approaches and their challenges and prospects.

We wish to thank the authors and the Jenny Stanford Publishing staff, who made this book possible.

Chapter 1

Modelling High-Capacity Nonlinear Transmission Systems

Hadrien Louchet, Nikolay Karelin, and André Richter

VPIphotonics, Berlin, 10587, Germany

Andre.Richte@VPIphotonics.com

1.1 Introduction

In 1966 Charles Kao and George Hockham [1] predicted that silica-based fibre could be a very effective medium for communication offering huge bandwidth and potentially very low loss compared to co-axial cables and copper wires despite of the very high attenuation of the first optical fibres (1000 dB/km). As early as 1970 a team at Corning [2] managed to produce about 100 m of fibre with characteristics close to the one predicted by Kao and Hockham[1]. Since then fibre fabrication

---

[1]The well-documented history of optical fibre can be found at http://opticalfibre-history.co.uk as well as in Hecht, Jeff. *City of Light: The Story of Fiber Optics*. Oxford University Press on Demand, 2004.

Optical Communication Systems: Limits and Possibilities
Edited by Andrew Ellis and Mariia Sorokina
Copyright © 2020 Jenny Stanford Publishing Pte. Ltd.
ISBN 978-981-4800-28-0 (Hardcover), 978-0-429-02780-2 (eBook)
www.jennystanford.com

and design have been improved thriving to design fibres with characteristics as close as possible to the ones of an ideal transmission channel with identity transfer function. All deviations from this *ideal channel* such as loss, chromatic dispersion (CD), polarization mode dispersion (PMD) and nonlinearities are usually called *transmission impairments*. Notable exceptions to this approach are *Soliton* systems (see Chapter 7) that explicitly make use of the dispersive and nonlinear nature of the optical fibre. Erbium-doped fibre amplifiers (EDFAs) [3–4] and dispersion compensating fibres [5] (DCF) have been introduced in the 1980s and 1990s, respectively, to compensate for the fibre attenuation and CD unlocking the optical fibre potential for long haul high-capacity transmission.

Nowadays the digital coherent optical technology, as introduced by Nortel Networks, allows to compensate all linear effects (CD, PMD, polarization rotation) at the receiver. Consequently, the remaining deviations from the *identity channel* are due to amplified spontaneous emission, ASE, modelled as additive white Gaussian noise, filtering effects and cross-talk resulting from optical add-drop multiplexers and complex interactions between linear and nonlinear propagation effects. While filtering and cross-talk effects can be modelled easily, accurate modelling of nonlinear interactions and their impact on the system performance remains an important challenge. This is especially true for high-capacity systems operating in the nonlinear regime and approaching or exceeding the nonlinear Shannon limit (see Chapter 5). Key aspects for accurate modelling of nonlinear effects in such systems are reviewed in this chapter.

1.2 Nonlinear Fibre Propagation: From Single to Multimode

1.2.1 Wave Equation

The propagation of an optical field, which is particular case of an electromagnetic wave, in an optical fibre is governed by Maxwell's equations. In a dielectric medium like fused silica, nonlinear effects are weak and can be treated as perturbation. In this framework, the wave equation takes the following form:

$$\nabla^2 \tilde{E} + \overline{n}^2 \frac{\omega^2}{c^2} \tilde{E} = 0, \tag{1.1}$$

where $\tilde{E}$ is the Fourier transform of the electrical field vector E expressed in the time domain, c is the light velocity in vacuum and $\overline{n}$ the (complex) refractive index of the waveguide. $\overline{n}$ is related to the medium relative permittivity ε_r and electric susceptibility χ_e as follows: $\overline{n} = \sqrt{\varepsilon_r} = \sqrt{1 + \chi_e}$. In an isotropic medium, i.e. when $\overline{n}$ is independent of the position or direction, the simplest solution for (1.1) is the single harmonic (i.e. monochromatic) plane wave:

$$E(t,z) = E_0 \exp(j(\omega_0 t - \overline{\beta}z)), \tag{1.2}$$

where z is distance along the direction of propagation, $\overline{\beta} \approx \overline{n}\omega/c$ is the complex propagation constant of the wave and ω_0 is the wave angular frequency. For a circular waveguide like the optical fibre the solution of (1.1) becomes [6]:

$$E(r,\varphi,z,t) = A(z,t)E^t(r,\varphi)\exp(j(\omega_0 t - \overline{\beta}z)), \tag{1.3}$$

where $E^t(r, \varphi)$ is the transverse field, i.e. the spatial distribution of the field in the waveguide cross-section, z the position along the fibre axis and $A(z, t)$ is the slowly varying complex envelope of the wave, describing, for instance, modulation or noise process affecting the monochromatic wave. The angular frequency is related the optical carrier wavelength in vacuum as $\omega_0 = 2\pi c/\lambda_0$.

In multimode fibres, several solutions (i.e. pairs of E^t and $\overline{\beta}$) exist (see Fig. 1.1) and can be well approximated as linearly polarized modes (*LP modes*). So-called *single mode fibres* are designed to support only a single *degenerate* solution, the fundamental mode LP_{01}, which presents an intensity profile similar to that of a Gaussian beam. The term *degenerate* relates to the fact that the optical field propagation can be represented as a superposition of two waves having the same propagation constant but orthogonal orientations for E (see Section 1.2.2.3) usually referred to as X and Y. Exact values for $\overline{\beta}$ depend on the refractive index profile of the fibre. The review of linear and nonlinear effects reported below is restricted to the LP_{01} mode but can be extended to higher-order modes.

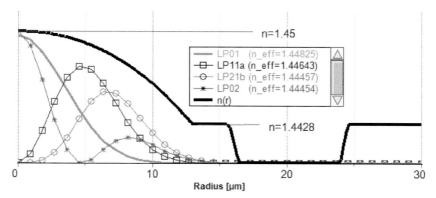

Figure 1.1 Refractive index profile of a few-mode fibre supporting four LP modes. The effective refractive indexes of the modes are indicated for 1550 nm.

1.2.2   Linear Propagation Effects

Here the term *linear* refers to fibre characteristics that remain independent of the propagating signal and therefore allow the propagation along the fibre to be described using a linear transfer function.

1.2.2.1   Loss

Optical waves propagating in optical fibre are attenuated due to scattering effects and molecule absorption. Rayleigh scattering is caused by intrinsic variation of the refractive index along the fibre. It is proportional to $1/\lambda^4$ and is the dominant loss effect at short wavelength (below 800 nm). Molecule absorption is mostly dominated by infrared absorption at wavelength above 1600 nm and by water ions at 975, 1250 and 1400 nm. Note that the absorption peaks due to water ions can be reduced via improved fabrication processes and are quasi-absent in low-water-peak or zero-water-peak fibres standardized as ITU-T G.652.C and G.652.D. In the framework of electromagnetic wave theory, this loss is described by the imaginary part of the complex refractive index, i.e. by the imaginary part of the complex the electric susceptibility χ_e:

$$\bar{n} = n + i\frac{\Im(\chi_e)}{2n} \text{ with } n = \Re(\chi_e)/2 \tag{1.4}$$

and with this, the complex propagation constant becomes

$$\bar{\beta} = n\omega/c + i\frac{\Im(\chi_e)}{2n} \cdot \frac{\omega}{c} = \beta + i\,\alpha/2,\tag{1.5}$$

where α is the fibre attenuation parameter, usually measured in dB/km as $\alpha[\text{dB/km}] = -10*\text{Log}(P_L/P_0)/L_{\text{km}}$, where P_0 and P_L are the transmitted and measured signal power after L km, respectively. Converting α in linear units gives $\alpha[1/\text{m}] = \alpha[\text{dB/m}]*10\,\log(e) \sim \alpha[\text{dB/km}]/1e3*4.434$. The factor $1/2$ in (1.5) is introduced as the attenuation coefficient α is defined for the optical power not for the optical field. The wavelength dependency of α is displayed in Fig. 1.2 for two fibre types. The so-called "optical wavelengths bands" are indicated as well.

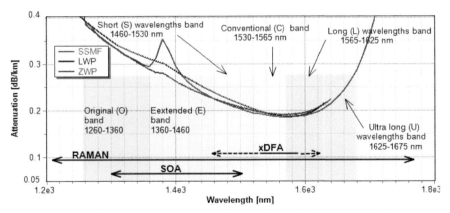

Figure 1.2 Fibre attenuation for standard single mode (ITU-T G652.A), low water peak (G652.B) and zero water peak fibres. The ITU-T wavelengths bands and operating range of Raman, erbium- and ytterbium-doped fibre and semiconductor optical amplifiers are indicated as well.

1.2.2.2 Chromatic dispersion

In silica fibre, $n \sim 1.45$, i.e. the signal travels at approximately 70% of the speed of light. Because of the fibre refractive index profile and material properties n is frequency-dependent. As a result, the signals propagating at different wavelengths travel at different speeds, an effect called chromatic dispersion. The frequency-dependency of the propagation constant can be expressed using the Taylor series expansion around the centre angular frequency of the considered signal, ω_0:

$$\beta(\omega) = \beta_0 + \beta_1(\omega - \omega_0) + \frac{1}{2}\beta_2(\omega - \omega_0)^2 + \frac{1}{3}\beta_3(\omega - \omega_0)^3 + \cdots \quad (1.6)$$

β_0 is a frequency-independent phase offset. β_1 refers to the zinverse of the group velocity v_g describing the speed of the modulated envelope, i.e. the speed at which pulses propagate through the fibre and therefore governs *inter-modal dispersion* in multimode fibres (see Section 1.2.6). β_2 is the group velocity dispersion (GVD) and accounts for the linear frequency dependency of v_g. GVD leads to pulse broadening or channels walk-off in wavelength division multiplexing (WDM) systems. β_3 is referred to as the slope of the GVD or the second order GVD. It plays a role in wide-band transmission where GVD is not constant (see Fig. 1.3) over a large frequency range.

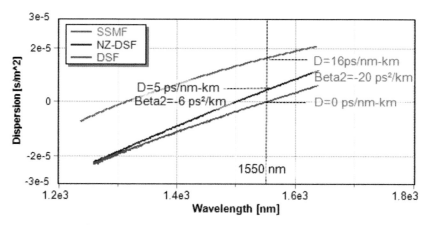

Figure 1.3 Chromatic dispersion of standard single mode (ITU-T G652), non-zero dispersion shifted (G655) and dispersion shifted (G653) fibres. Values for D and β_2 are indicated for 1550 nm.

Note that the terms *fibre dispersion* (D) and *dispersion slope* (S) describing the propagation constant dependency on wavelength rather than frequency are related to β_2 and β_3 as

$$D = \frac{\partial}{\partial\lambda}\frac{1}{v_g} = -\frac{2\pi C}{\lambda^2}\beta_2 \quad S = \frac{\partial D}{\partial\lambda} = \frac{(2\pi C)^2}{\lambda^4}\beta_3 - \frac{2D}{\lambda}. \quad (1.7)$$

Standard single mode fibre (ITU-G652) presents zero dispersion at approximately 1320 nm, while $\beta_2 \sim -20$ ps^2/km ($D \sim 16$ ps/nm-km) at around 1550 nm [7]. When the transmitted fibre distance is larger than the *dispersion length*, defined as $L_D = T_0^2/|\beta_2|$, which is the distance over which a pulse of width T_0 has broaden over one symbol interval, the propagation is considered to be in the *dispersive regime*. Note that not only the local dispersion but also the accumulated dispersion GVD$_{acc}$ = $\int \beta_2(z)\cdot\delta z$ play an important role for determining the signal propagation regime.

1.2.2.3 Birefringence

Under ideal conditions, the two degenerate modes of a single mode fibre have identical propagation constant $\beta(\omega)$ and signals propagating in these modes do not couple. However, small fluctuations in material anisotropy (e.g. refractive index variations) and small deviations from the cylindrical geometry make the fibre birefringent, i.e. degenerated LP modes have slightly different propagation constants [8]. Locally or for a short piece of fibre it is possible to define the fast and slow birefringence axes of the fibre (x and y in the following) corresponding to the orientation of the supported modes LP_{01X} and LP_{01Y} by adapting the propagation constants $\beta_{X/Y}$ as follows:

$$\beta_{0,X/Y} = \beta_0 \pm \Delta\beta_0 \text{ with } \Delta\beta_0 = \frac{\pi}{2L_b} \tag{1.8}$$

L_b is called the fibre *beat length* defined as the distance for which the accumulated phase difference between the two modes reaches 2π, i.e. where a polarized monochromatic signal retrieves its initial polarization state. L_b is typically between 1 and 20 m for transmission fibres [8, 9].

Fibre birefringence varies along the fibre due to asymmetries introduced during the fabrication process already and also in time due other sources such as vibrations, bending, twisting and temperature. Therefore, not only $\Delta\beta_0$ but also the orientation of the fibre birefringence axes (x and y) evolve randomly along the fibre. This leads to mixing between orthogonal modes as well as a random dispersion phenomenon called polarization-mode dispersion (PMD). For longer fibre spans random birefringence is modelled by adjusting the propagation constant by introducing $\beta_{1X/Y} = \beta_1 \pm \Delta\beta_1$ with

$$\Delta\beta_1 = \frac{D_{PMD}}{2\sqrt{2L_c}}. \tag{1.9}$$

L_c is called the *fibre correlation length* characterising the power transfer between the two orthogonal polarisations resulting from random coupling processes. For transmission fibres, typical values for L_c are between 30 and 300 m [10]. D_{PMD} is the *fibre PMD coefficient*, which ranges from 1 ps/km$^{1/2}$ for old fibres (installed prior to the 1990s) down to 0.01 ps/km$^{1/2}$ for recent ones (installed after 2010). It is related to the mean of the differential group delay (DGD), $\Delta\tau$, between the fastest and slowest polarization of a fibre of length L as follows:

$$D_{PMD} = \frac{\langle \Delta\tau \rangle}{\sqrt{L}} \tag{1.10}$$

$\Delta\tau$ varies randomly in time (and thus in optical frequency) and presents a probability density function close to the Maxwellian distribution [10] (see Fig. 1.4). PMD leads to pulse broadening and possibly to system penalty and outage in direct and differential detection-based transmission systems [11, 12]. In digital coherent optical transmission, slowly varying PMD-induced changes of DGD and signal state of polarisation (SOP) lead to channel crosstalk that can be corrected using an digital multiple input multiple output (MIMO) filter [13]. Therefore, in coherent systems the main impact of fibre birefringence is due to the interaction of PMD and nonlinear propagation effects leading to fast changing channel cross-talk, to penalty for nonlinearities compensation schemes such as digital-back-propagation and to polarization-dependent gain when distributed Raman amplification (DRA) is employed [122].

Note: Besides amplification-induced noise, fibre birefringence is the only fully stochastic process taking place during fibre propagation.

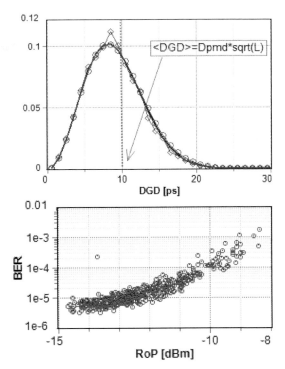

Figure 1.4 (Top) DGD distribution for a 100 km fibre with $D_{PMD} = 10$ ps/km$^{1/2}$. (Bottom) BER of a 28 Gbaud DP-16QAM vs. received optical power after 200 km span with distributed Raman amplification (0.6 W polarized backward pump) for 400 fibre birefringence profiles ($D_{PMD} = 0.7$ ps/km$^{1/2}$).

1.2.3 Nonlinear Propagation Effects

The term *nonlinear* refers to the wave equation that becomes nonlinear when the transmission medium characteristics become signal-dependent. The induced polarization, i.e. the polarization resulting from the applied optical field E of the electrical dipoles constituting the dielectric, can be written as

$$P = \varepsilon_0 (\chi_e^{(1)} \cdot E + \chi_e^{(2)} : EE + \chi_e^{(3)} \vdots EEE + ...) = P_L + P_{NL}. \qquad (1.11)$$

In silica fibres, $P_L = \varepsilon_0 (\chi_e^{(1)} \cdot E)$ is the dominant contribution to P, while $\chi_e^{(2)}$ is zero because of the symmetry of the SiO$_2$

molecule such that nonlinear effects are governed by the third-order susceptibility $\chi_e^{(3)}$ of the fused silica. The response of the silica medium to the applied field is of electronic and molecular origin [14, 15]. The electronic response is very fast (~60 fs) and can be considered as instantaneous if the variations of the signal envelope $A(z,t)$ are slower than 1 ps. The delayed molecular response is related to the Raman and Brillouin scattering processes. The impact of stimulated Brillouin scattering (SBS) is neglected in the present chapter as its threshold (~2 mW/ 20 MHz = 1e^{-10} Watt/Hz) is well above the power spectral density of modulated signals (~1e^{-14} to 1e^{-12} W/Hz). Considering both electronic and molecular responses of the media, (1.4) can be rewritten as:

$$n = n_0 + n_2((1 - f_r)|E|^2 + f_r \int h_r(\tau)|E(z,t-\tau)|^2 \, \delta\tau) \tag{1.12}$$

$$\text{with } n_0 = \Re(\chi_e)/2 \quad \text{and} \quad n_2 = \frac{3}{8n}\Re(\chi_e^{(3)}), \tag{1.13}$$

where n_2 is the *fibre nonlinear refractive* index, and h_r and f_r represent the normalized response and fractional contribution of the delayed Raman response, respectively. n_2 is approximately 2.4–2.6e–20 m^2/W, f_r is typically 17–19% and the characteristic time of h_r is 30 fs [15] in silica fibres. The dependency of the material refractive index on the signal intensity due to the electronic contribution is called the Kerr effect. It induces an intensity-dependent phase shift to the signal, the impact of which will be discussed in Section 1.4.

The impact of the molecular contribution is more complex to describe as it affects both real and imaginary parts of the refractive index $\bar{n}(\omega)$ expressed in frequency and leads therefore to power-dependent phase modulation (molecular contribution to cross-phase modulation, XPM) and power transfer (Raman scattering). Raman-induced power transfer is governed by the imaginary part of $H_r(\omega)$, the Fourier transform of $h_r(t)$. The power transfer taking place frequency components f_k and f_i is governed by the Raman-gain, g_r:

$$g_r(f_k, f_i) = 2\gamma f_r \Im(H_r(f_k - f_i)) \tag{1.14}$$

Signals at long wavelengths (Stokes waves) are pumped by signals at shorter wavelength (anti-Stokes waves) with maximal power transfer (Raman peak) occurring at frequency difference of about 12 THz. Similarly to n, α also exhibits a power dependency called *two-photon absorption*. However, this effect is very small in silica fibre and usually neglected.

1.2.4 The Scalar Nonlinear Schrödinger Equation

The nonlinear Schrödinger equation (NLSE) describes that birefringent-independent unidirectional propagation of light in single-mode fibre. With $U(z, t) = A(z, t)^{*}\exp(i(\beta_0 - \beta_1 \cdot \omega_0)z - \omega_0 t)$, where the slowly varying amplitude $A(z, t)$ of the optical field E is expressed in the frame moving at the group velocity (e.g. following the pulse), the NLSE can be written as follows [16]:

$$\frac{\partial U(z,t)}{\partial z} = [\hat{L} + \hat{N}]U(z,t) \tag{1.15}$$

$$\hat{L} = -\frac{\alpha}{2} + j\frac{\beta_2}{2}\frac{\partial^2}{\partial t^2} + \frac{\beta_3}{6}\frac{\partial^3}{\partial t^3} \tag{1.16}$$

$$\hat{N} = -j\gamma(1 - f_r)|U(z,t)|^2 - j\gamma f_r \int h_r(\tau)|U(z,t-\tau)|^2 \delta\tau \tag{1.17}$$

with γ being the *fibre nonlinear coefficient* defined as

$$\gamma = \frac{2\pi n_2}{\lambda A_{\text{eff}}}. \tag{1.18}$$

A_{eff} is the fibre effective core area describing the wave intensity distribution in the fibre cross-section ($I = P/A_{\text{eff}}$). Note that A_{eff} and therefore γ is slightly frequency-dependent. When two different frequencies interact A_{eff} could be estimated as follows:

$$A_{\text{eff}}(\omega_i, \omega_j) = \frac{\int\int |E^t(\omega_i)|^2 \, dxdy \int\int |E^t(\omega_j)|^2 \, dxdy}{\int\int |E^t(\omega_i)|^2 |E^t(\omega_j)|^2 \, dxdy}, \tag{1.19}$$

where $E^t(\omega)$ is the transverse field of the fundamental mode at frequency ω. For modern transmission fibres A_{eff} is typically in the range of 50 to 100 μm^2 but can also be as low as 20 μm^2 for DCF fibre. At 1550 nm the resulting fibre nonlinear coefficient γ is approximately 1–1.5 W^{-1} km^{-1} for transmission fibres and up to 3 W^{-1} km^{-1} for DCF [17]. When the fibre length is larger than the fibre *nonlinear length* defined as $L_{\text{NL}} = 1/(\gamma P_0)$, where P_0 is the signal peak power at the fibre input, the propagation is said to be in the *nonlinear regime*. This is the case for most of all high-capacity systems requiring large signal powers.

1.2.5 The Manakov-PMD Equation

In their papers from 1982 [18, 19], Botineau and Stolen have characterized the impact of signal polarization on the nonlinear Kerr effect. Based on these results, Menyuk derived a vectorial form of the NLSE called the *coupled nonlinear Schrödinger equations* (CNLSE) accounting for the impact of PMD on the linear and nonlinear signal propagation [20, 21]:

$$\frac{\partial \vec{U}(z,t)}{\partial z} = [\hat{L} + \hat{N}]\vec{U}(z,t) \quad \text{with} \quad \vec{U} = \begin{bmatrix} U_x \\ U_y \end{bmatrix} \tag{1.20}$$

$$\hat{L} = -\frac{\alpha}{2} + j\frac{\beta_2}{2}\frac{\partial^2}{\partial t^2} + \frac{\beta_3}{6}\frac{\partial^3}{\partial t^3} \pm \left[\Delta\beta_0(\omega) + j\Delta\beta_1(\omega)\frac{\partial}{\partial t} \right]\hat{\Sigma}(z) \tag{1.21}$$

$$\hat{N} = -j\gamma\left[|\vec{U}|^2 - \frac{1}{3}\left(\vec{U}^* \hat{\sigma}_2 \vec{U}\right)\hat{\sigma}_2 \right],$$

where $*$ is the transpose conjugate and σ_j are the Pauli matrices. Σ is given by $\Sigma = \sigma_3 \cos(2\theta) + \sigma_1 \sin(2\theta)$ with $\theta(z)$ being the orientation angle of the fibre local birefringence axis. As $\theta(z)$ randomly changes along the fibre and in time, it can only be described statistically. In numerical simulation this can be accomplished using the *coarse-step* approach (see Section 1.3.2.3). Ignoring the random rotation of the local birefringence axis exaggerates the interaction between PMD and nonlinear effects especially for cross-polarization modulation, XpolM (see Fig. 1.5 and Section 1.4.6).

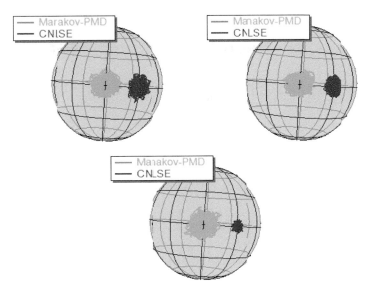

Figure 1.5 Time-domain Poincaré representation illustrating XpolM-induced depolarization of a CW signal after propagation over 20 × 80 km SSMF-based non-DM link according to the CNLSE (constant birefringence) and Manakov-PMD equation (random birefringence) for D_{PMD} = 0.005 (left), 0.05 (centre) and 0.5 ps/km$^{1/2}$ (right). The CW signal is launched in the X polarization while the pump signal is a depolarized 28 GB signal with 6 dBm input power.

For constant birefringence and assuming linear signal polarization, the nonlinear terms reduces to $\hat{N} = -j\gamma \left[|U_x, y|^2 + \frac{2}{3}|U_{y,x}|^2 \right]$. The factor 2/3 accounts for linear birefringence [18, 19], which is a good approximation for a *short* piece of transmission fibre. Even if the CNLSE describes correctly the signal evolution in presence of PMD, it is not suitable to describe optical transmission using numerical simulation. The reason for this is that contrary to the NLSE, where the nonlinear operator is proportional to the signal power $|U|^2$, it is proportional to $\frac{1}{3}(\vec{U^*}\hat{\sigma}_2 \vec{U})\hat{\sigma}_2$ in the CNLSE. This term is dependent on the signal phase, which varies very rapidly (even in the slowly varying envelope framework) because of the $\Delta\beta_0$ term in the linear operator. As discussed in Section 1.2.2.3, $\Delta\beta_0$ is related to L_B which can be as low as one meter. Therefore much smaller steps (~cm scale) would be required to solve the CNLSE using the common numerical techniques, such as the split-step Fourier approach (see Section 1.3.2.1). However, it is possible to express the CNLSE by another form where the nonlinear operator consists of two terms:

$$\hat{N} = -j\gamma\frac{8}{9}|\vec{U}|^2 - j\gamma\underbrace{\frac{1}{9}\left(\frac{1}{3}(\vec{U}^*\hat{\sigma}_2\vec{U})\hat{\sigma}_2 - |\vec{U}|^2\right)}_{\text{nonlinear PMD}} \approx -j\gamma\frac{8}{9}|\vec{U}|^2 \qquad (1.22)$$

Menyuk showed [22] that the second term referred to as *nonlinear PMD* vanishes after uniform averaging over the Poincare sphere. The averaging takes place in random birefringent fibre when the nonlinear propagation transmission length is much larger than L_B and L_C. With this, the nonlinear operator depends only on the signal power (and not field). Further, $\Delta\beta_0$ can be removed from the linear operator $\hat{L}$, as this term only leads to a constant phase shift, which is not relevant for propagation and does not affect the nonlinear operator. Summarizing, the linear and nonlinear operator of the so-called *Manakov-PMD equation* are given by

$$\hat{L} = -\frac{\alpha}{2} + j\frac{\beta_2}{2}\frac{\partial^2}{\partial t^2} + \frac{\beta_3}{6}\frac{\partial^3}{\partial t^3} \pm j\Delta\beta_1(\omega)\frac{\partial}{\partial t}\hat{\Sigma}(z) \qquad (1.23)$$

$$\hat{N} = -j\gamma\frac{8}{9}|\vec{U}|^2$$

Furthermore, it is necessary to consider the polarization dependency of the molecular contribution to the nonlinear response of the material. Since the nonlinear interaction between orthogonal polarization is very low compared to the one between co-polarized signals ($h_{r\perp} \ll h_{r//}$), usually it can be neglected and $\hat{N}$ is adjusted as follows:

$$\hat{N} \approx -j\gamma\frac{8}{9}(1-f_r)|\vec{U}(t)|^2 - j\gamma f_r \int \delta\tau(h_{r//}(\tau) - h_{r\perp}(\tau))|\vec{U}(t-\tau)|^2$$

$$(1.24)$$

1.2.6 Extension to SDM Systems Using Multimode Fibre

In the case of multimode fibres where the index contrast between core and cladding is low (weakly guided assumption), the supported modes are well approximated by LP modes with the following transverse field $E^t(r,\varphi)$:

$$E^t(r,\varphi) = \psi_{l,m}(r)\begin{Bmatrix} \cos l\phi \\ \sin l\phi \end{Bmatrix}, \tag{1.25}$$

where $\psi_{l,m}(r)$ contains the radial dependence [23]. Each $\{l,m\}$ combination with $l \geq 0$ and $m \geq 1$ corresponds to one LP mode. The number of supported modes depends on the refractive index profile of the fibre and on the wavelength. The modes transverse field and propagation constant are the solutions of the wave eigenvalue equation obtained by replacing (1.3) in (1.1) and can be found numerically using a mode solver [24] or estimated experimentally.

When $l \geq 1$, four degenerated modes (with same β) exist for a LP mode: two spatial modes with different radial dependency each supporting two polarizations. Each LP mode presents a different linear propagation constant β_{lm} leading to inter-modal dispersion, while CD and the velocity difference between degenerated modes (modal birefringence) lead to intra-modal dispersion. Attenuation also depends on the fibre design and can vary between LP modes [23]. Nevertheless, main sources of mode-dependent-loss in classical multimode fibre are non-ideal splices and fibre bending, which affects stronger higher order mode that propagate in the cladding. Similarly to single mode fibres, nonlinear effects affect the signal propagation in multimode fibres: noting $U_i(z, t)$ the slowly varying envelope of the i-th degenerated LP mode containing two polarizations the system of coupled Schrödinger equations for a multimode fibre supporting n degenerate LP modes has been derived by Poletti [25]. Expressed in a moving frame with group velocity v_{gr} and neglecting the molecular contribution to the nonlinear effect, it is expressed as [26]:

$$\frac{\partial \vec{U}_i}{\partial z} = j(\beta_{0,i} - \beta_r)\vec{U}_i + \left(\beta_{1,i} - \frac{1}{v_{gr}}\right)\frac{\partial \vec{U}_i}{\partial t} - j\frac{\beta_{2,i}}{2}\frac{\partial^2 \vec{U}_i}{\partial t^2} - \alpha_i \vec{U}_i \tag{1.26}$$

$$+ j\sum_{lmn} f_{lmni} \frac{\gamma}{3}\left[\left(\vec{U}_n^T \vec{U}_m\right)\vec{U}_l^* + 2\left(\vec{U}_i^* \vec{U}_m\right)\vec{U}_i\right],$$

where β_r is the reference propagation constant (usually $\beta_1 = \beta_{01}$) and $\beta_{n,i}$ the n-th frequency derivative of β_i at the central angular frequency of the signal (see Section 1.2.2.2). γ is the fibre nonlinear coefficient defined for the fundamental mode in Section 1.2.4 and f_{lmni} are the nonlinear coupling factors between spatial modes [26]:

$$f_{lmni} = \frac{\iint E_l^{t*} E_m^t E_n^t E_i^{t*} \, dr d\varphi}{\sqrt{\iint |E_l^t|^2 \iint |E_m^t|^2 \iint |E_n^t|^2 \iint |E_i^t|^2}} \qquad (1.27)$$

The system of coupled equations described in (1.26) is extremely challenging to solve in practice because the nonlinear terms depends on the phase difference between modes which evolves rapidly due to modal dispersion. Using the split-step Fourier algorithm, this would require the use of extremely small steps (~mm). However, fibre birefringence leads to random and uncorrelated rotation of the state of polarization of the LP modes along the fibre averaging the cross-terms in (1.26) to zero after a certain length (modes correlation length). Under this condition the nonlinear term in (1.26) becomes independent of the phase relation between modes and varies much slower. The resulting system of coupled Manakov equations describing the propagation of LP mode in the fibre in the weakly coupling regime becomes [26]:

$$\frac{\partial \vec{U}_i(z,t)}{\partial z} = [\hat{L}_i + \hat{N}_i]\vec{U}_i(z,t)$$

$$\hat{L}_i = j(\beta_{0,i} - \beta_r) + \left(\beta_{1,i} - \frac{1}{v_{gr}}\right)\frac{\partial}{\partial t} - j\frac{\beta_{2,i}}{2}\frac{\partial^2}{\partial t^2} - \alpha_i \qquad (1.28)$$

$$\hat{N}_i = j\gamma\left(\frac{8}{9}f_{iiii}|\vec{U}_i|^2 + \sum_m \frac{4}{3}f_{mmii}|\vec{U}_m|^2\right)$$

Fibre imperfections also lead to linear random coupling between modes, which should be accounted for in the modelling. Furthermore, Mecozzi et al. [27] showed that the propagation equations reported in (1.28) can be simplified in presence of coupling. These aspects will be discussed in Section 1.3.3.

1.3   Solving the Manakov-PMD Equation

1.3.1   Signal Representations

Time-resolved or time-averaged representations can be used to describe the signal propagation along an optical link [28, 29]. In the classical time-resolved representation, the signal propagation is described by updating the slowly varying envelope of the optical field U along the link according to the Schrödinger or Manakov-PMD equations. $U(t,z)$ is represented over a certain duration (time-window, T_w) and discretized at a certain sample-rate, f_s. The sample-rate is the bandwidth of the discrete signal spectrum and $\Delta t = 1/f_s$ the time-resolution. Similarly, the frequency resolution, i.e. spacing between the spectral components of the discrete signal is $\Delta f = 1/T_w$. An accurate representation of the signal envelope requires a large sample-rate (to ensure, for instance, that the Nyquist–Shannon sampling criterion is met and that FWM products can be properly generated). Gathering robust statistics (e.g. long bit sequences for accurate BER estimation) or accurately representing "slow" effects (e.g. laser noise processes) requires a long time-window. The number of complex samples used to describe U is equal to $p \times$ time-window $\times$ sample-rate, p being equal to 1 or 2 depending on whether the scalar or vector (for PMD) representation is used. When the amount of samples is too large, computational issues (e.g. limited RAM[2] of CPU or GPU) may prevent from solving efficiently the Manakov-PMD equation. In such case it is possible to switch from a *single* to a *multiple frequency bands* signal representation [28], where the Manakov-PMD equation is solved for each frequency band separately while nonlinear cross-terms are introduced to account for inter-band (Kerr, Raman) interactions [30]. In the time-averaged [31] or *parameterized* representation U is described using signal characteristics that are assumed constant (or averaged) over the time-window. For instance, this includes central wavelength, power, bandwidth, OSNR, state-of-polarization, accumulated GVD, accumulated DGD, accumulated nonlinear phase shift. These characteristics and their uncertainties can be tracked along the transmission line to

---

[2]An optical sample (1 polarization) requires 16 bytes memory for 64 bits (double) floating point arithmetic.

characterize the signal propagation or analysed at the receiver side to estimate the system performance according to certain design rules (required received power, required OSNR, maximal accumulated dispersion). This approach is used, for instance, in commercial link engineering solutions such as VPItransmissionMaker Optical Systems (see Fig. 1.6) and VPIlinkConfigurator. Tracking of these parameters along the link as well as the knowledge of the signal characteristics can be used to estimate independently from each other the impact of CD, SPM, XPM or FWM [33–37] or the aggregate nonlinear interference noise (NLIN) [38] at the receiver side using analytical or semi-analytical models such as the Gaussian or Extended Gaussian Noise models [39, 40].

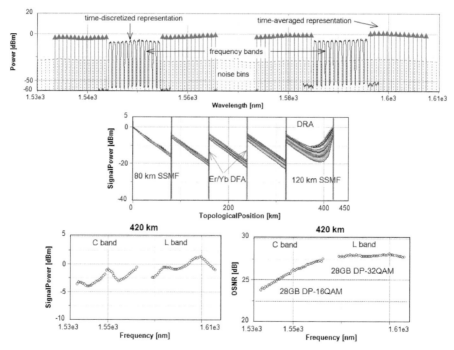

Figure 1.6 (Top) spectrum of WDM signal (C+L band) modelled using time-discretized and -averaged representations. (Middle, bottom) tracking of signal characteristics along the link.

The time-averaged representation is computationally very efficient and can be used to predict roughly the performance.

It allows, for instance, to model slow dynamic effects, which can be assumed to be constant over a single time-window, such as the evolution of EDFA transients in meshed-networks [41]. Further representing ASE noise using the parameterized representation is also very useful as its particular waveform is governed by its statistical characteristics (e.g. variance, SOP, spectral shaping). Of course the time-average representation does not account for the pattern-dependency of effects such as (i) FWM and (ii) XPM or for the stochastic nature of the interaction of nonlinear and polarization effects and is therefore limited to the prediction of the *expected* or *average* system performance. In addition the validity domain of most analytical models is restricted to specific modulation formats [33–36] or link design (for instance, long, non-dispersion managed links are assumed in [39, 40]).

Note that it is possible to combine different signal representations in simulation: In systems with distributed Raman amplification, for instance, the pump can be described using the time-averaged approach and the probe (e.g. WDM channels) using the time-resolved one. While the calculations of the Raman-induced power transfers and Raman spontaneous emission is not affected by this simplification, effects such as noise transfer between pump and probe are neglected. We restrict the following discussion to the time-resolved representation of the signal. Detailed discussion of system modelling using the parameterized representation can be found, for instance, in [42].

1.3.2 Numerical Methods

Noting $\hat{D} = \hat{N} + \hat{L}$ the combination of the linear and nonlinear operator, the general solution of the Manakov equation is given as $x(L,t) = x(0,t) \cdot \exp(\hat{D}\delta z)$. Even if the operators $\hat{N}$ and $\hat{L}$ are analytically integrable, there are not commutative and thus the full propagation operator $\hat{D}$ is usually not integrable [43]. For this reason, there is no general solution for the Manakov equation for arbitrary links and propagated signals.

1.3.2.1 The split-step (Fourier) method

The split-step approach [44] addresses the problem of solving differential equations of the form $\partial x(z,t)/\partial z = (\hat{A} + \hat{B})x(z,t)$ where individual solutions exist for the operators $\hat{A}$ and $\hat{B}$. This is the case for the NLSE and Manakov-PMD equations, where simple solutions exist for $\hat{L}$ and $\hat{N}$ in the frequency and time domain, respectively:

$$U(z,\omega) \xrightarrow{\hat{L}} U(0,\omega)e^{-j\frac{\beta_2}{2}\omega^2 z}$$

$$U(z,t) \xrightarrow{\hat{N}} U(0,t)e^{-j\gamma|U(0,t)|^2 z} \tag{1.29}$$

When considered over a small step Δz, $\hat{D} = \hat{N} + \hat{L}$ can assumed to be constant and a closed-form solution is given as

$$x(z + \Delta z,t) = \exp\left(\int_l^{l+\Delta z} \hat{D}(l)\delta l\right) \cdot x(z,t)$$

$$\approx \exp(\hat{D}(z)\Delta z) \cdot x(z,t) + O(\Delta z^2)$$

$$\approx \exp(\hat{L}\Delta z) \cdot x(z,t) \cdot \exp(\hat{N}\Delta z) + O(\Delta z^2)$$

$$\approx \exp(\hat{L}/2 \cdot \Delta z) \cdot [x(z,t) \cdot \exp(\hat{N}\Delta z)] \cdot \exp(\hat{L}/2 \cdot \Delta z) + O(\Delta z^3) \tag{1.30}$$

The limit $O(\Delta z^2)$ of the truncation error can be found by developing the exponential term using Taylor series expansion and reduces to $O(\Delta z^3)$ by using the symmetric split-step approach. Note that splitting the nonlinear operator instead of the linear operator leads to a different error term but with same accuracy $O(\Delta z^3)$. As mentioned above applying $\hat{L}$ and $\hat{N}$ separately requires simple multiplication in the frequency and time domain, respectively. Switching between the time and frequency can be performed very efficiently using the *Fast Fourier Transform* (FFT) as described in Fig. 1.7. This implementation of the split-step approach is called the *split-step Fourier* method (SSF). Note that the split-step method can be efficiently solved and dramatically sped up by taking advantages of modern graphics processing units (GPU) [45–47].

In principle, switching from the time to frequency domain using the FFT requires either finite or periodic signals. This assumption can lead to numerical errors like spectral leakage or generation of erroneous spectral components. This is the case for random but non-ergodic signals (e.g. in presence of slow power transients) or for periodic signals when the considered time window is not a multiple of the signal period. The time-domain split-step (TDSS) method [48, 49] is an alternative to the SSF method where the linear operator is modelled using a finite impulse response (FIR) filter. The advantage of the TDSS method is that it enables to process signals sequentially (in smaller blocks) relaxing the memory requirements of the simulation. However, the TDSS approach suffers from several drawbacks: First, it is much slower than the SSF method because the convolution operation required to apply the FIR filter to the signal is much slower than the FFT-multiplication operation even when recursive all-pass filtering is replaced by the more efficient FFT-based convolution. In addition, the zero-padding and truncation requirements become excessive for modern uncompensated transmission. On the contrary, periodic boundaries conditions provide great advantages for the simulation of WDM systems, as walk-off and therefore the resulting XPM between WDM channels can be modelled realistically without the need of simulating a long time window. Another limitation of the TDSS is that CD and PMD are not modelled accurately in the time domain especially at high frequencies, which can lead to numerical errors when simulating WDM systems. State-of-the-art photonic simulators enable to combine both approaches, by applying an overlap-add process on successive signal blocks avoiding the periodicity condition and enabling the stitching of successive signal blocks in a longer sequence (see Fig. 1.8).

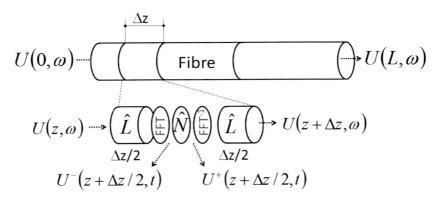

Figure 1.7 Principle of the split-step Fourier method.

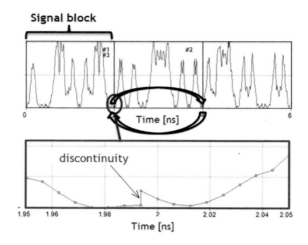

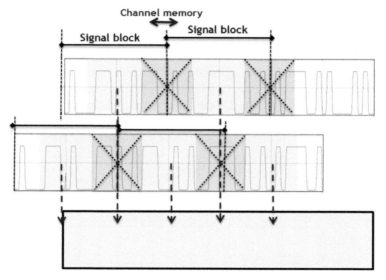

Figure 1.8 (Top) FFT-induced signal discontinuities in multiple blocks simulation. (Bottom) Principle of overlap-add based stitching of multiple signal blocks.

1.3.2.2   Step-size control

The choice of Δz of is essential to achieve an optimum trade-off between computation time and accuracy. It is usually adaptive, i.e. changes along the fibre to accommodate for the intensity of nonlinearities changing with the signal power. In numerical simulation Δz can be controlled using the following approaches:

Constant step-size: For single channel transmission Δz should be set much shorter than the *dispersion length* or the shortest *channel walk-off length* (see Section 1.4.3) in WDM systems. Unless chosen very carefully, the *constant step-size* approach is either inefficient or inaccurate and can even lead to numerical artefacts [50].

Logarithmic step-size: Δz is chosen so that the amount of *averaged* SPM or XPM-induced phase shift or generated FWM-components [50] is constant in each step. In other words, the step size increases as the signal power and consequently the nonlinear effects, decrease due to fibre attenuation. This follows the profile of the averaged power along the fibre which evolves in a logarithmic manner when lumped amplification is used.

Nonlinear Phase Change [51]: Δz is determined by preventing the nonlinear phase shift (proportional to *instantaneous* optical power) to exceed a pre-defined value, $\Delta\varphi_{\mathrm{NL}}$. The *Nonlinear Phase Change* method is a popular approach when modelling WDM systems where most of the Kerr-induced distortions are governed by phase change (SPM, XPM). Common values for $\Delta\varphi_{\mathrm{NL}}$ are around 0.01°.

$$\Delta z_{\varphi} < \frac{\Delta\varphi_{\mathrm{NL}}}{\gamma \max(|U(t)|^2)} \tag{1.31}$$

Local error method [52]: Δz is controlled according to the local error method (LEM) by maintaining the relative error ε between the field calculated using Δz and $\Delta z/2$ within the range $[\delta/2;\ \delta]$. δ is the target local error (typical value 1e-3) and ε is defined as

$$\varepsilon = \sqrt{\frac{\int |U_{\mathrm{f}}(t,z) - U_{\mathrm{c}}(t,z)|^2\ \delta t}{\int |U_f(t,z)|^2\ \delta t}}, \tag{1.32}$$

where U_f and U_c are the field computed with the *fine* ($\Delta z/2$) or *coarse* (Δz) step-size, respectively. The LEM is usually more efficient (i.e. require fewer steps) than other methods when high accuracy is required. However, due to a larger number of computations per step, low-accuracy simulations can be slightly slower. Note that the LEM tends to make larger errors in low-power regions of the spectrum and can therefore underestimate the build-up of FWM-products.

1.3.2.3 The coarse-step model

Because Kerr and Raman effects are polarization-dependent, it is important to model the impact of PMD during nonlinear propagation and not only to account for it at the end of the link, e.g. through the use of PMD-emulator (see [53], for instance). A standard technique for simulating polarization-dependent propagation effects in fibres is the *coarse step* method [21, 54], which approximates the continuous variations of birefringence by a series of many short polarization sections with constant birefringence. At the end of each section, the polarization state of the optical field is scattered onto a new point on the Poincaré sphere (see Fig. 1.9). Because full depolarization is assumed, the length of each scattering section should be longer than the fibre correlation length, L_c. Ideally, section lengths should be randomly chosen to avoid periodic artefacts in the frequency domain and model properly higher order PMD [55, 56]. Noting $z_{\text{scatt},i}$ the length of the i-th scattering section, the difference of the propagation constant of the fast and slow axes in this section should be set to

$$\Delta\beta_{1,i} = \pm\frac{D_{\text{PMD}}}{\sqrt{z_{\text{scatt},i}}}, \tag{1.33}$$

where D_{PMD} is the fibre PMD coefficient introduced earlier. For long transmission distances (>100 km), typical values for $z_{\text{scatt},i}$ are around 1 km with a variance of $\sim 1–10\%$.

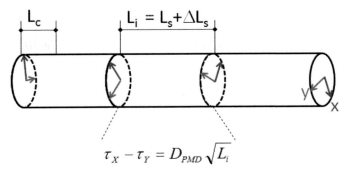

Figure 1.9 Coarse-step approach as implemented in VPItransmissionMaker Optical Systems. The orientation of the slow and fast axes is modified at the beginning of each new scattering section (polarization scattering).

1.3.3 Simulation Framework for SDM Systems

Considering ideal multimode fibres, modes propagate independently from each other in the linear regime. Therefore, the width of the fibre impulse response is proportional to the difference of the highest and lowest modal group delays per unit length multiplied by the fibre length. In real multimode fibres, however, imperfections of the fibre geometry lead to random coupling between different modes. In this case, the width of the impulse response scales with the square root of distance [57–59]. Modes that have similar propagation constants couple strongly; usually on a length scale between 100–300 m, they become fully coupled [59]. In graded-index fibres, LP modes with the same *prime* (or *principal*) index $G = l + 2m + 1$ (*l* and *m* are the radial and azimuthal mode index) usually have similar propagation constants [23, 60, 61] and can be sorted into a *mode group* (see Fig. 1.10). The random coupling between modes belonging to different mode groups is weaker and its length scale is much longer than the length scale over which strong intragroup coupling occurs [62]. Important causes for intergroup coupling are fibre bending, non-ideal splices and flaws in the fibre structure. In step- and graded-index multimode fibres, close values of propagation constant (and strong mode coupling) are observed between degenerated groups of modes having identical {*l,m*} indices.

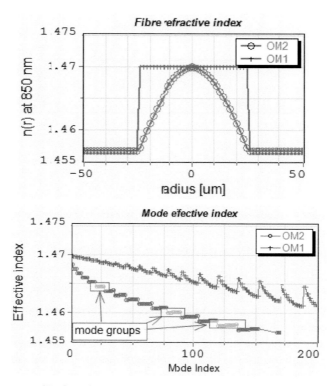

Figure 1.10 (Top) Refractive index profile of OM1 and OM2 multimode fibres. (Bottom) Effective index of the supported modes. Note the well-defined mode-groups in the case of the OM2 (graded-index) fibre.

Similarly to the *coarse-step* model for single mode fibres (see Section 1.3.2.3), mode coupling can be emulated by random mode mixing at discrete fibre locations. The corresponding scattering matrixes are assumed unitary, i.e. mode coupling is assumed lossless as fibres are not the main source of mode-dependent attenuation in typical fibre optic transmission systems [59]. The mixing process within mode groups is performed at intervals much longer than the correlation length of the intragroup coupling. As mode mixing is ignored between discrete coupling events, long section lengths will lead to an overestimation of the accumulated DGD. Therefore scaling of the DGD per unit length is required. This technique is similar to the coarse step method used to emulate PMD in single-mode fibres [21]. In presence of intra-mode group coupling, the *adjusted* propagation constant for the LP mode {*l,m*} for a section of length L_i is

$$\beta_{lm}(f) = \beta_{mean}^{G_i}(f) + (\beta_{lm}^{solver}(f) - \beta_{mean}^{G_i}(f))\frac{\sqrt{L_{MG}}}{\sqrt{L_i}}$$

$$\underbrace{\pm \frac{\tau_{MG}\sqrt{L_{MG}}}{2\sqrt{L_i}}2\pi(f - f_r)}_{spatial\, mode} \underbrace{\pm \frac{D_{PMD}}{2\sqrt{L_i}}2\pi(f - f_r)}_{polarization}, \quad (1.34)$$

where β_{lm}^{solver} is the actual mode propagation constant (as calculated by the mode solver) and $\beta_{mean}^{G_i}(f)$ the average of the mode propagation constants of group G_i. L_{MG} is the intragroup correlation length, τ_{MG} in s/m is the local birefringence of the spatially degenerate modes $LP_{lm,a}/LP_{lm,b}$ and $D_{PMD} = \Delta\beta_{1l,m} \cdot L_c^{1/2}$ (see Section 1.2) measured at the reference frequency f_r.

In presence of mode coupling, the dimension of the system of coupled equations reported in (1.26) and (1.28) can be further simplified [27]: Assuming that full intragroup coupling is achieved on a length scale much shorter than the length scale of nonlinear interactions, nonlinear effects can be averaged over all modes of a mode group:

$$\frac{\partial \vec{A}_i(z,t)}{\partial z} = [\hat{L}_i + \hat{N}_i]\vec{A}_i(z,t)$$

$$\hat{L}_i = j(B_{0,i} - B_r) + \left(B_{1,i} - \frac{1}{v_{gr}}\right)\frac{\partial}{\partial t} - j\frac{B_{2,i}}{2}\frac{\partial^2}{\partial t^2} - \alpha_i$$

$$\hat{N}_i = -j\gamma\left(k_{ai}|\vec{A}_a|^2 + \cdots + k_{wi}|\vec{A}_w|^2\right)\vec{A}_i \quad (1.35)$$

The vector A_i in the equations contains the fields of all spatial modes belonging to the mode group i. $\vec{A}_i = [U_X^1, U_Y^1, U_X^{1+1}, U_Y^{1+1}, ..., U_X^p, U_Y^p,]$. Its power is given by $\left|\vec{A}_i\right|^2 = +\left|U_X^1\right|^2 + \left|U_Y^1\right|^2 + \cdots + \left|U_X^p\right|^2 + \left|U_Y^p\right|^2$. The propagation constant of the individual modes is given by (1.34). γ is the fibre nonlinear parameter defined for the fundamental mode in (1.18). The nonlinear factors k_{ij} account for nonlinear correction terms due to the averaging over the strongly coupled modes within the mode groups G_i and G_j. In particular, k_{ii} considers the averaging within the same mode group G_i. This is similar to single mode fibres, where $k_{11} = 8/9$. The k_{ij} factors are given as [27]

$$k_{ij} = \sum_{k,m \in M_i \cup M_j} \sum_{l \in M_i} \sum_{h \in M_j} f_{lhkm} \frac{\delta_{hk}\delta_{lm} + \delta_{hm}\delta_{lk}}{|M_i|\,(|M_j| + \delta_{ij})}. \tag{1.36}$$

M_i and M_j represent the sets of mode indices for the modes that belong to the mode groups labelled by i and j, respectively. The indices k, m can take all values in the union set $M_i \cup M_j$. $|M_i|$ and $|M_j|$ denote the number of modes in G_i and G_j, respectively (taking into account polarizations). The coefficients f_{lhkm} are defined through overlap integrals of the modal distributions, as defined in Section 1.2.6. Equations (1.35) and (1.36) enable to model inter-modal nonlinear effects such as cross-mode modulation (XMM, [63]) and inter-modal four-wave mixing (imFWM [64], see Fig. 1.11).

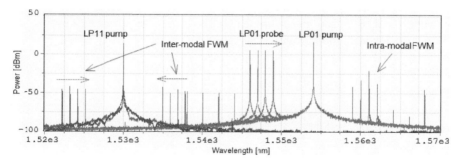

Figure 1.11 Simulated inter-modal four-wave mixing as experimentally demonstrated in [64].

Besides a realistic and efficient multimode fibre model, the modelling of multimode-based transmission systems requires a complete simulation framework [65], including a mode solver to calculate the spatial distribution and propagation characteristics of the modes supported by a specific fibre, an overlap integral solver to calculate the coupling between optical source and device or between different multimode devices and a doped-multimode fibre model to support the design of fibre amplifiers for SDM applications.

1.4 Accurate Modelling of System-Level Nonlinear Impairments

From the system design point of view, it is often useful to distinguish between the contributions of different channels in WDM systems. For this purpose the slowly varying amplitude of the signal $U(z,t)$ can be written as sum of contributions from different WDM channels

$$\left|\vec{U}(z,t)\right|^2 = \sum \vec{U}_k(z,t)\exp(2\pi\Delta f_{k-i}) = \sum \vec{V}_k(t,z), \qquad (1.37)$$

where U_k is the envelope of the k-th channel, $k = i$ is the channel of interest (*probe*) and Δf_{k-i} the frequency spacing between the i-th and k-th channels. Traditional WDM systems use a fixed frequency grid (@ 25, 50 and 100 GHz standardized by the ITU-T) while modern systems employ a Flex-Grid approach enabling the spectral optimization when using transceivers with different baud-rates and modulation formats. Using this notation and ignoring the Raman and polarization effect, the nonlinear part of the Manakov-PMD equation becomes

$$\frac{\partial V_i(z,t)}{\partial z} = i\gamma\frac{8}{9}\left(\underbrace{|V_i|^2 V_i}_{\text{SPM}} + \underbrace{2\sum_{k\neq i}|V_k|^2 V_i}_{\text{XPM}} + \underbrace{\sum_{k\neq i}\sum_{l\neq k}V_k V_l^* V_{i-k+l}}_{\text{FWM}}\right). \qquad (1.38)$$

The individual terms are discussed in the following sections.

1.4.1 Self-Phase Modulation

The first term involves only the probe channel and is therefore referred to as *self-phase modulation* (SPM). In amplitude-modulation (AM) transmission the nonlinearity-induced phase modulation (PM) is directly proportional to the amplitude of the transmitted symbols $|V_i(t)|^2$. This additional PM typically leads to spectral broadening. However, SPM-induced system impairments are mostly due to the interaction of SPM and CD: Indeed under the influence of CD, phase (-or frequency) modulation is turned into to amplitude modulation, a phenomenon called FM-AM conversion [66, 67] (see Fig. 1.12). Dispersion compensation can only mitigate a fraction of this FM-AM conversion process as SPM is distributed along the line and not localized at the beginning of the span.

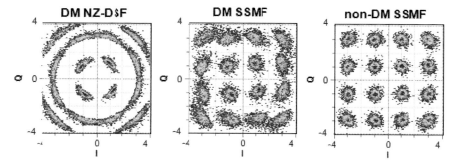

Figure 1.12 Impact of SPM on Nyquist-shaped 28 Gbaud DP-16QAM signal (2 dBm input power) after 400 km of NZ-DSF-based (left) or SSMF-based (centre) dispersion managed link (1% residual acc. dispersion per span), and after 400 km of SSMF-based non dispersion managed link (right). Note the increase of FM-AM conversion with accumulated dispersion.

1.4.2 Intra-Channel Cross-Phase Modulation and Four-Wave Mixing

SPM may not be limited to a single pulse (symbol) but can include the contributions of neighbouring symbols that broaden and overlap due to chromatic dispersion. This takes place in so-called *dispersive systems* where $L > L_D$ or for multi-span transmissions when the amount of accumulated dispersion during the nonlinear propagation, $GVD_{ACC,NL}$ is larger than T_0^2, T_0 being the symbol duration. The phase modulation due to the collision with neighbouring symbol waveforms is referred to as intra-channel cross phase modulation (iXPM). It usually leads to timing jitter. Intra-channel four wave mixing (iFWM) refers to the case when new frequency components are generated by the nonlinear interaction of three leading to additional amplitude noise (for more details see XPM and FWM sections below). iXPM and iFWM are therefore pattern- or modulation-dependent as they depend on the transmitted message and modulation format. When the accumulated chromatic dispersion $GVD_{acc} \gg T_0^2$ many symbols overlap and the pattern-dependency is averaged. The magnitude of the resulting distortion is usually reduced compared to less dispersive systems and looks similar to additive white Gaussian Noise (AWGN). The system is then referred to as *quasi-linear* [68]. It is important to notice that in quasi-linear transmissions, neighbouring symbols interact with the same symbols and that therefore iXPM-induced distortions become correlated in time [69].

1.4.3   Cross-Phase Modulation (XPM)

The second term in (1.38) describes cross-phase modulation (XPM). The *probe* channel is phase-modulated by the waveform of a neighbouring channel $|U_k(t)|^2$ (referred to as *pump*) and is therefore dependent on the pump modulation or *pattern*. Because of CD, WDM channels travel at different velocities, determined by the distance required for one pulse travelling at frequency ω_0 to overtake another pulse travelling at frequency ω_1. This distance is called *walk-off length*, and approximated by $L_W = T_0/|\beta_1(\omega_0) - \beta_1(\omega_1)| \sim T_0/|\Delta\lambda D|$.

When $GVD_{ACC,NL} > T_0/|\Delta\omega|$ ($\Delta\omega$ being the angular frequency spacing of the probe and pump), a symbol collides with multiple symbols from the pump. Similarly to SPM and iXPM, XPM leads to time jitter but also to amplitude distortion due to FM-AM conversion. It is useful to distinguish between transmission regimes (see Fig. 1.13):

When $GVD_{ACC,NL} \gg T_0/|\Delta\omega|$ the collision process is fast and a symbol interacts with many symbols from the pump. Similarly to iXPM, nonlinear collisions with multiple pulses lead to an averaging and thus to a reduction of XPM-induced impairments. For this reason the XPM effect decreases with the channel spacing and dispersion and presents a low-pass response [70].

In addition, when $L_W \ll L_D$ the symbols shapes remain the same during many collisions: Neighbouring symbols in the probe channel interact therefore with more or less the same pump symbols and XPM-induced distortions become time-correlated [69]. This aspect can be beneficial for DSP-based mitigation of XPM-induced jitter [71]. Furthermore, there is the molecule contribution to XPM. It is due to the real part of the delayed Raman response, which can be estimated from the imaginary part (Raman gain) using Kramers–Kronig relations. The molecular contribution to XPM leads to a slight reduction of the effective XPM in the system [72].

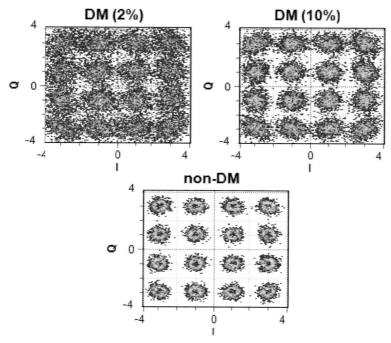

Figure 1.13 Impact of XPM for Nyquist-shaped 28 Gbaud DP-16QAM signal (central of 9 50 GHz spaced channels) after 800 km of SSMF with 2, 10 or 100% residual accumulated dispersion per span. Channel input power is 0 dBm, ASE noise and SPM of the probe channel are not considered.

1.4.4   Four-Wave Mixing (FWM)

The third term in (1.38) describes the four-wave-mixing (FWM) effect where three frequency components at f_i, f_j and f_k generate another component at frequency $f_{ijk} = f_i + f_j - f_k$ with $j \neq k$, such that energy and momentum conservation principles are satisfied. Degenerate FWM (two freq. components give raise to a third one) is possible for $f_{ik} = f_i + f_i - f_k$ or $f_{ik} = f_i - f_i + f_k$. iFWM refers to the case when all 3 frequency components belong to the same channel.

According to (1.38) FWM is phase-sensitive, i.e. generated frequency components depend on the relative phase ($\Delta\varphi_{ijk} = \varphi_i + \varphi_j - \varphi_k$) of the interacting waves. If the relative phase varies in time or in distance (e.g. due to chromatic dispersion), the frequency components generated at f_{ijk} add incoherently and averaged out. The impact of FWM can accumulate over long distance if the so-called a phase matching condition is met [73]. In WDM systems the FWM efficiency is given by [37]

$$\eta \approx \frac{\alpha^2}{\alpha^2 + (\beta_2 \Delta\omega_{ch}^2)^2},$$
(1.39)

which reduces with increasing the channel spacing $\Delta\omega_{ch}$ and chromatic dispersion, β_2. As stressed before, high local and accumulated dispersion has been shown beneficial to reduce distortions due to SPM, XPM and FWM [74]. For this reason, and also since in-line dispersion compensation requires additional EDFAs and therefore reduces the achievable OSNR and increases costs, most long-haul systems installed after 2010 are non-dispersion managed (non-DM). As a consequence FWM efficiency is very low in modern transmission system and most of the signal-signal nonlinear interactions are dominated by XPM, iXPM and iFWM.

1.4.5   Signal-Noise Interaction

In long-haul systems where ASE noise accumulates along the link, the contribution of ASE noise to nonlinear effects may become non-negligible. In this case the slowly varying envelope of the optical field can be written: $U(z, t) = U_s(z, t) + n(z, t)$ where U_s is the signal modulation and n the ASE noise components usually modelled as AWGN. The nonlinear operators for the signal and ASE become

$$\frac{\partial U_s(z,t)}{\partial z} = i\gamma \left(\underbrace{|U_s|^2\, U_s}_{\text{SPM+XPM+FWM}} + \underbrace{2\,|n|^2\, U_s}_{\text{NLPN}} \right)$$

$$\frac{\partial n(z,t)}{\partial z} = i\gamma \left(\underbrace{|n|^2\, n}_{\text{small}} + \underbrace{2\,|U_s|^2\, n}_{\text{PM}} + \underbrace{U_s n^* U_s}_{\text{FWM}} + \underbrace{n U_s^* n}_{\text{small}} \right). \qquad (1.40)$$

The second nonlinear term in the signal equation is referred to as *nonlinear phase noise* or *Gordon Mollenauer effect* [75]. The term noted as FWM (also called parametric process) is responsible for ASE amplification and therefore affects the OSNR. All other terms are small or do not change the magnitude of the noise power and therefore not relevant. Because of the stochastic nature of the ASE, signal-noise nonlinear interactions lead to stochastic impairments that present long temporal correlation [76]. Because of its parametric nature, the impact of nonlinear phase noise is similar to FWM and typically negligible in non-DM transmission.

1.4.6   Cross-Polarization Modulation

In order to investigate polarization-dependent nonlinear effects, the Manakov equation in vector form as expressed in (1.23) is considered. Writing the envelope of the propagating signal as sum of *probe* (S) and *pump* (P) contributions $U = P + S$ and expressing the polarization states of the pump in a referential moving with the instantaneous state of polarization probe, $P = P_\perp + P_\parallel$ the Manakov-PMD equation becomes [77]

$$\frac{\partial S(z,t)}{\partial z} = i\frac{8}{9}\gamma \left(\left(\underbrace{|S|^2}_{\text{SPM}} + \underbrace{|P|^2 + |P_\parallel|^2}_{\text{XPM}} \right) \cdot S + \underbrace{(P_\parallel^* \cdot S) \cdot P_\perp}_{\text{XpolM}} \right). \qquad (1.41)$$

The first nonlinear term affects only the phase of *S* whereas the second one is pointing in the orthogonal direction $(P_\perp)$ and therefore describes the depolarization of the signal, an effect referred to as cross-polarization modulation, XpolM [77–81].

Figure 1.14 illustrates relative amount of XPM and XpolM on a probe for different polarizations of pump.

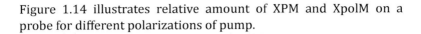

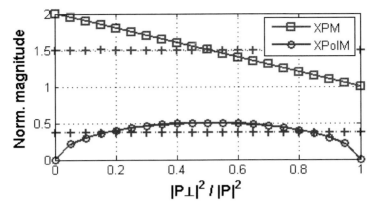

Figure 1.14 Normalized magnitude of XPM and XpolM terms as defined in (1.41), with '+' mean magnitude (averaged over Poincare sphere) (reproduced by permission of © 2011 The Institute of Electrical and Electronics Engineers).

XpolM-induced depolarization does not affect direct-detection systems, while it slightly affects the sensitivity of differential-detection receivers. XpolM is of particular relevance for dual-polarization (DP) channels because the induced depolarization leads to cross-talk between the two tributaries of the DP channel—an effect which is too fast to be mitigated by digital adaptive equalizers. XpolM is particularly strong when the pump consists of linearly polarized channels (e.g. OOK or DPSK channels) because the pump SOP varies *slowly* magnifying the depolarization effects (see Fig. 1.15).

On the contrary when the pump consists of DP channels, the XpolM effect averages out rapidly [80]. Due to the stochastic nature of fibre birefringence, the polarization of the pump and probe evolve randomly along the fibre. Therefore, the XpolM effect is random and cannot be compensated. In addition, the SOP of the pump is pattern dependent and changes rapidly compared to the envelope of the probe when the pumping occurs from many different channels. Therefore, XpolM-induced depolarization is often only weakly correlated in time.

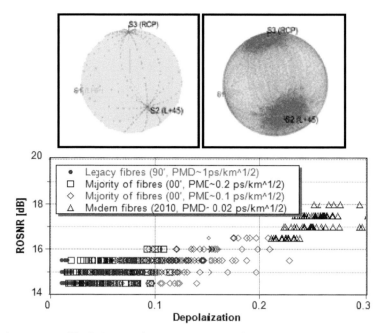

Figure 1.15 (Top) Poincaré representation of a DP-QPSK signal before (far left) and after transmission. (Bottom) required OSNR for 112 Gb/s DP-QPSK vs. depolarization, after co-propagation with 10 × 10 G NRZ channels (3 dBm input power, 100 GHz spacing) over 1000 km of dispersion-managed link.

1.4.7 Stimulated Raman Scattering (SRS)

We consider now Raman-induced distortions and use further the probe/pump representation for the optical field. Neglecting all other effects, the Manakov equation for the probe becomes

$$\frac{\partial S(z,t)}{\partial z} = \frac{g_r(f_i - f_k)}{2}|P_{\parallel}(z,t)|^2 S(z,t). \qquad (1.42)$$

In principle, the power transfer between two frequency components is pattern-dependent and also depends on their relative SOP, which leads to random and a time-dependent power-transfer. However, the Raman-induced power transfer being negligible for frequency spacing below 6 THz, the pump envelope and SOP evolve very rapidly in a frame moving with the probe—especially in the presence of large CD and PMD. Consequently, the power transfer depends only on the mean power of probe and pump and becomes deterministic. Note that stochastic polarization-dependent gain (PDG) can be observed for the case where the pumps are polarized (non-DP channels or a polarized CW light used pump for distributed Raman amplification).

1.4.8   The Nature of the Nonlinear Interference Noise

Kerr-induced signal distortions can be described as an additive noise term referred to as nonlinear-induced interference noise (NLIN) [38]. While popular models [39, 40] assume the NLIN to be similar to an AWGN in non-dispersion managed systems, careful analysis shows that the NLIN characteristics can deviate from the ones of the AWGN [69, 82]. This is especially the case for XPM- and iXPM-induced jitter and phase noise is correlated in time. The non-white nature of the NLIN has a proven impact of the system performance: The coloured nature of the XPM-induced distortions enables their mitigation through simple averaging such as the one performed in digital carrier phase recovery [71], which would not be possible in the case of an AWGN. On the other side the non-while nature and eventually non-Gaussian [83, 84] distribution of the NLIN is expected to degrade the performance of FEC schemes. This is illustrated in Fig. 1.16, where the Pre-FEC and Post-FEC performance of a Low Density Parity Check (LDPC) code (soft decision detection with 20 iteration of the MIN-SUM[3] decoder) is reported versus the channel input power in the case of eleven 28 Gbaud Nyquist shaped DP-16QAM channels (66 GHz spacing) propagation over a non-DM link consisting of 15 × 80 km G652-based spans. These results show that the expected FEC performance deviates from the one assuming an AWGN-approximation in presence of SPM and XPM.

---

[3]Particular implementation of the belief-propagation decoding.

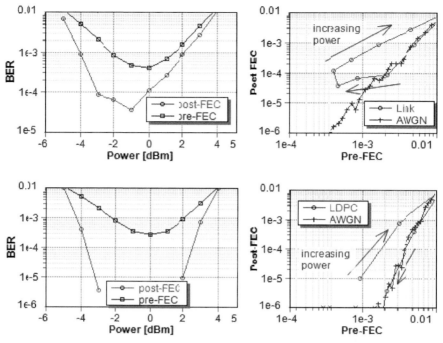

Figure 1.16 Performance of LPDC codes vs. channel input power after 800 km of non-DM transmission (11 × 56 Gbaud DP-16QAM, 66 GHz channel spacing). 1200 (top) and 4161 (bottom) bits frame length with ~20% overhead, both decoded with MIN-SUM algorithm (20 iterations). The expected performance deviates from the AWGN performance in presence of SPM and XPM.

1.5 Guidelines for Modelling High-Capacity Nonlinear Systems

1.5.1 Overview of System Performance Criteria

Besides the bit-error-rate (BER), which is often presented as the ultimate figure of merit for the performance of digital communication systems, various other criteria are employed to describe the quality of a digital optical transmission system. We restrict the following review to criteria that can be employed to describe the performance of digital coherent transmission systems.

1.5.1.1 Bit-error-rate

The BER is defined as the ratio of wrongly detected bits E to the total number of bits N transmitted during a certain observation interval. Because of slowly varying stochastic transmission effects such as PMD in single mode fibres, mode coupling in few-mode fibres [12, 85–87], or laser phase noise and resulting rare events such as cycle-slips, the transmission channel may be non-ergodic and the BER may fluctuate in time. The performance of such systems is better characterized by the *system outage probability*, which is the probability that the BER exceeds a given limit characterizing the quasi-error free operation during a certain time. This limit depends on the transport protocol and ranges from 1e-7 for DVBT to 1e-12 for 100 G Ethernet IEEE 802.3bx, while a limit of 1e-15 is targeted for future 400 G standard. In practice, the system should be designed to ensure operation below a certain *outage probability*. For instance, the so-called *four nines* or 99.99% availability corresponds to 52 min, 36 sec downtime per year. In addition, a system should be deployed with a certain *OSNR margin* (see Section 1.5.1.3) to ensure that *outage probability* is maintained if the system characteristics (OSNR, insertion loss) degrade in time, for instance, as a consequence of components aging or substitution. This margin should be at least 2–3 dB for terrestrial transmission where components can be easily replaced, but could be much larger for sub-sea transmission links.

1.5.1.2 Signal-to-noise ratio

One of the most elementary performance criteria is the ratio of signal power to noise power that can be expressed in different forms:

- *Signal-to-noise ratio* (SNR) is the ratio of the signal and noise power. When expressed in the electrical domain, the signal and noise power are measured within the same bandwidth.

- *Optical signal-to-noise ratio* (OSRN): Optical counterpart of the electrical *SNR* with the notable difference that the total signal power is considered while the noise power is measured in a particular bandwidth (usually 0.1 nm = 12.5 GHz).

- *Energy per bit to noise power spectral density ratio (E_b/N_0):* The energy per bit is expressed in Joules and the power spectral density of the noise in W/Hz (= Joules). E_b/N_0 is related to the SNR as follows:

$$\text{SNR} = E_b/N_0 \frac{f_b}{B}, \tag{1.43}$$

  where f_b is the data-rate (bits/s) of the signal and B the channel bandwidth (Hz). Contrary to *SNR*, E_b/N_0 enables comparing the performance of different signals independently of their bandwidth.

- *Effective Q-factor (Q_{eff}):* The *Q-factor* is a measure of the SNR for binary signals. It is defined as $Q = (\mu_1 - \mu_0)/(\sigma_1 + \sigma_0) = \sqrt{\text{SNR}}$, where $\mu_{1,0}$, is the mean value of the marks/spaces voltages or currents, and $\sigma_{1,0}$ their standard deviation. While the *Q*-factor is limited to binary signals only, it is possible to define an *effective Q-factor* for non-binary modulation when the signal BER is known:

$$\text{BER} = \frac{1}{2}\text{erfc}\left(\frac{Q_{eff}}{\sqrt{2}}\right) \tag{1.44}$$

  with erfc being the complementary error function. Q_{eff} values of 6 and 7 correspond to BER of 1e-9 and 1e-12, respectively. For binary signals in presence of additive white Gaussian noise (AWGN) and when inter-symbol interferences are negligible Q and Q_{eff} are equal.

1.5.1.3 System penalty and system margin

For system design purposes, it is important to characterize the impact of a particular effect on the system performance, for instance, the Kerr-induced impairments due to co-propagating channels. Looking directly at the BER is not always convenient as the magnitude of its variation depends not only on the impairments but also on the reference BER level. A common alternative criterion is the OSNR penalty [88] that describes the amount of optical noise that should be removed from the system in order to obtain the same performance as when the considered effect is absent. The OSNR penalty is expressed in dB:

$$OSNR_p = ROSNR - ROSNR_{ref},\qquad(1.45)$$

where $ROSNR_{ref}$ (resp. ROSNR) are the OSNR (in dB) required to achieve a desired performance (usually the quasi-error free BER or the BER threshold of a particular FEC code) in absence (resp. presence) of this effect. An example is shown in Fig. 1.17.

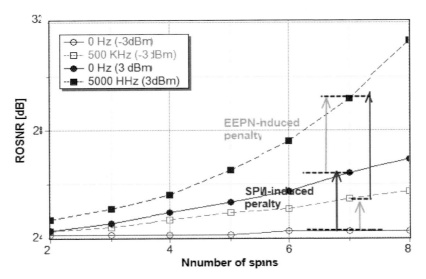

Figure 1.17 Required OSNR for 56 Gbaud DP-16QAM signal vs. number of non-DM spans (80 km SSMF) for different input powers and LO linewidths. Increased input power leads to Kerr-induced penalty while the LO linewidth combined with electronic CD compensation leads to EEPN. Note that OSNR penalties are not additive.

The OSNR penalty is a popular criterion because the impact of arbitrary transmission impairment is translated in terms of an *equivalent* amount of optical noise, enabling a simple understanding of their impact on the system performance. For instance, 1 and 3 dB OSNR penalty are roughly equivalent to a reduction of, respectively 25% and 50% of the transmission distance[4]. The OSNR margin is defined as the difference between the ROSNR and the actual OSNR of the system (see Fig. 1.18).

---

[4]Note that OSNR penalties from different impairments can usually not be added directly together, as it would be the case if all of these impairments can be regarded as AWGN.

Penalty and margin can be defined similarly for the effective Q-factor expressed in dB $Q_{eff\_dB} = 20.\log10(Q_{eff})$.

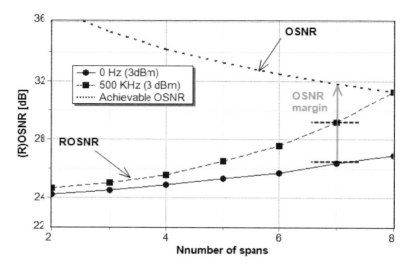

Figure 1.18 Achievable OSNR, required OSNR and resulting OSNR margin vs. number of spans for the 3 dBm input power case in Fig. 1.16. with an ideal (solid line) or realistic (dashed line) LO.

1.5.1.4 Error-vector magnitude

Similarly to the Q-factor for binary signals, the error-vector magnitude (EVM) describes a measure of the SNR for quadrature amplitude modulated (QAM) signals, which is evaluated by comparing N symbols before and after transmission:

$$\text{EVM} = \sqrt{\frac{\frac{1}{N}\sum_{i=1}^{N}|\vec{r}_i - \vec{t}_i|^2}{\frac{1}{N}\sum_{i=1}^{N}|\vec{t}_i|^2}} \approx \sqrt{\frac{P_{error}}{P_{ref}}} = \sqrt{\frac{1}{\text{SNR}}}, \tag{1.46}$$

where r and t are the vectors describing the received and transmitted symbols, respectively. Like for the Q-factor, there is a direct relationship between EVM and BER in an AWGN channel (no ISI, Gaussian distributed noise). For large square

m-QAM ($m \gg 2$) constellations and assuming raised cosine pulses, this relationship is given as [89, 90]

$$\text{BER} \approx \frac{4(1-m^{-1/2})}{\log_2(m)} Q\left(\sqrt{\frac{3}{(m-1)\text{EVM}^2}}\right), \quad Q(x) = \frac{1}{2}\text{erfc}\left(\frac{x}{\sqrt{2}}\right). \quad (1.47)$$

The accuracy of the estimate of the EVM depends on the number of considered symbols N:

$$\sigma_{\text{EVM}} \propto \frac{1}{2}\frac{\text{EVM}}{\sqrt{N}}. \tag{1.48}$$

1.5.2 Estimating the Bit-Error-Rate

In numerical simulations, the BER can be estimated using different methods [91, 92]. In the following section, we restrict the discussion to techniques that can be employed for digital coherent transmission using spectral efficient modulation.

1.5.2.1 Error-counting

Error counting (also called Monte-Carlo simulation) consists of estimating the BER by counting the errors. The main advantage of this method is that no particular assumption on the transmission channel (e.g. AWGN noise, filter bandwidth, receiver structure) is required to estimate the BER. It is very reliable, provided that enough errors are detected. Indeed, for a large number of errors ($E > 10$), the $(1 - \alpha)*100\%$ confidence interval for the number of detected errors E is given by [93]

$$\hat{E} = E \mp z_{\alpha/2}/\sqrt{E} \quad \text{with} \quad z_{\alpha/2} = \sigma\sqrt{2}\text{erf}^{-1}(\alpha-1), \quad \alpha \in [0;0.5], (1.49)$$

where $z_{\alpha/2}$ is 1.96 and 2.576 for $\alpha = 0.05$ and $\alpha = 0.01$, respectively. Assuming $E = 10$ and $N = 1e10$, the 95% confidence interval is [3.8e-10; 1.6e-9].

Note: 13 errors are necessary to ensure a 99% confidence interval when using the error counting method.

The main disadvantage of this method is that it requires many bits to be simulated in order to detect errors at low BER. Therefore, it is in general not practical to estimate BER below 1e-9 using numerical simulation.[5]

Techniques such as multi-canonical Monte-Carlo (MMC) [94, 95] simulation and importance sampling [96, 97] can be used to estimate low BER regions using the error-counting method. Both techniques rely on the same principle, which consists of biasing the statistics of the noise or other transmission effects (e.g., induced distortions [98]) such that errors happen more frequently. These errors are then weighted to estimate the performance of the unbiased system:

$$\mathrm{BER}_{\mathrm{IS}} = \frac{1}{N}\sum_{i=1}^{N} E(b_i) \cdot w(b_i) \quad w(b_i) = \frac{p\left(E(b_i)\right)}{p^*\left(E(b_i)\right)} \tag{1.50}$$

where $p(E(b_i))$ is the probability of the error on bit i to occur in the in the system of interest and p^* its probability in the biased system. In multi-canonical Monte-Carlo simulations, p and p^* are estimated numerically using dedicated algorithms [99, 100]. This offers the advantage that no a priori knowledge (assumption, model) about the system is required. However, the algorithms are time-consuming and the speed-up compared to standard error-counting is limited especially when Forward Error Correction (FEC) is considered [94]. In the importance sampling technique, p and p^* (or equivalently w) are estimated using a semi-analytical approach, requiring a precise model (or a good guess) of the error mechanisms in the considered system as well as a careful choice of the bias configuration. On the other side, this approach is much faster than the MMC one and can be applied in complex systems including DSP and FEC [101].

1.5.2.2 BER estimation techniques

The pre-FEC BER can be derived from the probability density function (PDFs) of the received symbols. Analytical models [102–105] predicting the PDF of the received symbols in direct or differential detection systems are of limited use in coherent

---

[5]Even with a real-time BER tester, 27 h are necessary to detect 10 errors at a BER of 1e-15 for a 100G channel.

systems. Indeed these techniques require an accurate analytical model of the receiver, which is usually not available when DSP is employed. If an ideal DSP-unit is assumed, the coherent receiver is equivalent to a linear filter. Under this assumption, the optical noise remains Gaussian distributed after detection and processing and the symbol-error-rate (SER) can be estimated by assuming the Normal distribution for the signal distortion, for instance, using the union-bound approximation [106]:

$$\text{SER} \leq \sum_{k=1}^{K} O_k \sum_{j \neq k}^{K} \frac{1}{2} \text{erfc} \left(\frac{1}{2} \frac{d_{kj}}{\sqrt{2}\sigma_{kj}} \right), \tag{1.51}$$

where M is the number of symbols in the constellation; O_k is the occurrence probability of symbol k; d_{kj} is the Euclidian distance between symbols k and j and σ_{kj} is the estimated average noise variance along k-j quadrature. The union-bound is a very good approximation for SER below 1e-3. In order to accurately estimate the noise variance σ_{kj} it is necessary to separate deterministic distortions (e.g. pattern effects induced by nonlinearities and filtering) from the noise. When the propagations of signal and noise cannot be considered separately this can be achieved by transmitting N times the same (short) sequence but using different noise realizations and computing the covariance matrix of the extracted noise [95].

1.5.3 Estimating System Average Performance and Outage Probability

Key aspects to ensure realistic system modelling and accurate performance prediction are reviewed in the following section.

1.5.3.1 System-level components modelling

Accurate modelling of key components such as optical sources, modulators, amplifiers, filters and detectors is of particular importance as components limitations and noise ultimately govern the system performance. Traditional components modelling, for example rate equations model for semiconductor lasers and EDFAs, requires the knowledge of the internal structure and a physical description of the components. Besides these accurate but complex *physical* models, much simpler *black-box* models emulating the components behaviour can be used for system-level simulations [32]. These models usually reproduce measurable components characteristics that are important on the system level and do not require a precise knowledge of the internal structure of the device. Therefore, they are also referred to as *data-sheets* or *behavioural* models. For instance, such characteristics include the linewidth and RIN of lasers and the gain and noise figure of EDFAs. Contrary to the *physical* models, *black-box* models require the a priori knowledge of the components characteristics affecting the system performance and for that reason should be employed with appropriate care.

In digital coherent transmissions, particular attention should be paid to the modelling of transmitter limitations: even if particular effects such as *I-Q* phase or amplitude imbalance at the modulator can be compensated at the receiver side, the equalization process with linear filter will be accompanied by noise enhancement. The transmitter performance is mainly impaired by

- optical source noise (laser RIN and phase noise or ASE of integrated semi-conductor optical amplifiers);
- noise and nonlinear response (compression) of the driver amplifiers;
- bandwidth limitation and imbalance (phase and amplitude) as well as parasitic effects in the modulator structure (e.g. electro-optical effects in silicon modulator [107]);
- bandwidth limitation and distortions (differential nonlinearities, gain and offset error, jitter, noise) of the digital-to-analogue converter (DAC) leading to a reduction of the effective resolution measured in *effective number of bits*, ENoB. Note that the ENoB is not the ultimate criterion to describe the performance of DAC or ADC as different combinations of distortions leading to the same ENoB may have a different impact on the system performance.

Most imperfections of the analogue coherent front-end can be mitigated in the digital receiver side, such that the receiver performance is finally limited by the following characteristics:

- bandwidth and noise characteristics of the photodiode and transimpedance amplifiers;

- sampling speed, bandwidth and resolution of the ADC;
- linewidth of the local oscillator which can lead to equalization enhanced phase noise (EEPN) when CD compensation is performed digitally at the receiver. The EEPN-induced signal variance σ_{EEPN}^2 grows linearly with the signal baud-rate, B_r, the local oscillator linewidth $\Delta\nu$ and the amount of CD compensated digitally at the receiver [108].

$$\sigma_{\text{EEPN}}^2 \approx 2\pi \cdot B_\text{r} \cdot \Delta\nu \cdot |\beta_2 \cdot L| \tag{1.52}$$

1.5.3.2 Transmission link modelling

Accurate modelling of the signal propagation requires precise specification of the fibre characteristics (attenuation, chromatic dispersion, nonlinear coefficient). When considering WDM systems, the wavelength dependency of the fibre GVD and attenuation coefficient as well as of the EDFA noise figure and gain should be accounted for. In principle the accurate estimation of nonlinear impairments requires extensive simulation as these effects result from multiple interactions between deterministic (signal pattern, dispersion, Kerr) and random (birefringence, noise) processes. However, the number of degrees of freedom and the resulting modelling effort can be substantially reduced depending on the considered system:

In [112], Bononi et al. showed that intra- and inter-channel XPM (see Sections 1.4.2 and 1.4.3) are the limiting nonlinear impairments in homogeneous non-dispersion managed transmission over standard single mode fibre (ITU-G652) and their impact in terms of OSNR penalty is independent of PMD. In addition since signal-noise interactions can be neglected, the optical noise can be added to the signal after *deterministic* propagation at the receiver side. This allows the generation of long signal sequences by adding different realizations of optical noise and LO phase noise to the signal that is obtained at the output of the link to perform, for instance, Monte-Carlo simulations (see Section 1.5.2.1).

Of course, the simulated sequence should be long enough to account correctly for the statistics of pattern-dependent effects such as iXPM and XPM (see Section 1.5.3.3). Since the impact of XPM depends on the fibre birefringence [113], multiple realizations of the birefringence profiles of the fibres constituting the link are required to estimate the statistical variation of the system performance. Such a statistical analysis is discussed in 1.5.3.4.

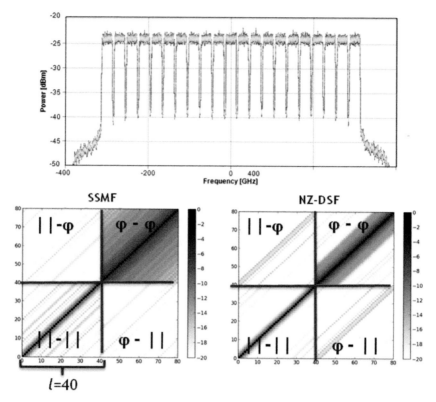

Figure 1.19 Covariance of Kerr-induced distortions for 21 × 100 G DP-QPSK @33 GHz after propagation over 2000 km of non-DM SSMF (Bottom left) or NZ-DSF (Bottom right) spans.

When distributed Raman amplification is employed, the following aspects should be considered:

- For the case that the Raman pump is not fully depolarized, the relative evolution of the pump and signal SOP should be accounted for because the Raman-induced gain is polarization-dependent. This requires a statistical analysis of the system performance (gain, nonlinearities) assembled over different fibre birefringence profiles.

- When a forward pumping scheme is used, noise processes of the pump laser must be modelled accurately to account for the RIN transfer phenomena between pump and signal [109]. This effect can be ignored in backward pumping schemes where the walk-off between pump and waveform is very short. Possibly reflections due to splices and connectors, as well as Brillouin and Rayleigh scattering processes should be modelled in order to account for additional noise sources leading to the reduction of OSNR [110] and sources of multi-path interferences [111] distorting the signal.

1.5.3.3 Deterministic propagation

In order to account correctly for pattern-dependent transmission effects such as SPM or XPM the transmitted signal should be chosen correctly to ensure gathering robust statistics. For instance, pseudo-random bits sequences (PRBS) can be used [114]. A PRBS of order N is a periodic sequence of length 2^N-1 and includes all possible combinations of N bits expect the sequence consisting of N zeros[6]. In simulations De Bruijn sequences [115] can be used. They are counterparts of PRBS sequences but include the all-zeros state and therefore are of length 2^N. De Bruijn sequences can be generated for a non-binary alphabet to support QAM signal and are in this case of length L^N where L is the size of the alphabet and N the order of the sequence. Such sequences comprise all possible transitions between N symbols, and present therefore optimal statistical properties to investigate pattern-dependent effects. Ideally N should be chosen larger than the channel memory, which includes the transmitter, transmission link and digital equalizer. For linear channels the system memory is usually limited to a few symbols. The channel memory is typically governed by the carrier phase recovery algorithm. For nonlinear systems, where the nonlinear propagation is combined with large amounts of CD or short walk-off lengths between WDM channels (see Section 1.2.2.2), the effective channel memory can be up several tens of symbols. The channel memory and N can be

---

[6]PRBS generators are implemented using linear feedback shift-registers with XOR gates in hardware. The all-zeros state is illegal because the register would remain locked-up in this state.

estimated by computing the covariance of the signal distortions [101].

1.5.3.4 Modelling stochastic propagation effects

Even if intra- and inter-channel effects can appear random in dispersive transmission or when a large number of WDM channels are involved, they are eventually governed by the Manakov equation, and remain deterministic processes as long as the PMD effect on the channels waveform is negligible. When accumulated DGD is large compared to the symbol duration, SPM and XPM-induced signal distortions the system performance varies with the actual birefringence profile of the fibres. This effect can affect the performance of nonlinear mitigation techniques such as optical phase conjugation [116] and digital back propagation (DBP) [117]. Similarly, the relative state-of-polarization of the WDM channels and the resulting XpolM effect depend on the random and time-varying fibre birefringence profile.

In order to capture the statistics of these random transmission effects, different realizations of the fibres birefringence profiles have to be simulated. For this, the following parameters of the fibre coarse-step model (see Section 1.3.2.3) can be varied:

- The orientation of the fast and slow axes in successive coarse-step sections. It should follow a circular uniform distribution, unless biasing is desired, for instance, when importance sampling is employed to target specific birefringence profiles.
- The length of the individual scattering sections. For instance, 1 km ± Δl, Δl being a random variable following the normal distribution with 100 m variance. Note that in the coarse step approach the mean section length should be larger than the fibre correlation length.

Simulated time-windows are usually too short to capture the statistics of slow noise processes in lasers that can lead to impairments such as cycle slips. To capture this behaviour, the system should be simulated multiple times (using Monte-Carlo or importance sampling) applying different realization of the lasers noise processes.

Extrapolating the outcome of statistical modelling can be used to derive the system outage probability. However, such a Monte-Carlo approach requires a large number of realizations to gather robust statistics about the system failure. Alternatively, importance sampling be used by (1) identifying the configurations possibly leading to an outage and (2) exploring the system performance for these configurations and weighting their contribution to the system outage.

For instance, it has been shown in [117] that the performance of DBP is affected by PMD and that this penalty grows with the DGD of the link as illustrated in Fig. 1.20. In order to explore the system outage probability it is possible to select the links with large DGD, and weight this outage by the probability of this DGD to take place.

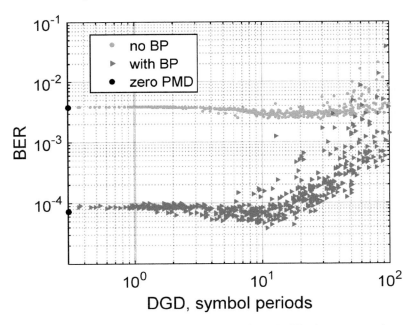

Figure 1.20 BER vs. link DGD for a 28 Gbaud DP-QPSK after propagation over a 1600 km non-DM link in nonlinear regime (5 dBM input power) with (blue triangles) and without (yellow dots) DBP. Hundreds of different birefringence profiles have been simulated while D_{PMD} has been varied between 0.05 and 2 ps/km$^{1/2}$.

1.6 Summary and Outlook

The modelling of transmission systems operating close to the nonlinear Shannon limit requires particular care as nonlinear propagation effects become the source of dominant transmission impairments, especially for systems employing techniques that should mitigate those effects.

In this chapter, the origins of linear and nonlinear effects in optical fibre are reviewed before the equations governing the signal propagation in single and multi-mode fibres are introduced. Simulation techniques for solving efficiently and accurately these equations are then described and the nature of Kerr-induced signal distortions in digital coherent optical transmission systems is discussed. In particular it is shown that the Kerr-induced nonlinear interference noise is correlated and that its impact on the system performance can strongly deviate from the one of additive white Gaussian noise (AWGN). In addition, it is shown that the random nature of fibre birefringence impacts the system performance, which therefore requires statistical analysis to be performed. This leads the authors to the conclusion that the compensation of nonlinear effects, required to operate beyond the nonlinear Shannon limit, is ultimately limited by the random birefringent nature of the fibre.

References

1. Kao, K. C., and Hockham, G. A. (1966). Dielectric-fibre surface waveguides for optical frequencies, *Electrical Eng., Proc. Inst.,* **113.7**, pp. 1151–1158.

2. Keck, D. B. (2000). Optical fiber spans 30 years, *Lightwave,* **17.8**, pp. 78–82.

3. Mears, R. J., Reekie, L., Jauncey, I. M, and Payne, D. N. (1987). Low-noise Erbium-doped fibre amplifier at 1.54 μm, *Electron. Lett.,* **23**, pp. 1026–1028.

4. Desurvire, E., Simpson, J., and Becker, P. C. (1987). High-gain erbium-doped traveling-wave fiber amplifier, *Opt. Lett.,* **12**, pp. 888–890.

5. Dugan, J. M., Price, A. J., Ramadan, M., Wolf, D. L., Murphy, E. F., Antos, A. J., Smith, D. K., and Hall, D. W. (1992). All-optical, fiber-based 1550 nm dispersion compensation in a 10 Gb/s,

150 km transmission experiment over 1310 nm optimized fiber, *Proceedings OFC 1992*, pp. 63–66.

6. Agrawal, G. P. (2012). *Nonlinear Fiber Optics*, 5th ed. (John Wiley & Sons).

7. ITU-T, telecommunication standardization sector of ITU-T, ITU-T G.652 (11/2009).

8. Kogelnik, H., and Jopson, R. M. (2002). *Optical Fiber Communication IVB*, eds. Kaminow, I. P., and Li, T., Chapter 15, polarization mode dispersion (Academic Press).

9. Galtarossa, A., Palmieri, L., Schiano, M., and Tambosso, T. (2000). (2000). Measurements of beat length and perturbation length in long single-mode fibers, *Opt. Lett.*, **25**, pp. 384–386.

10. Palmieri, L., Galtarossa, A., Schiano, M., and Tambosso, T. (2001). PMD in single-mode fibers: Measurments of local birefringence correlation length, *Proceedings OFC 2001*, paper ThA2.

11. Sunnerud, H., Karlsson, M., and Andrekson, P. A. (2001). A comparison between NRZ and RZ data formats with respect to PMD-induced system degradation, *Proceedings OFC 2001*, pp. WT3–WT3.

12. Bülow, H. (1998). System outage probability due to first-and second-order PMD, *Photonics Technol. Lett.*, **10**, pp. 696–698.

13. Savory, S. J. (2008). Digital filters for coherent optical receivers, *Opt. Express*, **16**, pp. 804–817.

14. Stolen, R. H., and Ashkin. A. (1973). Optical Kerr effect in glass waveguide, *Appl. Phys. Lett.*, **22**, pp. 294–296.

15. Stolen, R. H., Tomlinson, W. J., Haus, H. A., and Gordon, J. P. (1989). Raman response function of silica-core fibers, *JOSA B*, **6**, pp. 1159–1166.

16. Hasegawa, A., and Tappert, F. (1973). Transmission of stationary nonlinear optical pulses in dispersive dielectric fibers. I. Anomalous dispersion, *Appl. Phys. Lett.*, **23**, pp. 142–144.

17. Vinegoni, C., Wegmuller, M., and Gisin, N. (2001). Measurements of the nonlinear coefficient of standard, SMF, DSF, and DCF fibers using a self-aligned interferometer and a Faraday mirror, *IEEE Photonics Technol. Lett.*, **13**, pp. 1337–1339.

18. Botineau, J., and Stolen, R. H. (1982). Effect of polarization on spectral broadening in optical fibers, *JOSA*, **72**, pp. 1592–1596.

19. Stolen, R. H., Botineau, J., and Ashkin, A. (1982). Intensity discrimination of optical pulses with birefringent fibers, *Opt. Lett.*, **7**, pp. 512–514.

20. Menyuk, C. R. (1987). Nonlinear pulse propagation in birefringent optical fibers, *J. Quantum Electron.*, **23**, pp. 174–176.

21. Marcuse, D., Menyuk, C. R., and Wai, P. K. A. (1997). Application of the Manakov-PMD equation to studies of signal propagation in optical fibers with randomly varying birefringence, *J. Lightwave Technol.*, **15**, pp. 1735–1745.

22. Menyuk, C. R. (2005). *Polarization Mode Dispersion* Chapter, Interaction of nonlinearity and polarization mode dispersion (Springer New York) pp. 126–132.

23. Gloge, D. (1971). Weakly guiding fibers, *Appl. Opt.*, **10**, pp. 2252–2258.

24. VPIphotonics (2016). VPItransmissionMaker Optical Systems: *SolverFiberMM Module Reference Manual.*

25. Poletti, F., and Horak, P. (2008). Description of ultrashort pulse propagation in multimode optical fibers, *JOSA B,* **25**, pp. 1645–1654.

26. Mumtaz, S., Essiambre, R.-J., and Agrawal, G. P. (2013). Nonlinear propagation in multimode and multicore fibers: Generalization of the Manakov equations, *J. Lightwave Technol.*, **31**, pp. 398–406.

27. Mecozzi, A., Antonelli, C., and Shtaif, M. (2012). Coupled Manakov equations in multimode fibers with strongly coupled groups of modes, *Opt. Express*, **20**, pp. 23436–23441.

28. Lowery, A. J. (2002). *Optical Fiber Communication IVB*, eds. Kaminow, I. P., and Li, T., Chapter 12, Photonic simulation tools (Academic Press).

29. Lowery, A., Lenzmann, O., Koltchanov, I., Moosburger, R., Freund, R., Richter, A., Georgi, S., Breuer, D., and Hamster, H. (2000). Multiple signal representation simulation of photonic devices, systems, and networks, *J. Selecl. Top. Quantum Electronics,* **6**, pp. 282–296.

30. VPIphotonics (2016). VPItransmissionMaker Optical Systems: *UniversalFiber, Module Reference Manual.*

31. Yu, Y., Reimer, W., Grigoryan, V. S., and Menyuk, C. R. (2000). A mean field approach for simulating wavelength-division multiplexed systems, *Photonics Technol. Lett.*, **12**, pp. 443–445.

32. Richter, A. (2011). *WDM Systems and Networks: Modeling, Simulation, Design and Engineering*, eds. Antoniades, N. N., Ellinas, G., and Roudas, I., Chapter 2, Computer modeling of transport layer effects (Springer Science & Business Media) pp. 13–61.

33. Djupsjobacka, A., Berntson, A., and Martensson, J. (2008). A method to calculate PMD-induced eye-opening penalty and signal outage for RZ-modulated signal formats, *J. Lightwave Technol.*, **26**, pp. 3186–3189.

34. Elbers, J. P., Faerbert, A., Scheerer, C., Glingener, C., and Fischer, G. (2000). Reduced model to describe SPM-limited fiber transmission in dispersion-managed lightwave systems, *J. Selected Topics Quantum Electronics*, **6**, pp. 273–281.

35. Cartaxo, A. V. T. (1999). Cross-phase modulation in intensity modulation-direct detection WDM systems with multiple optical amplifiers and dispersion compensators, *J. Lightwave Technol.*, **17**, pp. 178–190.

36. Marki, C. F., Alic, N., Gross, M., Papen, G., Esener, S., and Radic, S. (2006). Performance of NRZ and Duobinary modulation formats in rayleigh and ASE-dominated dense optical links, *Proc OFC 2006*, paper OFD6.

37. Inoue, K. (1992). Four-wave mixing in an optical fiber in the zero-dispersion wavelength region, *J. Lightwave Technol.*, **10**, pp. 1553–1561.

38. Dar, R., Feder, M., Mecozzi, A., and Shtaif, M. (2014). Accumulation of nonlinear interference noise in fiber-optic systems, *Opt. Express*, **22**, 14199–14211.

39. Poggiolini, P. (2012). The GN model of non-linear propagation in uncompensated coherent optical systems, *J. Lightwave Technol.*, **30**, pp. 3857–3879.

40. Carena, A., Bosco, G., Curri, V., Jiang, Y., Poggiolini, P., and Forghieri, F. (2010). EGN model of non-linear fiber propagation, *Opt. Express*, **22**, pp. 16335–16362.

41. Richter, A., Karelin, N., Louchet, H., Koltchanov, I., and Farina, J. (2010). Dynamic events in optical networks—emulation and performance impact analysis, *Proceedings IEEE Military Communication Conference (MILCOM)*, pp. 1083–1087.

42. VPIphotonics (2016). *VPItransmissionMaker Optical Systems: User's Manual*, Chapter 5: Advanced Signal Modes and Representations.

43. Rieznik, A., Tolisano, T., Callegari, F. A., Grosz, D. F., and Fragnito, H. L. (2005). Uncertainty relation for the optimization of optical-fiber transmission systems simulations, *Opt. Express*, **13**, pp. 3822–3834.

44. Stoffa, P. L., Fokkema, J. T., Freire, R. M. D., and Kessinger, W. P. (1990). Split-step Fourier migration, *Geophysics,* **55.4**, pp. 410–421.

45. Pachnicke, S., Chachaj, A., Helf, M., and Krummrich, P. M. (2010). Fast parallel simulation of fiber optical communication systems accelerated by a graphics processing unit, *Proc 12th International Conference on Transparent Optical Networks (ICTON)*.

46. Hellerbrand, S., and Hanik, N. (2010). Fast implementation of the split-step Fourier method using a graphics processing unit, *Proceedings OFC 2010*, paper OTuD7.

47. Karelin, N., Shkred, G., Simonov, A., Mingaleev, S., Koltchanov, I., and Richter, A. (2014). Parallel simulations of optical communication systems *Proc 14th International Conference on Transparent Optical Networks* (*ICTON*).

48. Carena, A., Curri, V., Gaudino, R., Poggiolini, P., and Benedetto, S. (1997). A time-domain optical transmission system simulation package accounting for nonlinear and polarization-related effects in fiber, *IEEE J. Selected Areas Commun.*, **15**, pp. 751–765.

49. Scarmozzino, R. (2008). *Optical Fiber Communication VB*, eds. Kaminow, I. P., Li, T., and Willner, A. E. T., Chapter 20, simulation tools for devices, systems, and networks (Academic Press).

50. Bosco, G., Carena, A., Curri, V., Gaudino, R., Poggiolini, P., and Benedetto, S. (2000). Suppression of spurious tones induced by the split-step method in fiber systems simulation, *IEEE Photonics Technol. Lett.*, **12**(5), pp. 489–491.

51. Tkach, R. W., Chraplyvy, A. R., Forghieri, F., Gnauck, A. H., and Derosier, R. M. (1995). Four-photon mixing and high speed WDM systems, *J. Lightwave Technol.*, **13**, pp. 841–849.

52. Sinkin, O. V., Holzlöhner, R., Zweck, J., Menyuk, C. R. (2003). Optimization of the split-step Fourier method in modeling optical-fibre communications systems, *J. Lightwave Technol.*, **21**, pp. 61–68.

53. Kogelnik, H., Nelson, L. E., Gordon, J. P., and Jopson, R. M. (2000). Jones matrix for second-order polarization mode dispersion, *Opt. Lett.*, **25**, pp. 19–21.

54. Kogelnik, H., Nelson, L. E., and Gordon, J. P. (2003). Emulation and inversion of polarization-mode dispersion (Invited Tutorial), *J. Lightwave Technol.*, **21**, pp. 61–68.

55. Prola, C. H., Da Silva, J. P., Dal Forno, A. O., Passy, R., Von der Weid, J. P., and Gisin, N. (1997). PMD emulators and signal distortion in 2.48-Gb/s IM-DD lightwave systems, *IEEE Photonics Technol. Lett.*, **9**, pp. 842–844.

56. Dal Forno, A. O., Paradisi, A., Passy, R., and Von der Weid, J. P. (2000). Experimental and theoretical modeling of polarization-mode dispersion in single-mode fibers, *IEEE Photonics Technol. Lett.*, **12**, pp. 296–298.

57. Ho, K.-P., and Kahn, J. M. (2011). Statistics of group delays in multimode fiber with strong mode coupling, *J. Lightwave Technol.*, **29**, pp. 3119–3128.

58. Juarez, A. A., Bunge, C. A., Warm, S., and Petermann, K. (2012). Perspectives of principal mode transmission in mode-division-multiplex operation, *Opt. Express*, **20**, pp. 13810–13824.

59. Ho, K.-P., and Kahn, J. M. (2013). Linear propagation effects in mode-division multiplexing systems, *J. Lightwave Technol.*, **32**, pp. 614–628.

60. Olshansky, R. (1975). Mode coupling effects in graded-index optical fibers, *Appl. Opt.*, **14**(4), pp. 935–945.

61. Marcuse, D. (1991). *Theory of Dielectric Optical Waveguides*, 2nd ed. (Academic Press).

62. Kawakami, S., and Tanji, H. (1983). Evolution of power distribution in graded-index fibres, *Electronics Lett.*, **19**, pp. 100–102.

63. Kroushkov, D. I., Rademacher, G., and Petermann, K. (2013). Cross mode modulation in multimode fibers, *Opt. Lett.*, **38**(10), pp. 1642–1644.

64. Essiambre, R.-J., Mestre, M. A., Ryf, R., Gnauck, A. H., Tkach, R. W., Chraplyvy, A. R., Sun, Y., Jiang, X., and Lingle, R. (2013). Experimental investigation of inter-modal four-wave mixing in few-mode fibers, *IEEE Photonics Technol. Lett.*, **25**(6), pp. 539–542.

65. Karelin, N., Louchet, H., Kroushkov, D., Uvarov, A., Mingaleev, S., Koltchanov, I., and Richter, A. (2016). Modeling and design framework for SDM transmission systems, *Proc 17th International Conference on Transparent Optical Networks (ICTON)*.

66. Bochove, E. J., de Carvalho, E. M., and Ripper Filho, J. E. (1981). FM–AM conversion by material dispersion in an optical fiber, *Opt. Lett.*, **6**(2), 58–60.

67. Petermann, K. (1990). FM–AM noise conversion in dispersive single-mode fibre transmission lines, *Electronics Lett.*, **26**, pp. 2097–2098.

68. Essiambre, R. J. (2002). *Optical Fiber Communication IVB*, eds. Kaminow, I. P., and Li, T., Chapter 6, *Pseudo-linear transmission of high-speed TDM signals*: 40 and 160 Gb/s (Academic Press).

69. Dar, R., Feder, M., Mecozzi, A., and Shtaif, M. (2013). Properties of nonlinear noise in long, dispersion-uncompensated fiber links, *Opt. Express*, **21**, pp. 25685–25699.

70. Ho, K. P., Kong, E. T. P., Chan, L. Y., Chen, L. K., and Tong, F. (1999). Analysis and measurement of root-mean-squared bandwidth of cross-

phase-modulation-induced spectral broadening, *Photonics Technol. Lett.*, **11**, pp. 1126–1128.

71. Fehenberger, T., Yankov, M. P., Barletta, L., and Hanik, N. (2015). Compensation of XPM interference by blind tracking of the nonlinear phase in WDM systems with QAM input, *Proceedings ECOC'15*.

72. Höök, A. (1992). Influence of stimulated Raman scattering on cross-phase modulation between waves in optical fibers, *Opt. Lett.*, **17**, pp. 115–117.

73. Inoue, K., and Hiromu, T. (1992). Wavelength conversion experiment using fiber four-wave mixing, *Photonics Technol. Lett.*, **4**, pp. 69–72.

74. Bayvel, P., and Killey, R. (2002). *Optical Fiber Communication IVB*, eds. Kaminow, I. P., and Li, T., Chapter 13, Nonlinear optical effects in WDM transmissions (Academic Press).

75. Ho, K.-P. (2005). *Phase-Modulated Optical Communication Systems* (Springer), pp. 143–188.

76. Ho, K. P., and Wang, H. C. (2006). Effect of dispersion on nonlinear phase noise, *Opt. Lett.*, **31**, pp. 2109–2111.

77. Louchet, H., Richter, A., Koltchanov, I., Mingaleev, S., Karelin, N., and Kuzmin, K. (2011). Comparison of XPM and XpolM-induced impairments in mixed 10G–100G transmission, *Proceedings 13th International Conference on Transparent Optical Networks (ICTON)* paper.

78. Karlsson, M., and Sunnerud, H. (2006). Effects of nonlinearities on PMD-induced system impairments, *J. Lightwave Technol.*, **24**, pp. 4127–4137.

79. Winter, M., Setti, D., and Petermann, K. (2010). Cross-polarization modulation in polarization-division multiplex transmission, *Photonics Technol. Lett.*, **22**, pp. 538–540.

80. Bononi, A., Serena, P., and Rossi, N. (2010). Nonlinear signal–noise interactions in dispersion-managed links with various modulation formats, *Opt. Fiber Technol.*, **16**, pp. 73–85.

81. Winter, M., Bunge, C. A., Setti, D., and Petermann, K. (2009). A statistical treatment of cross-polarization modulation in DWDM systems, *J. Lightwave Technol.*, **27**, pp. 3739–3751.

82. Fan, Y., Dou, L., Tao, Z., Li, L., Oda, S., Hoshida, T., and Rasmussen, J. C. (2012). Modulation format dependent phase noise caused by intra-channel nonlinearity, *Proceedings ECOC' 12*, paper We.2.C.3.

83. Tanimura, T., Koganei, Y., Nakashima, H., Hoshida, T., and Rasmussen, J. C. (2014). Soft decision forward error correction over nonlinear transmission of 1-Tb/s superchannel, *Proceedings ECOC'14*.

84. Louchet, H., Koltchanov, I., Richter, A., Ye, Y., Zhou, E., Binh, L. N., and Zhang, S. (2012). Generalized scaling laws for Kerr-induced signal distortions in nondispersion-managed systems, *Photonics Technol. Lett.*, **18**(24), pp. 1618–1620.

85. Nelson, L. E., Nielsen, T. N., and Kogelnik, H. (2001). Observation of PMD-induced coherent crosstalk in polarization-multiplexed transmission, *Photonics Technol. Lett.*, **13**(7), pp. 738–740.

86. Winzer, P. J., and Foschini. G. J. (2011). MIMO capacities and outage probabilities in spatially multiplexed optical transport systems, *Opt. Express*, **19.17**, pp. 16680–16696.

87. Vassilieva, O., Inwoong, K., and Motoyoshi, S. (2012). Statistical analysis of the interplay between nonlinear and PDL effects in coherent polarization multiplexed systems, *Proc, ECOC 2012*, paper We.3.C.4.

88. Rongqing, H., and O'Sullivan, M. (2009). *Fiber Optic Measurement Techniques* (Academic Press).

89. Shafik, R. A., Rahman, M. S., and Islam, A. R. (2006). On the extended relationships among EVM, BER and SNR as performance metrics, *Proceedings ICECE'06*, pp. 408–411.

90. Nebendahl, B., et al. (2012). Quality metrics in optical modulation analysis: EVM and its relation to Q-factor, OSNR, and BER, *Proceedings Asia Communications and Photonics Conference*. paper AF3G-2.

91. Jeruchim, M. C. (1984). Techniques for estimating the bit error rate in the simulation of digital communication systems, *Selected Areas Commun. IEEE J.*, **2.1**, pp. 153–170.

92. Louchet, H., Kuzmin, K., Koltchanov, I., and Richter, A. (2009). BER estimation for multilevel modulation formats, *Proceedings Asia Communications and Photonics*, pp. 763218–763218.

93. Snedecor, G. W., and Cochran, W. G. (1989). *Statistical Methods*, 8th ed. (Iowa State University Press).

94. Holzlöhner, R., Mahadevan, A., Menyuk, C. R., Morris, J. M., and Zweck, J. (2005). Evaluation of the very low BER of FEC codes using dual adaptive importance sampling, *Commun. Lett.*, **9**(2), pp. 163–165.

95. Holzlöhner, R., and Menyuk, C. R. (2003). Use of multicanonical Monte Carlo simulations to obtain accurate bit error rates in optical communications systems, *Opt. Lett.*, **28**(20), pp. 1894–1896.

96. Bohdanowicz, A. (2001). On efficient BER evaluation of digital communication systems via importance sampling, *Proceedings SCVT 2001 (8th Symposium on Communications and Vehicular Technology)*, pp. 61–67.

97. Romano, G., Drago, A., and Ciuonzo, D. (2011). Sub-optimal importance sampling for fast simulation of linear block codes over BSC channels, *Proceedings Wireless Communication Systems (ISWCS), 2011 8th International Symposium on Wireless Communication Systems*, pp. 141–145.

98. Biondini, G., Kath, W. L., and Menyuk, C. R. (2002). Importance sampling for polarization-mode dispersion, *Photonics Technol. Lett.*, **14**(3), pp. 310–331.

99. Bononi, A., and Rusch, L. A. (2011). *Impact of Nonlinearities on Fiber Optic Communications,* ed. Kumar, S., Chapter 1, Multicanonical Monte Carlo for simulation of optical links (Springer New York).

100. Bononi, A., Rusch, L. A., Ghazisaeidi, A., Vacondio, F., and Rossi, N. (2009). A fresh look at multicanonical Monte Carlo from a telecom perspective, *Proceedings GLOBECOM 2009*.

101. Louchet, H., and Richter, A. (2016). Performance estimation of digital coherent systems with FEC under non-AWGN channel statistics, *Proceedings OFC '16,* paper M2A-1.

102. Schwartz, M., Bennet, W. R., and Stein, S. (1966). *Communication Systems and Techniques* (McGraw-Hill, New York) Appendix B.

103. Richter, A., Koltchanov, I., Myslivets, E., Khilo, A., Shkred, G., and Freund, R. (2005). Optimization of multi-pump Raman amplifier, *Proceedings OFC'05*, paper NTuB4.

104. Forestieri, E., and Secondini, M. (2009). On the error probability evaluation in lightwave systems with optical amplification, *J. Lightwave Technol.*, **27**, pp. 706–717.

105. Bosco, G., and Poggiolini, P. (2005). The impact of receiver imperfections on the Performance of optical direct-Detection DPSK, *J. Lightwave Technol.*, **23**, pp. 842–848.

106. Benedetto, S., and Biglieri, E. (1999). *Principles of Digital Transmission: With Wireless Applications* (Kluwer, New York) p. 190.

107. Goi, K., Oka, A., Kusaka, H., Terada, Y., Ogawa, K., Liow, T., Tu, X., Lo, G., and Kwong, D. (2014). Low-loss high-speed silicon IQ modulator for QPSK/DQPSK in C and L bands, *Optics Express*, **22**, pp. 10703–10709.

108. Kakkar, A., Navarro, J. R., Schatz, R., Louchet, H., Pang, X., Ozolins, O., Jacobsen, G., and Popov, S. (2015). Comprehensive study of equalization-enhanced phase noise in Coherent optical systems, *J. Lightwave Technol.*, **33**, pp. 4834–4841.

109. Fludger, C. R. S., Handerek, V., and Mears, R. J. (2001). Pump to signal RIN transfer in Raman fiber amplifiers, *J. Lightwave Technol.*, **19**, p. 1140.

110. Hansen, P. B., Eskildsen, L., Stentz, A., Strasser, T., Judkins, J., DeMarco, J., Pedrazzani, R., and DiGiovanni, D. (1998). Rayleigh scattering limitations in distributed Raman pre-amplifiers, *Photonics Technol. Lett.*, **10**.1, pp. 159–161.

111. Lewis, S. A. E., Chernikov, S. V., and Taylor, J. R. (2010). Characterization of double Rayleigh scatter noise in Raman amplifiers, *Photonics Technol. Lett.*, **12**.5, pp. 528–530.

112. Bononi, A., Rossi, N., and Serena, P. (2011). Transmission limitations due to fiber nonlinearity, *Proceedings OFC'11*, paper OWO7.

113. Serena, P., Rossi, N., and Bononi, A. (2009). Nonlinear penalty reduction induced by PMD in 112 Gbit/s WDM PDM-QPSK coherent systems, *Proceedings ECOC'09*.

114. Wichmann, B. A., and Hill, I. D. (1982). Algorithm AS 183: An efficient and portable pseudo-random number generator, *J. R. Stat. Soc. Series C (Appl. Stat.)*, **31**(2), pp. 188–190.

115. Annexstein, F. S. (1997). Generating de Bruijn sequences: An efficient implementation, *IEEE Trans. Comput.*, **46**(2), pp. 198–200.

116. Ellis, A. D., McCarthy, M. E., Al-Khateeb, M. A. Z., and Sygletos, S. (2015). Capacity limits of systems employing multiple optical phase conjugators, *Opt. Express*, **23**(16), pp. 20381–20393.

117. Goroshko, K., Louchet, H., and Richter, A. (2016). Fundamental limitations of digital back propagation due to polarization mode dispersion, *Proceedings Asia Communications and Photonics Conference 2016*, paper ASu3F-5.

Chapter 2

Basic Optical Fiber Nonlinear Limits

Mohammad Ahmad Zaki Al-Khateeb, Abdallah Ali, and Andrew Ellis

Aston Institute of Photonic Technologies, Aston University,
Aston Triangle, Birmingham, B4 7ET, UK

alkhamaz@aston.ac.uk

In the past years, optical communication systems witnessed different developments in the fields of optical amplification, coherent receivers, digital signal processing (DSP), and high spectral efficiency modulation formats. However, these technologies have reached a point where limited improvements can be achieved due to physical considerations such as the properties of silica fibers, optical amplifiers bandwidth, power efficiency, and electronics bandwidth for real-time DSP. The limited potential improvements of optical transmission systems and the exponential growth in demand for digital communication (annual rate of 40% since 1999 [1]) have resulted in predictions of an optical capacity crunch [2, 3]. The intensive digital communication market

Optical Communication Systems: Limits and Possibilities
Edited by Andrew Ellis and Mariia Sorokina
Copyright © 2020 Jenny Stanford Publishing Pte. Ltd.
ISBN 978-981-4800-28-0 (Hardcover), 978-0-429-02780-2 (eBook)
www.jennystanford.com

competition has pushed Internet service providers to deploy optical fibers closer to the end users (Fiber to the Home, FTTH [4]) to deliver higher data rates, which leads to a higher pressure on the long-haul core networks to accommodate this demand for capacity.

Long-haul optical fiber transmission systems are designed to achieve a target distance reach and maximum capacity throughput of the fiber. During the design of long-haul optical links, both linear and nonlinear noise accumulation along the way from transmitter to the receiver must be taken into consideration; these noise terms will limit the capacity or distance reach of optical signals (with a resilient bit error rate [BER]). This limit has been commonly called the nonlinear Shannon limit [5, 6], which defines the digital capacity to optical bandwidth ratio (bps/Hz) as a function of the signal-to-noise ratio (SNR) of the received optical signals [7].

The maximum performance limits and distance reach of a long-haul optical transmission system is directly dependent on a trade-off between of the linear noise, generated from inline optical amplifiers, and nonlinear noise, generated from signal–signal interference due to optical fiber's nonlinearity. Discrete optical amplifiers (erbium-doped fiber amplifiers, EDFA [8], fiber optical parametric amplifiers, FOPA [9], discrete Raman amplifiers, DRA [10]) can be deployed in long-haul optical transmission system to compensate for the optical power loss as the signals propagate through the fiber, these amplifiers generate linear amplified spontaneous emission (ASE) noise that is dependent on the noise figure and the gain of the amplifier. The negative effect of linear noise on the quality of optical signals can be avoided by increasing the optical signals peak power; however, the susceptibility of the single-mode fiber will generate a signal power–dependent (noise-like) interference that grows cubically with the signal power; this type of noise is dependent on the properties of the fiber and the design of the optical transmission link. In order to understand the performance limits of optical fiber communication systems, we must first understand accumulation of those noise sources that are limiting the performance of the system.

In this chapter, we focus on the theoretical model that can predict the maximum performance and reach of discretely amplified long-haul coherent optical single-mode fiber transmission systems. The first section will discuss the theoretical description of the basic nonlinear Kerr effects of optical fiber and provide closed form equations that describe the accumulation of these nonlinear effects in single-span and periodically amplified multispan (dispersion unmanaged and dispersion managed) optical transmission systems. The second section of this chapter will build on the theoretical model that describes Kerr nonlinearities accumulation to derive an equation that can quantify the power spectral density of the nonlinear noise generated due to the collective signal–signal interactions and signal–ASE interactions. The theoretical models are derived with the assumption of using spectrally efficient modulated optical signals that use Nyquist wavelength division multiplexed (WDM) superchannels with negligible channel spacing [11] and densely spaced orthogonal frequency division multiplexed (OFDM) [12–14]. The final section of this chapter will identify the theoretical description of the received SNR of the optical signals after crossing long-haul discretely amplified optical transmission system (with and without ideal signal–signal nonlinearity compensation). The quantification of the received SNR will provide a tool that allows system designers to evaluate the Q factor of the coherently received signals at a given distance, choose the modulation format with highest spectral efficiency while ensuring resilient data transmission, and identify the highest distance reach that can be achieved in a particular transmission system.

2.1　Nonlinear Behavior of Optical Fibers

In literature, the nonlinear interference effects of the optical fiber were categorized into three types [15]: four wave mixing (FWM), cross phase modulation (XPM), and self-phase modulation (SPM). The three types of interference were used to distinguish the intra-channel, near inter-channel, and far inter-channel interference. In general, the collective effect generated from the three types of nonlinearities is called nonlinear Kerr effects [16], which involve the interaction between optical spectral components (w_P, w_Q, and w_R) to generate a new optical spectral component (at $w_K = w_P + w_Q - w_R$). If the frequencies of all three mixing optical components were identical ($w_P = w_Q = w_R$), then the resulting nonlinear optical field (at $w_K = w_P$) is referred to as SPM. If the frequencies of two of the mixing components were identical ($w_P = w_Q \neq w_R$), then the resulting nonlinear optical field (at $w_K = 2w_P - w_R$ and $w_K = 2w_R - w_P$) is referred to as XPM. Finally, if the frequencies of three of the mixing components were unique ($w_P = w_Q = w_R$), then the resulting nonlinear optical field (at all combinations of $w_K = w_P + w_Q - w_R$) is referred to as FWM [17]. The optical nonlinear Kerr effects produce a signal power–dependent noise-like interfering optical field fields that can diminish the performance of modulated optical signals, especially in broadband densely spaced Nyquist superchannels [18] or densely spaced OFDM systems [19].

The optical fiber is a lossy, dispersive, nonlinear transmission medium; the propagation of optical signals in optical fiber can be described by the nonlinear Schrödinger equation [10]. Since the bandwidth of optical communication signals is much smaller than the carrier frequency, the nonlinear Schrödinger equation can be simplified by assuming slowly varying wave envelope ($d^2E(w_K, z)/dz^2 \approx 0$), which then can be written as [23]

$$\frac{d}{dz}E_K(z) = -\frac{\alpha}{2}E_K(z) + i\frac{2\pi\omega_K D\chi}{nc}E_P(0)E_Q(0)E_R(0)^* \exp\left[\left(-\frac{3}{2}\alpha + i\Delta\beta\right)z\right],$$

(2.1)

where c is the speed of light, $E_X(z)$ is the optical field measured at fiber length z and located spectrally at w_X. α, χ, and n are, respectively, the loss coefficient, susceptibility, and refractive index of the optical fiber. $\Delta\beta = -\beta''(\omega_P - \omega_R)(\omega_Q - \omega_R)$ is the phase mismatching coefficient of the optical fiber which is dependent on the second-order propagation constant (β'') of the different mixing components. The term chromatic dispersion (D_c) is used in literature referring to β'', which can be described as $D_c = -\beta''c/(2\pi\lambda^2)$. The degeneracy factors (D) often arise from mathematically identical permutations of the fields in the description of certain Kerr phenomena, the factor $D = 1, 3, 6$ describes, respectively, the repetition factor of the occurrences of SPM, XPM, and FWM [20].

2.1.1 Kerr Nonlinear Effects in a Single-Span Transmission System

To calculate the quantity of the Kerr optical field generated by the end of a single span system, the first-order differential Eq. 2.1 should be solved in terms of $E_K(z)$ as [17], which yields

$$E_K(L) = i\frac{2\pi\omega_K D\chi}{nc} E_P(0)E_Q(0)E_R(0)^* \exp\left[\left(-\frac{1}{2}\alpha + i\beta_K\right)L\right] \quad (2.2)$$

$$\left[\frac{\exp[(-\alpha + i\Delta\beta)L] - 1}{-\alpha + i\Delta\beta}\right].$$

The total Kerr nonlinear power resulted from the optical field can be calculated by finding $|E_K(L)|^2$ divided by the effective area of the propagated mode in the fiber [17]. The resulting Kerr nonlinear power can be simplified into a closed form equation as [20–23]

$$P_K(L) = \left[\frac{1024\pi^6(D\chi)^2}{A_{eff}^2 n^4 \lambda^2 c^2} P_P(0)P_Q(0)P_R(0)\right]\left[\frac{\alpha^2 \exp[-\alpha L]L_{eff}^2}{\alpha^2 + \Delta\beta^2}\right] \quad (2.3)$$

$$\left[1 + \frac{4\exp[-\alpha L]\sin(\Delta\beta L/2)}{(1 - \exp[-\alpha L])^2}\right].$$

The first term in Eq. 2.3 represents the scaling effect of nonlinear factor ($\gamma = 96\pi^3\chi/A_{eff}n^2\lambda c$) of the optical fiber and the product of the mixing components powers, which suggests a cubic growth of Kerr resulting product as a function of the mixing components power (assuming they all have the same power). The second term shows that the Kerr nonlinear product depends on the total fiber loss, the effective length of the fiber $L_{eff} = |\exp(-\alpha L) - 1|/\alpha$, and the phase mismatching factor. The third term represents the coherent oscillation of the Kerr nonlinear products because of the phase mismatching accumulation during the propagation through the optical fiber.

Figure 2.1a shows a single span optical transmission system that deploy two types of optical fiber: non-zero dispersion shifted fiber (NZDSP) and standard single-mode fiber (SSMF).

Two CW lasers (the power of each one is 0 dBm) were transmitted through the span, and the spectral separation between the two CWs changes to observe the effect of phase mismatching on the power Kerr nonlinear product generated by the end of the fiber, as shown in Fig. 2.1b. The figure shows that XPM power has the highest value at strongly phase-matched mixing components (low frequency separation, $\Delta\beta \to 0$) for both types of fiber. As the frequency separation between CWs increases, the growth of phase mismatching $\Delta\beta$ between the CWs causes a degradation in the XPM product power which can be analytically concluded from the second term in Eq. 2.3. At weakly phase-matched regime ($\Delta\beta \gg 0$), the XPM power shows lower degradation slope in NZDSF (compared to SSMF) due to the fact that it has lower dispersion value (lower phase mismatch growth rate).

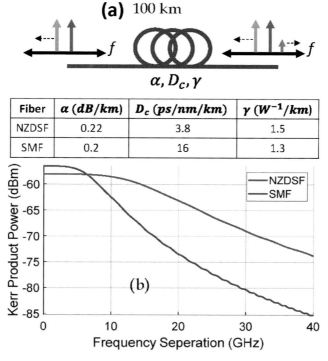

Figure 2.1 (a) Single span optical system transmitting two CW lasers (0 dBm each) through non-zero-dispersion fiber (NZDSF) or standard single mode fiber (SSMF). (b) The power of Kerr nonlinear product as a function of frequency separation between the mixing CWs.

2.1.2 Kerr Nonlinear Effects in a Multi-Span Transmission System

In a multi-span system, as shown in Fig. 2.2, an optical amplifier is located by the end of each span to compensate for the total optical power loss due to the attenuation factor (α) of the fiber. In a symmetric uniform optical transmission system, all the spans of the system have the same length (same insertion loss), and accordingly, all the inline amplifiers will provide similar gain ($G = \exp(\alpha L)$).

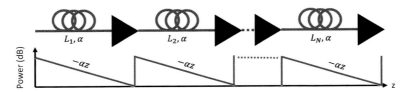

Figure 2.2 Discretely amplified long-haul optical transmission system.

As explained in the previous section, the Kerr nonlinear products field by the end of first span of a multispan transmission link can be described by Eq. 2.2. Figure 2.3 shows an optical transmission link and a conceptual diagram of Kerr nonlinear fields accumulation resulted from each span in the system. The mixing fields at the input of each span ($E_P(0)$, $E_Q(0)$, $E_R(0)$, and $E_K(0)$) will change their phase according to phase mismatching accumulation from previous spans, which explains the phase shift factor $\exp[i(\Delta\beta + \beta_K)(n-1)L]$ applied to the nonlinear Kerr field resulted from the n-th span. The nonlinear Kerr fields by the end of n-th span then propagate through the rest of the link, which results in a phase shift ($\exp[i\beta_K(N-n)L]$) due to the propagation constant of the Kerr nonlinear field. The Kerr nonlinear fields resulted from each span can be treated separately to analyze the accumulation (coherent addition [23–25]) of these fields. The nonlinear Kerr field generated in the n-th span and measured by the end of the transmission line can be written as [23]

$$E_K(L_n \to \text{end}) = i \frac{2\pi\omega_K D\chi}{nc} E_P(0)E_Q(0)E_R(0)^* \exp[i\beta_K NL] \quad (2.4)$$

$$\left[\frac{\exp[(-\alpha + i\Delta\beta)L] - 1}{-\alpha + i\Delta\beta} \right] \exp[i\Delta\beta(n-1)L].$$

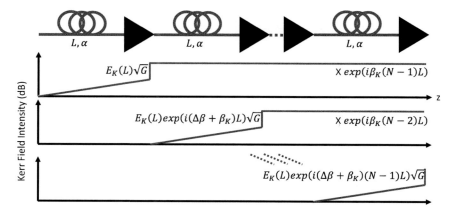

Figure 2.3 Kerr nonlinear field accumulation map for multi-span optical transmission system.

The total nonlinear Kerr field resulted by the end of the transmission link is simply the summation of all the Kerr fields generated from each span in the link, this summation can be written as

$$E_K(N) = i\frac{2\pi\omega_K D\chi}{nc} E_P(0)E_Q(0)E_R(0)^* \exp[i\beta_K NL]$$

$$\left[\frac{\exp[(-\alpha + i\Delta\beta)L] - 1}{-\alpha + i\Delta\beta}\right]\sum_{n=1}^{N}\exp[i\Delta\beta(n-1)L]. \qquad (2.5)$$

Accordingly, the total power of Kerr fields can be written as [21–23]

$$P_K(N) = \left[\frac{D^2\gamma^2}{9}P_P(0)P_Q(0)P_R(0)\right]\left[\frac{\alpha^2 L_{eff}^2}{\alpha^2 + \Delta\beta^2}\right]$$

$$\left[1 + \frac{4\exp[-\alpha L]\sin(\Delta\beta L/2)}{(1 - \exp[-\alpha L])^2}\right]\left[\frac{\sin^2(\Delta\beta NL/2)}{\sin^2(\Delta\beta NL/2)}\right]. \qquad (2.6)$$

Compared to Eq. 2.3, Eq. 2.6 introduces a new oscillating scaling term (the last term) which represents the quasi-phase matching effect of the cascaded amplifier chain. This fourth term is expressed for a uniform system with identical spans and is readily verified experimentally [26]. Figure 2.4 shows a 5-span system that transmits two CW lasers (0 dB each, with varying frequency separation) through two types of optical fiber: NZDSP and SSMF. In the strongly phase-matched region (low frequency separation between the mixing CWs, $\Delta\beta \to 0$), the power of nonlinear Kerr product scales proportional to N^2, the NZDSF system shows higher Kerr power due to the higher nonlinear factor of NZDSF compared to SSMF. At the weakly phase-matched region (high frequency separation between the mixing components, $\Delta\beta \gg 0$), the oscillating last term in Eq. 2.6 starts to degrade the power of the resulting Kerr power. The Kerr power oscillates with a frequency proportional to the squared value of the frequency separation between the mixing CWs. It can be seen that the NZDSF will oscillate in a slower pace and slower roll-off of the peak fringes compared to the SSMF system, this is due to the lower value of dispersion in NZDSF.

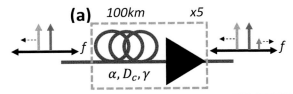

Fiber	α (dB/km)	D $(ps/nm/km)$	γ (W^{-1}/km)
NZDSF	0.22	3.8	1.5
SMF	0.2	16	1.3

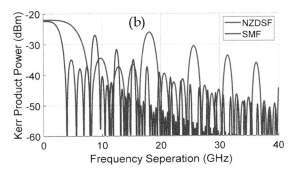

Figure 2.4 (a) 5-span optical system transmitting two CW lasers (0 dBm each) through NZDSP and SSMF. (b) Kerr nonlinear product power as a function of frequency separation between the mixing components.

For any discretely amplified system, Eq. 2.6 can be used to calculate the total nonlinear Kerr interfering products that negatively impact the quality of received optical signals by the end of an optical transmission system [18, 19]. A multispan that deploy low dispersive fiber, or uniform dispersion managed, spans were designed to use direct detection based receivers, in which the goal is to minimize the pulse broadening (dispersion accumulation) within the transmission system because of the inability to digitally compensate for pulse broadening. In dispersion-managed system, the low dispersion accumulation in the system tends to enhance the nonlinear Kerr effects by extending the phase matching bandwidths to a higher frequency separations, which can be seen from Figs. 2.1 and 2.4. Some dispersion-compensated transmission systems deploy optimized dispersion maps that result in zero total dispersion but maximizing the local dispersion accumulation within the transmission system; these dispersion maps can be realized by changing the location of the dispersion compensation fibers (DCF) [27] and optimizing the lengths of DCF [28]. In dispersion-managed transmission systems, the nonlinear Kerr effects generated from the system can be derived by following the approach in Fig. 2.3 and Eqs. 2.4–2.6. The total Kerr nonlinear field resulted in a multispan dispersion-managed system, each span contains two fibers: SSMF (length L_1) and DCF (length L_2), can be written as [25]

$$E_{\mathrm{K}}(N) = i\frac{2\pi\omega_{\mathrm{K}}D\chi}{nc}E_{\mathrm{P}}(0)E_{\mathrm{Q}}(0)E_{\mathrm{R}}(0)^* \exp[i\beta_{\mathrm{K}}NL] \tag{2.7}$$

$$\left[\frac{\exp[(-\alpha+i\Delta\beta)L]-1}{-\alpha+i\Delta\beta}\right]\sum_{n=1}^{N}\exp\left[i(n-1)\overline{\Delta\beta L}\right].$$

The power of the resulting Kerr nonlinear product can be written as

$$P_{\mathrm{K}}(N) = \left[\frac{D^2\gamma^2}{9}P_{\mathrm{P}}(0)P_{\mathrm{Q}}(0)P_{\mathrm{R}}(0)\right]\left[\frac{\alpha^2 L_{\mathrm{eff}}^2}{\alpha^2+\Delta\beta^2}\right] \tag{2.8}$$

$$\left[1+\frac{4\exp[-\alpha L]\sin(\Delta\beta L/2)}{(1-\exp[-\alpha L])^2}\right]\left[\frac{\sin^2(N\overline{\Delta\beta L}/2)}{\sin^2(\overline{\Delta\beta L}/2)}\right],$$

where $\overline{\Delta\beta L}$ is the total weighted phase mismatching:

$$\overline{\Delta\beta L} = -\left(\beta''_{SMF}L_1 + \beta''_{DCF}L_2\right)\left(\omega_P - \omega_R\right)\left(\omega_Q - \omega_R\right)$$
$$= -\beta''_{SMF}\delta(L_1 + L_2)\left(\omega_P - \omega_R\right)\left(\omega_Q - \omega_R\right), \tag{2.9}$$

where δ is the ratio of the residual dispersion resulted from the SSMF fiber. The previously described analytical approach proved to be accurate provided that $L_1 \gg L_2$, as the nonlinear Kerr products generated within the DCF (L_2) is negligible compared to the nonlinearities resulted in the SSMF (L_1), since the mixing components power entering the DCF will be attenuated by the insertion loss of the SSMF fiber. Figure 2.5a shows a 5-span dispersion-compensated system, each span consists of 100 km SSMF and varying length of DCF to show effect of residual dispersion ratios (δ). Figure 2.5b shows the resulting nonlinear Kerr product power as a function of frequency separation between the mixing CWs.

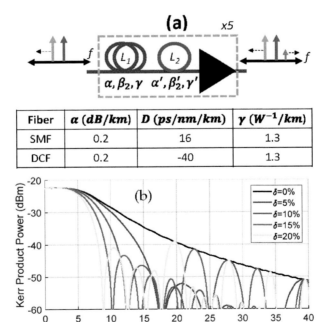

Fiber	α (dB/km)	D (ps/nm/km)	γ (W^{-1}/km)
SMF	0.2	16	1.3
DCF	0.2	-40	1.3

Figure 2.5 (a) 5-span optical system transmitting two CW lasers (0 dBm each) through SSMF dispersion compensated by DCF. (b) Kerr nonlinear product power as a function of frequency separation between the mixing components.

The DCF impacts the relative phases of the interacting signals, so the dispersion compensation ratio has a significant effect on the nonlinear Kerr power oscillation. It can be seen that fully dispersion-compensated (δ = 0%) transmission system shows nonlinear response similar to the single span system but scaled by N^2 (representing the fourth term in Eq. 2.8). At the weakly phase-matched region ($\Delta\beta$ >> 0), the nonlinear Kerr power resulted in a fully dispersion-compensated system is higher compared to dispersion-uncompensated system (δ = 100%, Fig. 2.4b), which suggests that a 0% residual dispersion ratio should be avoided. As the residual dispersion ratio increases, the phase mismatching oscillation along the system will start to show nonlinear Kerr fringes that gets closer to the response appearing in Fig. 2.4b.

2.2 Noise Accumulation Optical Transmission Systems

We have seen how, throughout the history of optical communications, it has been possible to develop accurate analytical predictions of the nonlinear Kerr products. These applied to single channel systems, with and without optical amplifiers, and to WDM systems with and without dispersion management. The performance of the coherently received optical signals from long-haul optical transmission system is dependent on the SNR value of those signals. SNR of the optical signal is degraded as the signals propagate through the transmission system because of the linear ASE noise accumulated from inline amplifiers, and the nonlinear noise generated from the Kerr effects of the optical fiber. Equation 2.8 provides the basis to calculate the nonlinear power spectral density of Kerr effects generated by the optical signals propagating through discretely amplified optical transmission system.

To fully exploit the available optical spectrum (that can be amplified by the inline amplifiers), the system designers must deploy orthogonal pulse shapes to minimize inter-channel interference [29] leading to concepts of OFDM and filter bank multicarrier [30, 31] or using signals with almost rectangular spectra. To calculate the nonlinear power spectral density resulted in OFDM optical signal, we must calculate the power of all the nonlinear Kerr products resulted from all the combinations of the subcarriers of the OFDM.

2.2.1 Total Nonlinear Kerr Noise

Wideband OFDM or densely spaced OFDM contains N_c subcarriers spectrally separated by Δf to form a total optical bandwidth $B_w = N_c \Delta f$. As OFDM passes through long-haul optical transmission system, the different subcarriers will congregate inter-subcarrier interference due to Kerr nonlinearities dominated by non-degenerate FWM which suggests degeneracy factor (D) of 6. For a long-haul optical transmission system that deploys relatively long span length (longer than 50 km) which imply that ($\exp(-\alpha L) \approx 0$ and $L_{eff}^2 = 1/\alpha^2$), surely this assumption is valid in a realistic and cost-efficient transmission system because of lower number of deployed amplifiers. Long span length can simplify FWM equation (that represents signal^2–signal interaction P_{sss}) and can be written as

$$P_{sss}(N) = 4\gamma^2 P_P P_Q P_R \left[\frac{1}{\alpha^2 + \Delta\beta^2} \right] \left[\frac{\sin^2(N\overline{\Delta\beta L}/2)}{\sin^2(\overline{\Delta\beta L}/2)} \right]. \qquad (2.10)$$

The spectral separation between the x-th and y-th subcarriers of an OFDM signal. Conventionally, all subcarriers of OFDM optical signal have the same power ($P = P_P = P_Q = P_R$) to achieve same performance for all the subcarrier. Therefore, Eq. 2.10 can be rewritten as

$$P_{sss}(N) = 4\gamma^2 P_s^3 \left[\frac{1}{\alpha^2 + (4\pi^2 \beta''_{SMF} \Delta f^2 [P-R][Q-R])^2} \right]$$
$$\left[\frac{\sin^2(2\pi^2 N \delta \beta''_{SMF} L \Delta f^2 [P-R][Q-R])}{\sin^2(2\pi^2 \delta \beta''_{SMF} L \Delta f^2 [P-R][Q-R])} \right] \qquad (2.11)$$

The spectral variables of Eq. 2.11 are the product of relative difference between the first two mixing subcarrier and a third reference subcarrier ($[P - R][Q - R]$). If we fix the spectral component f_R as a reference subcarrier that is spectrally located in the center of the OFDM populated spectrum, then we can calculate the maximum nonlinear Kerr products power since the maximum number of FWM products will aggregate in the center of the OFDM bandwidth, where these nonlinear interfering products will act as source of noise that degrades the performance of data carried on the subcarrier.

To calculate the total power of FWM products, we have to find the summation of the nonlinear Kerr products resulting from all the spectral compensation of subcarriers $[P - R]$ and $[Q - R]$, which we will call x and y since we have fixed R as reference subcarrier. As a result, the total nonlinear power generated from an OFDM signal can be written as

$$P_{\text{sss}}(N) = \frac{4\gamma^2 P_s^3}{2} \sum_{x=-N_c/2}^{N_c/2} \sum_{y=-N_c/2}^{N_c/2} \left[\frac{1}{\alpha^2 + (4\pi^2 \beta_{\text{SMF}}'' \Delta f^2 xy)^2} \right]$$
$$\left[\frac{\sin^2(2\pi^2 N \delta \beta_{\text{SMF}}'' L \Delta f^2 xy)}{\sin^2(2\pi^2 \delta \beta_{\text{SMF}}'' L \Delta f^2 xy)} \right]. \qquad (2.12)$$

The double summation of x and y will result in double counting of the combinations, which is fixed simply by dividing the total nonlinear power by a factor of 2. The summations of Eq. 2.12 can be converted to integrations over the bandwidth (B_w) of the OFDM signal, but this conversion can be valid only when the total FWM power is slowly changing as function of Δf, which enforces the following assumptions:

$$\Delta f << \left[\sqrt{\frac{\alpha}{4\pi^2 |\beta_{\text{SMF}}''|}} \approx 1.2\text{GHz} \right]$$

$$\Delta f << \left[\sqrt{\frac{1}{4\pi^2 \delta \beta_{\text{SMF}}'' L}} \approx \frac{177.3\text{GHz}}{\sqrt{\delta L}} \right] \qquad (2.13)$$

The previous assumptions are valid since the typical OFDM subcarrier separation is 100 MHz. As a result, Eq. 2.12 can be rewritten in the integration form as

$$P_{sss}(N) = \frac{2\gamma^2 P_s^3}{\Delta f^2} \int\limits_{-B_w/2}^{B_w/2} \int\limits_{-B_w/2}^{B_w/2} \left[\frac{1}{\alpha^2 + (4\pi^2 \beta''_{SMF} f_x f_y)^2} \right]$$

$$\left[\frac{\sin^2(2\pi^2 N \delta \beta''_{SMF} L f_x f_y)}{\sin^2(2\pi^2 \delta \beta''_{SMF} L f_x f_y)} \right] df_x df_y. \tag{2.14}$$

By dividing both sides of the equation by Δf, we can convert the powers (P) in the equation to power spectral density (I) of signals and FWM product ($I_s = P_s/\Delta f$ and $I_{sss} = P_{sss}/\Delta f$). Since the integration over the bandwidth of the OFDM is symmetric in reference to the central subcarrier, then the lower limit of the integrations can be changed to 0 the rescaling the equation by a factor of 4:

$$I_{sss}(N) = 8\gamma^2 I_s^3 \int\limits_{0}^{B_w/2} \int\limits_{0}^{B_w/2} \left[\frac{1}{\alpha^2 + (4\pi^2 \beta''_{SMF} f_x f_y)^2} \right]$$

$$\left[\frac{\sin^2(2\pi^2 N \delta \beta''_{SMF} L f_x f_y)}{\sin^2(2\pi^2 \delta \beta''_{SMF} L f_x f_y)} \right] df_x df_y \tag{2.15}$$

As explained in [19, Appendix A], the integration limits in Eq. 2.15 can be modified to simplify the solution of the double integration. When assuming that the OFDM subcarriers are spread over large bandwidth ($B_w > 50$ GHz), the enhancement factor (η_{sss}) of the Kerr nonlinear power spectral density can be written as (with the expansion of the last term inside the integration) [19]

$$\eta_{sss}(N) = \frac{I_{sss}(N)}{I_s^3} \approx 8\gamma^2 \int\limits_{B_0/2}^{B_w/2} \int\limits_{0}^{\infty} \left[\frac{1}{\alpha^2 + (4\pi^2 \beta''_{SMF} f_x f_y)^2} \right]$$

$$\left[N + 2\sum_{n=1}^{N-1} (N-n)\cos(4\pi^2 n \delta \beta''_{SMF} L f_x f_y) \right] df_x df_y, \tag{2.16}$$

where

$$B_0 = \frac{\alpha}{2\pi^2 |\beta''_{SMF}| B_w}. \tag{2.17}$$

Since $\cos(x) = \text{Re}[\exp(ix)]$, Eq. 2.16 can be rewritten as

$$\eta_{sss}(N) = 8\gamma^2 \int\limits_{B_0/2}^{B_w/2} \int\limits_{0}^{\infty} \left[\frac{1}{\alpha^2 + (4\pi^2 \beta''_{SMF} f_x f_y)^2} \right]$$

$$\mathrm{Re}\left[N + 2\sum_{n=1}^{N-1}(N-n)\exp(4\pi^2 in\delta\beta''_{SMF} Lf_x f_y) \right] df_x df_y. \quad (2.18)$$

If we use the complex functional analysis

$$\int\limits_{0}^{\infty} \frac{F(x)}{x^2 + a^2} dx = \frac{\pi}{2a} F(ia), \quad (2.19)$$

accordingly, then the inner integration of Eq. 2.18 can be solved to become

$$\eta_{sss}(N) = \frac{\gamma^2}{\pi\alpha|\beta''_{SMF}|} \left[N + 2\sum_{n=1}^{N-1}(N-n)\exp(-\alpha n\delta L) \right] \int\limits_{B_0/2}^{B_w/2} \frac{1}{f_y} df_y. \quad (2.20)$$

The second integration can be solved to form a logarithmic relationship with the bandwidth:

$$\eta_{sss}(N) = \frac{\gamma^2}{\pi\alpha|\beta''_{SMF}|} \left[N + 2\sum_{n=1}^{N-1}(N-n)\exp(-\alpha n\delta L) \right] \log\left(\frac{B_w}{B_0}\right) \quad (2.21)$$

And finally, the second term can be rewritten so that Eq. 2.21 becomes

$$\eta_{sss}(N) = \frac{\gamma^2}{\pi\alpha|\beta''_{SMF}|} \left[\frac{2e^{-\alpha\delta L}(N\sinh(\alpha\delta L) + e^{-\alpha\delta NL} - 1)}{(e^{-\alpha\delta L} - 1)^2} \right] \log\left(\frac{B_w}{B_0}\right), \quad (2.22)$$

where $\gamma = \gamma_0$ for a single polarization transmission, while $\gamma = 3\gamma_0/8$ for a dual polarization transmission system [32], γ_0 is the intrinsic nonlinear factor of the fiber. Note that strictly, the natural logarithms in Eq. 2.22 can be replaced with inverse hyperbolic sin [33] to have an accurate nonlinear power spectral density on OFDM signals that have total bandwidth lower than 20 GHz. For a system that is not fully loaded, noise contributions the integrals need to take into account gaps between the signals [33, 34]. For a totally uniform input spectrum, gain flat amplifiers (and noise) and wavelength-independent fiber loss, these noise spectral densities (with the exception of nonlinear noise) are approximately uniform within the signal band.

The second term of Eq. 2.22 illustrates the difference between dispersion-managed (δ < 1) and dispersion-unmanaged (δ = 1) SSMF optical transmission systems. It can be clearly seen that dispersion-managed systems will have an enhanced nonlinearity efficiency (compared to a dispersion-unmanaged system). This observation is consistent with the conclusions that we have made from Eq. 2.8 and Fig. 2.5. For a dispersion-unmanaged optical transmission system (δ = 1), the second term of Eq. 2.22 will simplify to be a scaling factor of N (assuming large span length, L > 50 km). On the other hand, fully dispersion managed (δ = 0), the enhancement factor will simplify to be a scaling factor of $(N + 1)/2$. Figure 2.6 shows the normalized enhancement factor ($[second\ term\ of\ 2.21]/N$) as a function of the residual dispersion (δ) in a dispersion-managed optical system. It can be seen from the figure, if the dispersion-managed SSMF transmission system has a residual dispersion ratio greater than 60%, then the resulting nonlinear power spectral density will be close to one of a dispersion-uncompensated optical system. As the residual dispersion goes below 40%, the nonlinear power spectral density starts to be enhanced (compared to dispersion-unmanaged system) until it reaches enhancement factor of $(N + 1)/2$ when the residual dispersion becomes 0%.

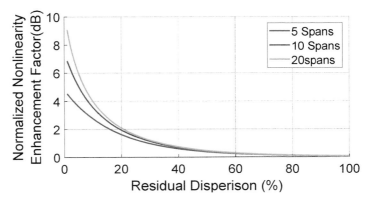

Figure 2.6 Normalized enhancement factor as a function of the residual dispersion ratio for a transmission system that uses SSMF fiber (α = 0.2 dB/km, L = 100 km).

2.2.2 Total Linear ASE Noise

It is well known that optical amplifiers, as it amplifies optical signals, add ASE noise to the optical signal across the amplification bandwidth, and there is no way of avoiding (or recovering) this noise. The ASE noise added by optical amplifiers is related to the quantum limit of those amplifiers. EDFA is one of the widely used optical amplifiers, where it uses optical fiber doped with rare earth element erbium, the fiber doping will lead to allow energy transfer from the pump to the signals. The total ASE noise generated from an EDFA (that has a gain of G) can be written as (for a dual polarization transmission system)

$$I_{ASE} = (G[NF] - 1)h\upsilon,\tag{2.23}$$

where h is the Planck constant, υ is the light frequency, and NF is the noise figure and can be written as (in a quantum-beat limited regime) [35]

$$NF = \frac{SNR_{out}}{SNR_{in}} = \frac{2n_{sp}(G-1)}{G} + \frac{1}{G},\tag{2.24}$$

where ni_{sp} is the spontaneous emission factor (ideally = 1, NF = 2 [3 dB]). If the EDFA is used in a single polarization transmission system, then I_{ASE} is scaled back to half of its value. In long-haul optical transmission systems, the total gain of each EDFA located by the end of each span compensates for the total span loss $(\exp(\alpha L))$. As a result, the total linear noise resulted in a system of N spans would be (NI_{ASE}). Ideal EDFAs maintain their full inversion state $(n_{sp} = 1)$ which suggests that the accumulated ASE noise in a transmission system is always lower when deploying a shorter span length (lower G). On the other hand, the nonlinear interference among signals propagating through the fiber increases with the deployment of shorter span length. As a result, there is a trade off optimization can be made by changing the span length which leads to significant capacity improvements [36].

Other types of discrete amplifiers try to find other techniques to reduce NF, one of the best examples are parametric phase-sensitive amplifiers (PSA), which can reach a noise figure in the order of −1.9 dB [37] leading to the enhancement of transmission reach or capacity of the system. While PSAs can achieve significant improvements, implementing multilevel PSA (for higher order QAM modulation formats) can involve a complex setup [38] and lacks the stability of the EDFA.

2.2.3 Total Signal-ASE Nonlinear Noise

The enhancement factor of the nonlinear Kerr power spectral density introduced in Eq. 2.22 results from the signal–signal interaction. These interactions can be fully recovered either in the electronic domain (digital back propagation DBP [39]) or in the optical domain (optical phase conjugation OPC [40]). Nonlinearity compensation of signal–signal interactions can be realized by DBP if the coherent receiver is able to coherently receive the full bandwidth (B_w) optical signals so that it can perform full-field backpropagation in the digital domain. On the other hand, full signal–signal nonlinearity compensation using OPC can be realized by placing mid-link OPC or equally spaced multi-OPC in a link that satisfies optical power profile symmetry in reference to the point of conjugation [41–44].

DBP or OPC assisted links can mitigate the deterministic Kerr effects leaving the system (that uses polarization multiplexed modulated signals) limited by the nondeterministic nonlinear Kerr effects of signal–signal interaction due to the influence of polarization mode dispersion (PMD) [45, 46]. If the transmission system uses optical fiber spans that has high PMD factor, then the improvements (in the nonlinear regime) of DBP-assisted system can be limited due to the stochastic randomness of the signal–signal interactions which cannot be predicted by the DBP [45]. Mid-link and multi-OPC (deployed in power and dispersion symmetric system) can show higher improvements (compared to DBP system) since it reduces the signal–signal polarization decorrelation length resulted by the PMD [46]. The influence of PMD can be avoided by using either low PMD or polarization maintaining (PM) fibers, which will unveil another nondeterministic Kerr effect due to the signal–ASE nonlinear Kerr interactions (known as the Gordon–Mollenauer effect [47] or parametric noise amplification [41]). In this chapter, we will concentrate on the signal–ASE nonlinear Kerr interaction limit in full-field DBP-assisted systems, where we believe it is the ultimate nonlinear limit of digital nonlinearity compensated systems.

FWM describes the interaction between three OFDM subcarriers (f_P, f_Q, and f_R) to generate a fourth tone at ($f_K = f_P + f_Q - f_R$) leading to subcarrier interference. In the case of a noise field superimposed on optical signal fields, the power spectral density generated by all combinations of the nonlinear Kerr interactions can be written as

$$I_{sss,ssn,snn,nnn}(n) = (I_P + I_p)(I_Q + I_q)(I_R + I_r)\eta_{sss}(n), \tag{2.24}$$

where I_Y is the power spectral density of the OFDM subcarrier Y, while I_y is the ASE noise power spectral density imposed under the Y-th OFDM subcarrier, $\eta_{sss}(n)$ is the Kerr nonlinear efficiency resulted by the propagation through n spans (described by Eq. 2.22). $I_{sss,ssn,snn,nnn}(n)$ is the total power spectral density of Kerr effects resulted by the superimposition of ASE noise on OFDM subcarriers which represents, respectively, signal–signal–signal, signal–signal–noise, signal–noise–noise, and noise–noise–noise interactions. If we expand Eq. 2.24, it will result in

$$\begin{bmatrix} I_{sss}(n) \\ I_{ssn}(n) \\ I_{snn}(n) \\ I_{nnn}(n) \end{bmatrix} = \begin{bmatrix} (I_pI_QI_R) \\ (I_pI_QI_r) + (I_pI_qI_R) + (I_pI_QI_R) \\ (I_pI_qI_r) + (I_pI_QI_r) + (I_pI_qI_R) \\ (I_pI_qI_r) \end{bmatrix} \eta(n)$$

$$= [I_s^3 + 3I_s^2 I_{ASE} + 3I_s I_{ASE}^2 + I_{ASE}^3]\eta(n). \tag{2.25}$$

The first row in the squared bracket of Eq. 2.25 represents signal–signal–signal Kerr power spectral density (as in Section 2.2.1), which can be fully compensated by nonlinear compensation techniques (if we assume zero PMD effects). The second row represents Kerr nonlinear power spectral density resulted from signal–signal–noise interaction [41, 45, 48]. The third and fourth rows represent the power spectral density of Kerr effects results, respectively, from signal–noise–noise interactions and noise–noise–noise interactions and may usually be neglected. If we assume that the ASE noise power spectral density is uniform across the amplification bandwidth and all of the OFDM subcarriers have the same power, then, we can conclude that the signal- signal-noise

Kerr power spectral density has extra degeneracy factor of 3 multiplied by the Kerr nonlinearity efficiency (described in Eq. 2.22). This Kerr nonlinear power spectral density efficiency for OFDM signal I_s copropagating with ASE noise I_{ASE} along n spans can be written as

$$I_{ssn}^{1st\text{-}order}(n) = 3I_s^2 I_{ASE}\eta_{sss}(n) \tag{2.26}$$

To analyze the accumulation of first-order signal–ASE nonlinear Kerr interactions (the first-order distinction will be clarified later in the paragraph), we must follow signal copropagating with ASE noise added by each EDFA all the way from transmitter to receiver, and the backpropagation process in the digital domain. Figure 2.7 shows transmission system and the accumulation of signal–signal–noise nonlinear Kerr power spectral density. From the figure, we can see that the signal–signal–signal nonlinear Kerr interactions can be fully recovered using DBP (assuming ideal DBP); this also applies to the signal–ASE (of the first EDFA) nonlinear Kerr interactions since OFDM subcarriers and ASE noise emitted from the first EDFA co-propagate and co-backpropagated (in DBP) through the same fiber distance. The ASE noise generated from the second amplifier co-propagate with signal for three spans and co-backpropagate for four spans; this will result in an over-compensation of the signal–ASE interactions equivalent to one span ($\eta(1)$), we will call this interaction the first-order signal–ASE interaction (represented by filled triangles in Fig. 2.7). The ASE noise generated from the third amplifier will co-propagate with signals over two spans and co-backpropagate for four spans; this will result in an over-compensation of the signal–ASE interactions to result in first-order interactions equivalent to two spans ($\eta(2)$). Furthermore, by the end of the first overcompensated span, the first-order signal–ASE interaction will also interact with OFDM signals over the second overcompensated span to result in second-order signal–ASE interactions (represented by black arrows in Fig. 2.7). The second-order signal–ASE interaction can be written (for n overcompensated spans) as [49]

$$I_{ssn}^{2nd\_Order}(n) = 3I_s^2 [3I_s^2 I_{ASE}\eta_{sss}(1)]\sum_{i=1}^{n-1}\eta_{sss}(n) \tag{2.27}$$

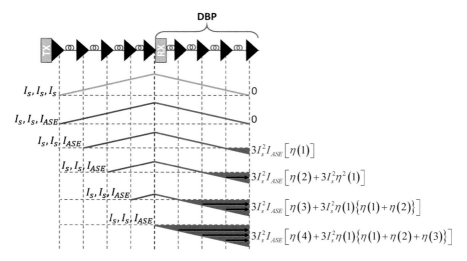

Figure 2.7 Kerr nonlinear power spectral density of signal-signal-noise interaction accumulation in a multi-span optical transmission system.

The second-order signal–ASE interactions result from a seed of first-order signal–ASE products interacting with OFDM subcarriers and it must be taken into account since it (the first-order signal ASE interactions) can reach level of power higher than the power spectral density of the ASE noise resulted by a single EDFA [49]. The second-order signal–ASE interactions (as it can be seen from Eq. 2.27) grows quartically as a function of the power spectral density of the OFDM signals. The ASE noise added by the fourth and fifth EDFAs results in signal–ASE interactions (first and second order) due to the overcompensation of the DBP to the optical fields for three and four spans, respectively, which will cause three and four span overcompensations leading to generate first-order signal–ASE interactions ($I_{\text{ssn}}^{(\text{1st-order})}(3)$ and $I_{\text{ssn}}^{(\text{1st-order})}(4)$) and second-order signal–ASE interactions ($I_{\text{ssn}}^{(\text{2nd-order})}(3)$ and $I_{\text{ssn}}^{(\text{2nd-order})}(3)$).

The summation of all signal–ASE nonlinear Kerr power spectral densities (including both first and second order) can be mathematically formulated for a system with N spans and $N + 1$ EDFAs (extra EDFA to boost the optical signals from the transmitter), which results in the following total enhancement factor [49]:

$$\eta_{ssn}(N) = \frac{I_{ssn}(N)}{3I_s^2 I_{ASE}} = \sum_{n=1}^{N} \eta_{sss}(n) + 3I_s^2 \eta_{sss}(1) \sum_{n=1}^{N} \sum_{i=1}^{n-1} \eta_{sss}(i) \qquad (2.28)$$

where the first term represents the enhancement factor of the first-order signal–ASE Kerr power spectral density, and the second term represents the enhancement factor of the second-order signal–ASE Kerr power spectral density. Equation 2.28 can be simplified for dispersion-uncompensated transmission systems that uses large span length, where the second term of Eq. 2.22 can be simplified to be equal to N. As a result, the efficiency signal–ASE power spectral efficiency can be simplified from the summation form to be

$$\eta_{ssn}(N) = \frac{\gamma^2 N(N+1)}{2\pi\alpha |\beta_{SMF}''|} \left[1 + \frac{\gamma^2 I_s^2 (N-1)}{\pi\alpha |\beta_{SMF}''|} \log\left(\frac{B_w}{B_0}\right) \right] \log\left(\frac{B_w}{B_0}\right) \qquad (2.29)$$

The equation shows that the first-order signal–ASE power spectral density grows quadratically with both the number of the amplifiers in the system and the OFDM signal power spectral density. On the other hand, the second-order signal–ASE power spectral density grow cubically with the number of spans and grows quartically as a function of the OFDM signal power spectral density.

2.3   Performance of Coherently Detected Optical Transmission Systems

The performance of optical modulated signals depends on the transmission design which defines the total noise generated along the system.

An optimized system would launch optical power into the system to achieve the highest possible SNR, by balancing between the nonlinear inter/intra-channel interference and ASE noise. Coherent detection allows the receiver to detect the in-phase and quadrature of the optical field on each polarization, the full optical field detection opens four dimensions that can carry bipolar multi-level modulation formats. The fact that the receiver can detect the quadrature's of the optical field allows for digital compensation of linear pulse broadening due to the dispersion of the optical fiber. Modulation formats of optical signals can have different spectral efficiencies (bps/Hz) by choosing a constellation with 2^M points [50], the constellations with higher number of points (higher M) requires higher SNR [7].

At the receiver side of the optical transmission system, the optical field contains modulated signals superimposed on linear optical ASE noise, nonlinear signal–signal–signal Kerr noise, and nonlinear signal–signal–noise Kerr noise. As a result, the SNR of the received optical signal at the end of N-span optical transmission system can be written as

$$\text{SNR} = \frac{I_s}{NI_{\text{ASE}} + I_s^3 \eta_{\text{sss}}(N) + [3I_s^2 I_{\text{ASE}} + 3I_s I_{\text{ASE}}^2 + I_{\text{ASE}}^3]\eta_{\text{ssn}}(N)} \quad (2.30)$$

The noise terms in denominator of Eq. 2.30 represents, respectively, the total linear ASE noise, the nonlinear signal–signal–signal, signal–signal–noise, signal–noise–noise, noise–noise–noise nonlinear Kerr interfering noise. The definition of noise terms in Eq. 2.30 has developed over the past two decades [48, 49, 51–53]. The signal–signal–signal nonlinear Kerr noise generated within a system that does not deploy nonlinearity compensation (DBP) is dominant over the signal–noise and noise–noise nonlinear limit $I_s^3 \eta_{\text{sss}}(N) \gg [3I_s^2 I_{\text{ASE}} + 3I_s I_{\text{ASE}}^2 + I_{\text{ASE}}^3]\eta_{\text{ssn}}(N)$. As a result, the third term of the denominator of Eq. 2.30 can be dropped. On the other hand, a DBP-assisted transmission system (deploying low PMD fiber) can eliminate the signal–signal–signal nonlinear Kerr noise $(I_s^3 \eta_{\text{sss}}(N) \approx 0)$. SNR for an optical system can be optimized by changing the optical power launched into the optical fiber, at low I_s levels SNR is considered to be operating in the linear regime $(NI_{\text{ASE}} \gg I_s^3 \eta_{\text{sss}}(N))$, while at high I_s levels SNR is considered to be operating in the nonlinear regime $(NI_{\text{ASE}} \ll I_s^3 \eta_{\text{sss}}(N))$.

The SNR calculation in Eq. 2.30 uses Eq. 2.24 to describe total ASE noise, Eq. 2.22 to describe the nonlinear Kerr enhancement factor (for nonlinearity uncompensated system), and Eqs. 2.28 and 2.29 to describe the signal–ASE nonlinear Kerr enhancement factor (for DBP-assisted systems). The assumptions that we have made for these equations were long span length ($L > 50$ km), large signal bandwidth ($B_w > 20$ GHz), zero PMD fiber, and full-field DBP (large coherent detection bandwidth) with large number of steps per span in the backpropagation process (> 1 step/km).

To verify Eq. 2.30, we have conducted a transmission system simulation using VPITransmissionMaker 9.5. The system transmits 8×28 Gbaud PM-QPSK through 12 spans (amplified by 6 dB noise figure EDFAs) of three common dispersion maps: SSMF (100 km spans), NZDSF (100 km spans), and SSMF+DCF (100 km spans), resulting in residual dispersion of 5% of the SSMF. All the fibers simulated in the previously described system have zero PMD parameter. The system uses conventional DSP at the receiver side to compensate for fiber's dispersion, as well as a full-field DBP with 120 steps/span in case of a nonlinearity compensated transmission system. Figure 2.8 shows the Q^2 factors of the received QPSK (which is equal to the SNR value) as a function of the launched signal power into the fiber, the figure compares theoretical prediction from Eq. 2.30 (lines) for the different system (different dispersion maps, DBP enabled or not) to the simulation results (dots).

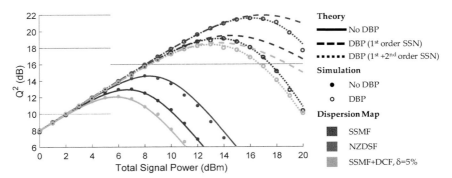

Figure 2.8 Simulated and predicted Q^2 as a function of the launched optical signal power to a lumped system with EDFA noise figure of 6 dB and passing 8×28 Gbaud PM-QPSK Nyquist WDM system over 12 spans of 100 km of fibers with different dispersion maps: SSMF, NZDSF, SSMF+DCP with 5% residual dispersion.

It can be seen from the figure that lower accumulated chromatic dispersion in the link will lead to a degradation in system performance both for the case of receiver compensating only for chromatic dispersion and for the case of receiver compensating for nonlinearities using DBP. The figure shows that full-field DBP can improve the Q^2 factor by 7 dB for the SSMF link, 6 dB for the NZDSF link, and 6.4 dB for the 95% dispersion-compensated SSMF. The improvements achieved by DBP are directly linked to the fact that DBP has fully recovered all the signal–signal nonlinear Kerr noise, but it is clear from the figure that the transmission system is still limited by the signal–signal–noise interactions (while signal–noise–noise and noise–noise–noise limit is still ignorable). The theoretical SNR calculation that considers both the first- and second-order signal–ASE interactions represents more accurate prediction of the nonlinear regime of the DBP-assisted system compared to the theoretical SNR calculations that consider only the first-order signal–signal–noise interactions. The second-order signal–signal–noise interactions cause a degradation of the SNR with a rate of I_s^3 (3 dB$_{SNR}$/dB$_{Pow}$). In general, Fig. 2.8 shows a good agreement between the simulation results and the theoretical predictions.

Figure 2.9 shows the Q^2 factor as a function of distance of the same system that was described before, but now we have fixed the signal power to the optimum power. The figure shows that the system that accumulates higher chromatic dispersion can achieve higher distance that a system with lower accumulation of fiber dispersion along the link. It can be seen that the Q^2 difference between the SSMF and the SSMF+DCF (with δ = 5%) starts with 2 dB at 1000 km distance, but this difference increases to almost 3 dB as the distance reaches 5000 km. This deviation can be explained by the fact that as the residual dispersion (δ) gets closer to 0%, the second term in Eq. 2.22 starts to scale the nonlinear power spectral density quadratically ($N(N + 1)/2$) instead of linearly (N) for dispersion-uncompensated transmission system (δ = 100%). For the DBP system, it can be seen that the theoretical prediction considering only the first-order signal–signal–noise interaction can be inaccurate, especially when passing through higher number of spans, but including the second-order signal–signal–noise interactions will get the predictions to be accurate representation of the simulation results. As the number of spans in a DBP-assisted transmission system increases, the Q^2 crosses the performance of a system that does not deploy DBP; this is directly related to the fact that the signal–signal–noise interactions grow in a quartic polynomial manner with the number of spans (as it can be seen in Eq. 2.29). Figures 2.8 and 2.9 illustrate the importance of considering the second-order signal–signal–noise Kerr nonlinear noise.

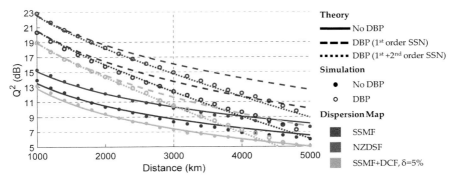

Figure 2.9 Simulated and predicted Q^2 as a function of distance (at optimum power identified by Fig. 2.9) of a discretely amplified system with EDFA noise figure of 6 dB and passing 8 × 28 Gbaud PM-QPSK Nyquist WDM system over 100 km spans of fibers with different dispersion maps: SSMF, NZDSF, SSMF+DCP with 5% residual dispersion.

In this chapter, we have reviewed the calculation of the maximum performance of discretely amplified coherent transmission system. By explaining the basic nonlinear Kerr effects of optical fiber, we were able to conceptualize the accumulation of the nonlinear noise generated in spectrally efficient modulated optical signals. The analysis of the different types of noise accumulation in a transmission system has helped to define the SNR of the received optical signals, which directly relates to the performance, reach, and capacity of the optical system. We have seen that dispersion-compensated optical transmission system tends to perform worse than the dispersion-uncompensated transmission systems, because the nonlinear Kerr effects shows higher efficiency over mixing bandwidth with lower accumulated dispersion. The basic nonlinear Shannon nonlinear limit can be expanded by compensating inter/intra-channel nonlinearities (using DBP); the transmission system will still be limited by the nondeterministic signal–signal–noise nonlinear interactions. The limits introduced in this chapter provide basic conceptual limits of the transmission system with and without nonlinearity compensation. We have made different assumptions in order to derive the closed formulas in the chapter. Practical implementations of transmission systems might deviate from the theoretical limits especially in DBP-assisted system, but those limits will still be valid once an ideal nonlinearity compensation system is implemented.

Acknowledgments

This work is partially funded by the Engineering and Physical Sciences Research Council (EPSRC) (EP/J017582/1-UNLOC, EP/L000091/1-PEACE).

References

1. A. D. Ellis, N. Mac Suibhne, D. Saad, and D. N. Payne, Communication networks beyond the capacity crunch, *Phil. Trans. R. Soc. A*, 374(2062), 20150191 (2016).

2. D. J. Richardson, Filling the light pipe, *Science*, 330(6002), 327–328 (2010).

3. H. Shinohara, Broadband access in Japan: Rapidly growing FTTH market, *IEEE Commun. Mag.*, 43(9), 72–78, Sept. (2005).

4. FTTH Handbook, Fibre to the Home Council Europe, Feb. (2016).

5. J.-C. Antona, Key technologies for Present and Future Optical Networks, presented at Topical Workshop on Electronics for Particle Physics Plenary Session 5, Paris, France, 21–25 September (2000).

6. A. D. Ellis, J. Zhao, and D. Cotter, Approaching the non-linear shannon limit, *J. Lightwave Technol.*, 28(4), 423–433 (2010).

7. C. E. Shannon, A mathematical theory of communication, *Bell Syst. Tech. J.*, **27**(379–423), 623–656 (1948).

8. D. N. Payne, L. Reekie, R. J. Mears, S. B. Poole, I. M. Jauncey, and J. T. Lin, Rare-earth doped single-mode fiber lasers, amplifiers, and devices, *Conference on Lasers and Electro-Optics*, G. Bjorklund, E. Hinkley, P. Moulton, and D. Pinnow, eds., OSA Technical Digest (Optical Society of America (1986), paper FN1.

9. R. Stolen and J. Bjorkholm, Parametric amplification and frequency conversion in optical fibers, *IEEE J. Quantum Electronics*, 18(7), 1062–1072, Jul (1982).

10. A. Hasegawa, Numerical study of optical soliton transmission amplified periodically by the stimulated Raman process, *Appl. Opt.*, 23, 3302–3309 (1984).

11. K. Igarashi, T. Tsuritani, I. Morita, Y. Tsuchida, K. Maeda, M. Tadakuma, and M. Suzuki, Super-Nyquist-WDM transmission over 7,326-km seven-core fiber with capacity-distance product of 1.03 Exabit/s km, *Opt. Express*, 22(2), 1220–1228 (2014).

12. T. Kan, K. Kasai, M. Yoshida, and M. Nakazawa, 42.3-Tbit/s, 18-Gbaud 64QAM WDM Coherent Transmission of 160 km over Full C-band using an Injection Locking Technique with a Spectral Efficiency of 9 bit/s/Hz, *Optical Fiber Communication Conference*, OSA Technical Digest Series (Optical Society of America, 2017), paper Th3F5.

13. D. Hillerkuss, R. Schmogrow, T. Schellinger, M. Jordan, M. Winter, G. Huber, T. Vallaitis, R. Bonk, P. Kleinow, F. Frey, M. Roeger, S. Koenig, A. Ludwig, A. Marculescu, J. Li, M. Hoh, M. Dreschmann, J. Meyer, S. Ben Ezra, N. Narkiss, B. Nebendahl, F. Parmigiani, P. Petropoulos, B. Resan, A. Oehler, K. Weingarten, T. Ellermeyer, J. Lutz, M. Moeller, M. Huebner, J. Becker, C. Koos, W. Freude, and J. Leuthold, 26 Tbit s-1 line-rate super-channel transmission utilizing all-optical fast Fourier transform processing, *Nat. Photonics*, 5(6), 364–371 (2011).

14. S. Chandrasekhar, X. Liu, B. Zhu, and D. W. Peckham. Transmission of a 1.2-Tb/s 24-carrier no-guard-interval coherent OFDM superchannel over 7200-km of ultra-large-area fiber. In Proceedings of 35th European Conference and Exhibition of Optical Communication (IEEE, 2009).

15. G. P. Agrawal, *Nonlinear Fiber Optics*, 2nd ed. (Academic Press, 1995).

16. R. H. Stolen and A. Ashkin, Optical Kerr effect in glass waveguide, *Appl. Phys. Lett.*, 22, 294 (1973).

17. K. O. Hill, D. C. Johnson, B. S. Kawasaki, and R. I. MacDonald. cw three-wave mixing in single-mode optical fibers, *J. Appl. Phys.*, 49(10), 5098–5106 (1978).

18. G. Bosco, V. Curri, A. Carena, P. Poggiolini, and F. Forghieri, On the performance of nyquist-WDM terabit superchannels based on PM-BPSK, PM-QPSK, PM-8QAM or PM-16QAM subcarriers, *J. Light. Technol.*, 29(1), 53–61 (2011).

19. X. Chen and W. Shieh, Closed-form expressions for nonlinear transmission performance of densely spaced coherent optical OFDM systems., *Opt. Express*, 18(18), 19039–19054 (2010).

20. N. Shibata, R. Braun, and R. Waarts, Phase-mismatch dependence of efficiency of wave generation through four-wave mixing in a single-mode optical fiber, *IEEE J. Quantum Electronics*, 23(7), 1205–1210, Jul (1987).

21. A. D. Ellis and W. A. Stallard, Four wave mixing in ultra long transmission systems incorporating linear amplifiers, in Proceedings of the IEE Colloquium on Non-Linear Effects in Fibre Communications (IEE, 1990), 6/1–6/4.

22. D. G. Schadt, Effect of amplifier spacing on four-wave mixing in multichannel coherent communications, *Electronics Lett.*, 27(20), 1805–1807 (1991).

23. K. Inoue, Phase-mismatching characteristic of four-wave mixing in fiber lines with multistage optical amplifiers, *Opt. Lett.*, 17, 801–803 (1992).

24. W. Zeiler, F. Di Pasquale, P. Bayvel, and J. E. Midwinter, Modeling of four-wave mixing and gain peaking in amplified WDM optical communication systems and networks, *J. Lightwave Technol.*, 14(9), 1933–1942 (1996).

25. K. Inoue and H. Toba, Fiber four-wave mixing in multi-amplifier systems with nonuniform chromatic dispersion, *J. Lightwave Technol.*, 13(1), 88–93 (1995).

26. C. Kurtzke, Suppression of fiber nonlinearities by appropriate dispersion management, *IEEE Photonics Technol. Lett.*, 5(10), 1250–1253 (1993).

27. M. E. Marhic, N. Kagi, T. K. Chiang, and L. G. Kazovsky, Optimizing the location of dispersion compensators in periodically amplified fiber links in the presence of third-order nonlinear effects, *IEEE Photonics Technol. Lett.*, **8**(1), 145–147 (1996).

28. K. Nakajima, M. Ohashi, K. Shiraki, T. Horiguchi, K. Kurokawa, and Y. Miyajima, Four-wave mixing suppression effect of dispersion distributed fibers, *J. Lightwave Technol.*, 17(10), 1814–1822 (1999).

29. R. R. Mosier and R. G. Clabaugh, Kineplex, a bandwidth-efficient binary transmission system, *Trans. Am. Inst. Electrical Eng. Part I: Commun. Electronics*, **76**(6), 723–728 (1958).

30. H. W. Chang, Synthesis of band-limited orthogonal signals for multichannel data transmission, *Bell System Tech. J.*, 45(10), 1775–1796 (1966).

31. B. Farhang-Boroujeny, OFDM Versus Filter Bank Multicarrier, *IEEE Signal Proc. Mag.*, 28(3), 92–112 (2011).

32. W. Shieh and X. Chen, Information spectral efficiency and launch power density limits due to fiber nonlinearity for coherent optical OFDM systems, *IEEE Photonics J.*, 3(2), 158–173, April (2011).

33. G. Bosco, A. Carena, V. Curri, P. Poggiolini, and F. Forghieri, Performance limits of nyquist-WDM and CO-OFDM in high-speed PM-QPSK systems, *IEEE Photonics Technol. Lett.*, **22**(15), 1129–1131 (2010).

34. P. Poggiolini, The GN model of non-linear propagation in uncompensated coherent optical systems, *J. Lightwave Technol.*, 30(24), 3857–3879, Dec.15 (2012).

35. E. Desurvire, *Erbium Doped Fiber Amplifiers: Principles and Applications*, John Wiley & Sons, N. Y. (1994).

36. N. J. Doran and A. D. Ellis, Minimising total energy requirements in amplified links by optimising amplifier spacing, *Opt. Express*, 22, 19810–19817 (2014).

37. Z. Tong, et al., Towards ultrasensitive optical links enabled by low-noise phase-sensitive amplifiers, *Nat. Photonics*, 5(7), 430–436 (2011).

38. S. L. I. Olsson, B. Corcoran, C. Lundström, T. A. Eriksson, M. Karlsson, and P. A. Andrekson, Phase-sensitive amplified transmission links for improved sensitivity and nonlinearity tolerance, *J. Lightwave Technol.*, 33(3), 710–721, Feb.1, 1 (2015).

39. X. Li, X. Chen, G. Goldfarb, E. Mateo, I. Kim, F. Yaman, and G. Li, Electronic post-compensation of WDM transmission impairments using coherent detection and digital signal processing, *Opt. Express*, 16, 880–888 (2008).

40. A. Yariv, D. Fekete, and D. M. Pepper, Compensation for channel dispersion by nonlinear optical phase conjugation, *Opt. Lett.*, 4, 52–54 (1979).

41. A. D. Ellis, M. E. McCarthy, M. A. Z. Al-Khateeb, and S. Sygletos, Capacity limits of systems employing multiple optical phase conjugators, *Opt. Express*, **23**, 20381–20393 (2015).

42. A. D. Ellis, M. E. McCarthy, M. A. Z. Al Khateeb, M. Sorokina, and N. J. Doran, Performance limits in optical communications due to fiber nonlinearity, *Adv. Opt. Photon.*, 9, 429–503 (2017).

43. A. D. Ellis, M. A. Z. Al Khateeb, and M. E. McCarthy, Impact of optical phase conjugation on the nonlinear shannon limit, *J. Lightwave Technol.*, 35(4), 792–798, Feb.15, 15 (2017).

44. A. D. Ellis et al., The impact of phase conjugation on the nonlinear-Shannon limit: The difference between optical and electrical phase

conjugation, *IEEE Summer Topicals Meeting Series (SUM), Nassau,* 209–210 (2015).

45. G. Gao, X. Chen, and W. Shieh, Influence of PMD on fiber nonlinearity compensation using digital back propagation, *Opt. Express*, **20**, 14406–14418 (2012).

46. M. E. McCarthy, M. A. Z. Al Khateeb, and A. D. Ellis, PMD Tolerant nonlinear compensation using in-line phase conjugation, *Opt. Express*, **24**(4) 3385–3392 (2016).

47. J. P. Gordon and L. F. Mollenauer, Phase noise in photonic communications systems using linear amplifiers, *Opt. Lett.*, 15, 1351–1353 (1990).

48. D. Rafique and A. D. Ellis, Impact of signal-ASE four-wave mixing on the effectiveness of digital back-propagation in 112 Gb/s PM-QPSK systems, *Opt. Express*, 19, 3449–3454 (2011).

49. M. A. Z. Al-Khateeb, M. McCarthy, C. Sánchez, and A. Ellis, Effect of second order signal–noise interactions in nonlinearity compensated optical transmission systems, *Opt. Lett.*, 41, 1849–1852 (2016).

50. V. Arya, I. Jacobs, Optical preamplifier receiver for spectrum sliced WDM, *J. Lightwave Technol.*, **15**(4), 576–583 (1997).

51. P. P. Mitra and J. B. Stark, Nonlinear limits to the information capacity of optical fibre communications, *Nature*, 411, 1027–1030 (2001).

52. K. S. Turitsyn, S. A. Derevyanko, I. V. Yurkevich, and S. K. Turitsyn, Information capacity of optical fiber channels with zero average dispersion, *Phys. Rev. Lett.*, 91, 203901–203904 (2003).

53. H. Louchet, A. Hodzic, and K. Petermann, Analytical model for the performance evaluation of DWDM transmission systems, *IEEE Photonics Technol. Lett.*, 15(9), 1219–1221 (2003).

Chapter 3

Fiber Nonlinearity Compensation: Performance Limits and Commercial Outlook

Danish Rafique

*Adva Optical Networking SE, Fraunhoferstr. 9a,
82152 Martinsried, Munich, Germany*

drafique@advaoptical.com

With recent advances in consumer technologies, optical networks are experiencing rapid exhaustion of transmission capacity [1–2]. Optical transport equipment vendors and network carriers have been challenged in terms of not only addressing exponentially increasing bandwidth demands but also stagnating revenue streams, leading to so-called *capacity-revenue crunch*. Consequently, upcoming optical communication products are expected to carefully trade off network equipment and operational expenditures with the overall system performance, and it is now more important than ever to utilize the installed network infrastructure to the fullest, before overhauling the entire network infrastructure [3].

Optical Communication Systems: Limits and Possibilities
Edited by Andrew Ellis and Mariia Sorokina
Copyright © 2020 Jenny Stanford Publishing Pte. Ltd.
ISBN 978-981-4800-28-0 (Hardcover), 978-0-429-02780-2 (eBook)
www.jennystanford.com

On the commercial end, successful deployments of multiple generations of 100 Gb/s polarization multiplexed quadrature phase shifted keying (PM-QPSK) line interfaces have been followed up by announcements of line systems incorporating polarization multiplexed quadrature amplitude modulated (PM-mQAM) transmission formats, supporting lines rates of 200 Gb/s, 400 Gb/s, and even 1 Tb/s [4–6]. However, these spectrally efficient modulation schemes are intrinsically limited in performance and transmission reach due to higher optical signal-to-noise ratio (OSNR) requirements [7]. For instance, PM-16QAM doubles the network capacity, compared to PM-QPSK, but at the cost of a 70% reach reduction. Consequently, from a commercial deployment viewpoint, modulation schemes beyond PM-16QAM are largely limited to short-reach transport applications [8].

In order to alleviate such spectral efficiency × reach restrictions, and meet the *capacity-revenue* sweet spot, fiber nonlinearity compensation (NLC) has emerged as one of the most vital approach, and a logical supplement to the currently commercial optical transmission products, as it allows for significant system performance improvements, with diminutive network footprint [9–10].

In this chapter, we discuss two key NLC techniques, namely, digital back propagation (DBP) and phase conjugation (PC). We establish ideal performance bounds and draw parallel for different configurations of these schemes, and finally offer industry perspective on application of NLC in commercially deployed optical transport networks.

3.1   Fiber Nonlinearity Compensation

Figure 3.1 depicts the basic concept of optical fiber NLC technique. The signal stream, $E(0, T)$, traversing through the optical fiber cable—termed as forward transmission, is impaired by multitude of physical transmission effects, including both linear effects like chromatic dispersion, and Kerr induced nonlinear fiber impairments. While today's commercial systems employ linear compensation (LC) techniques to sufficiently mitigate linear channel impairments, NLC is required to address nonlinear fiber distortions. The most straightforward, yet comprehensive, NLC approach to address such distortions is termed as backward propagation (Fig. 3.1), which processes the distorted signal with a channel response inverse to that of forward transmission, typically including both linear and nonlinear impairments. In case only nonlinear response is considered in NLC technique, linear distortions are treated separately in the system.

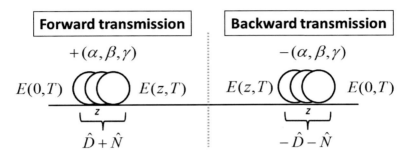

Figure 3.1 Conceptual diagram for forward (real) fiber transmission, and backward (virtual) fiber transmission. E, D, N, and α represent signal electric field, dispersion operator (includes β), fiber nonlinearity operator (includes γ), and fiber loss parameters [11].

Figure 3.2 shows a schematic diagram of optical transmission system, presenting several locations to place the optical, digital, or optoelectronic flavor of the NLC module. In a conventional system, without NLC, digitally processed electrical signal is mapped to modulation format of choice, and converted to analog domain using a digital to analog converter (DAC). The output signal stream is then used to modulate an in-phase and quadrature-phase carrier. For the optical part, a continuous wave laser is used as the light source for a dual-parallel nested Mach–Zehnder modulator (MZM) structure. The output of several such modules is optically multiplexed, resulting in a dense wavelength division multiplexed (DWDM) system. The DWDM signal is typically transmitted over a dispersion-unmanaged link, employing erbium doped fiber amplifier (EDFA)-based lumped signal amplification. Finally, at the receiver, the signal-of-interest is de-multiplexed using an integrated coherent receiver (ICR), digitized using an analog to digital converter (ADC). The linear transmission impairments are then digitally compensated using standard digital signal processing (DSP) algorithms. In the case of fiber nonlinearity compensation, in addition to the conventional setup, NLC may be employed in either digital or optical domain or as a combination of the two domains. Furthermore, NLC may be placed at the transmitter, within the link or at the receiver (see Fig. 3.2). Note that a combination of multiple modules may also be employed, termed as a hybrid NLC, as reported in [12]. In this chapter, we focus on the two most representative and widely investigated NLC techniques, i.e., digital back propagation and phase conjugation. While these two techniques can be placed in any number of positions and combinations, as shown in Fig. 3.2, we consider receiver-side and link placement for DBP and PC, respectively.

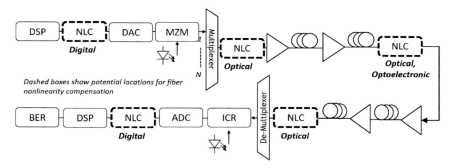

Figure 3.2 Schematic diagram of a coherent optical transmission system with DSP-enabled transmitter and receiver. Multiple potential locations of fiber nonlinearity compensation module are shown [10].

3.2   Digital Back Propagation

DBP is a widespread NLC solution, which inverses the fiber channel response by plainly back-propagating the distorted signals and launching them into a digital virtual fiber—as opposed to real fiber [13–16]. The virtual fiber operates on fiber parameters of opposite-sign to those in the transmission channel. In practice, this method can be implemented in a way similar to that of forward fiber channel modeling, using the nonlinear Schrödinger equation (NLSE). For back-propagation, this equation is numerically solved using the split-step Fourier method (SSFM)—see Chapter 2, where

α, β, and γ parameters are set to be the opposite values to those in the transmission fiber, and the optical amplifiers are replaced with digital attenuators. The SSFM representation for DBP may be given as

$$\frac{\partial E(z,t)}{\partial z} = (-\widehat{D} - \widehat{N})E(z,t). \tag{3.1}$$

Figure 3.3 shows a typical EDFA-based 3-span optical transmission system. It can be seen that during forward propagation, the signal is severally distorted by multitude of transmission effects—see constellation diagram at the receiver, whereas receiver-based DBP reverses the channel impairments, leading to signal quality similar to that of transmitter. The effects of signal distortion are also depicted qualitatively by signal power profile as a function of transmission, where broadening of transmission pulses is counteracted by the DBP method. Note that in the absence of noise, back-propagation method can fully compensate the linear and nonlinear fiber impairments to arbitrary precision, and is typically used to benchmark other NLC methods. However, typically, DBP-based NLC does not address non-deterministic channel impairments, arising from the interaction of Kerr nonlinearities with amplified spontaneous emission (ASE) noise generated in the optical amplifiers [17]. While these effects limit the maximum potential performance of DBP-based algorithms, more optimum detectors, e.g., maximum a posteriori (MAP) and Viterbi detectors, have been proposed in literature to minimize these effects [18–19].

As discussed in Chapter 1, DBP may be employed using asymmetric SSFM solution, where linear operator is followed by the nonlinear operator or vice versa, or symmetric SSFM solution, where the linear operator is split in two halves, placed around the nonlinear operator. Figure 3.4a shows the symmetric approach, where dispersion is calculated in two steps each accounting for half of the accumulated dispersion per step. At each step, channel gain is applied, while forward transmission amplification is reversed on per span basis (Fig. 3.4b). The entire transmission link is traversed by repeating these steps, as shown in Fig. 3.4c. This leads to distributed compensation of channel impairments, effectively reconstructing the transmission channel with precise linear and nonlinear steps, albeit with signs opposite to that of transmission. This is in contrast to traditional channel impairment compensation, for instance, dispersion compensation only, which acts on the whole system at once.

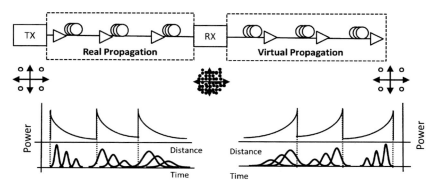

Figure 3.3 Typical EDFA-based lumped amplified transmission system employing digital back propagation. Qualitative representation of pulse broadening as a function of transmission distance and DBP-based compensation.

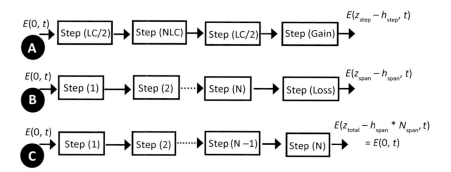

Figure 3.4 Split step Fourier method-based solution of NLSE for digital back propagation. (a) Per DBP step, (b) per span, and (c) per link. E, z, h, and t represent signal electric field, step/span distance, step/span size, and time instant, respectively [11].

Note that digital back-propagation necessitates full knowledge of forward physical channel response and signal power evolution at the receiver. However, in practical cases, these system parameters are not precisely known, and some of these effects are even time-varying in reconfigurable systems—necessitating the need for an adaptive algorithm for DBP optimization. The most straightforward approach is to iteratively apply a range of values for critical system parameters, including span length, dispersion, and nonlinear parameters and choose the most optimum configuration. However, this brute force approach is highly cumbersome and not a favorable solution for practical systems. On the other hand, the use of linear filters for chromatic dispersion approximation is an accepted practice and may reduce one degree of complexity for adaptive DBP. More concretely adaptive optimization of DBP parameters with limited link knowledge has been shown to optimally work for DBP-based optical transmission systems using steepest decent algorithms, etc. [20–22].

The DBP approach may broadly be divided in to two categories: (1) single-channel NLC (SC-NLC) or narrowband NLC and (2) multichannel NLC (MC-NLC) or wideband NLC. SC-NLC operates on a single channel-of-interest and allows for full compensation of intra-channel fiber nonlinearities (Fig. 3.5a). On the other hand, MC-NLC enables compensation of inter-channel effects, in addition to intra-channel nonlinear fiber impairments. MC-NLC may manifest itself either in form of single-receiver detection (Fig. 3.5b)—necessitating wideband optical and electronic components, compensating for first-order inter-channel nonlinearities from immediate channel neighbors; or phase-locked multireceiver detection (Fig. 3.5c), allowing total field reconstruction and ideal deterministic impairment mitigation. Note that a phase independent MC-NLC (Fig. 3.5c) may also compensate for all deterministic channel impairments, except four wave mixing-based nonlinear sums and products.

Note that today's commercial optical and electronic subsystems can only support –3 dB bandwidth up to ~30–40 GHz, limiting the use-cases for single-receiver-based MC-NLC. Furthermore, the application of back-propagation in WDM systems is dependent on the system architecture. Figure 3.6a shows a point-to-point link, where all the deterministic inter-channel nonlinear effects like cross-phase modulation and four-wave mixing are compensable using MC-NLC (Fig. 3.5c). On the other hand, in a typical meshed network (Fig. 3.6b), application of digital back-propagation is limited since co-propagating traffic may get added and/or dropped at various reconfigurable optical add drop multiplexers (ROADM) nodes, and consequently the receiver may not have access to all the channels which originated at the same network node.

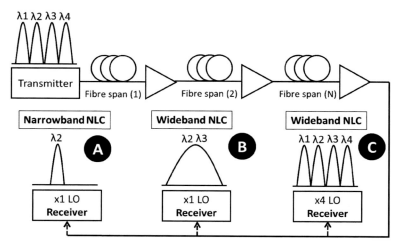

Figure 3.5 Various configurations of digital back propagation. (a) SC-NLC: Narrowband NLC for intra-channel nonlinearity compensation. (b) Wideband NLC for limited inter-channel nonlinearity compensation. (c) MC-NLC: wideband NLC for full inter-channel nonlinearity compensation.

In particular, two potential key issues could be the following: First, at a particular node, even in case the receiver may detect a few channels out of many originating at the same point, the effect of digital back propagation may be substantially minimized, or even incorrect. Second, the traffic added or dropped at intermediate nodes has not traversed the full path from original node to the receiver, and therefore when the target signal employs digital back-propagation at the receiver, spurious nonlinear terms might be produced. Both uncompensated and spurious nonlinearities may further deteriorate the received signal of interest, particularly if the added and/or dropped channels are next to the desired channel, or even worse, if the channels are added or dropped near the drop node of the target channel.

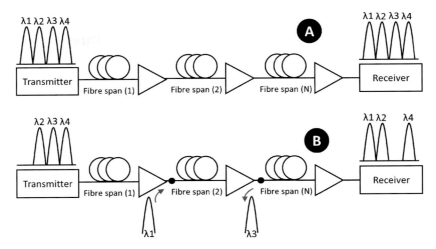

Figure 3.6 Application of digital back propagation in different network topologies. (a) Point-to-point transmission. (b) Meshed network transmission. Dark circles represent network nodes [11].

Consequently, application of digital back-propagation to compensate inter-channel nonlinearities is largely limited in dynamic networks. However, a potential approach could be to employ DBP aware routing and wavelength assignment algorithms, allowing channel co-propagation requirement as an additional system optimization criterion. Nonetheless, both from network topology and complexity viewpoint, typically SC-NLC is considered as the most viable NLC approach [10].

3.2.1   DBP Performance Scaling

In this section, we report the performance gains from DBP-based NLC for PM-mQAM modulation formats over a dispersion-unmanaged transmission link. In particular, we establish performance limits of LC, SC-NLC and MC-NLC (multireceiver) techniques, as described earlier. Figure 3.7a depicts the Q-factor as a function of launch power per channel for repeater less transmission of WDM 100 Gb/s PM-256QAM system. It can be clearly seen that compensation of nonlinear fiber impairments, using MC-NLC, enables substantial performance improvements, compared to both LC and SC-NLC methods. On the other hand,

SC-NLC is seen to enable miniscule performance improvement, compared to LC-based system. This behavior may be attributed to signal strongly limited by inter-channel fiber nonlinearities– not compensable by SC-NLC for formats with multiple signaling levels, such that intra-channel effects are not dominant.

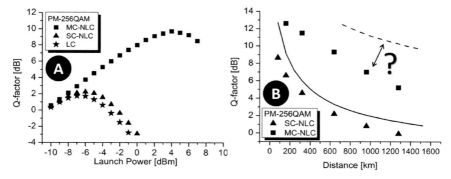

Figure 3.7 (a) Q-factor versus fiber launch power for 100 Gb/s PM-256QAM transmission system. LC (Stars), SC-NLC (up-triangle), and MC-NLC (squares) [23]. (b) Q-factor as a function of transmission distance. Solid line: Inter-channel theoretical limit (XPM+FWM). Dashed line: Inter-channel theoretical limit (XPM).

Nonetheless, the three curves follow the typical system optimization behavior, where the Q-factor peaks at an optimum launch power, referred to as nonlinear threshold (NLT). This trend is a well-known consequence of uncompensated or partially compensated fiber nonlinearities for an LC or a SC-NLC-based system, respectively. On the other hand, for an ideal MC-NLC-based system one would expect all the deterministic impairments to be fully compensated, and Q-factor either growing as a linear function of input power or saturating beyond a certain breakeven point between OSNR and nonlinearity limited regions. However, as it seems, a NLT can still be observed after MC-NLC, alluding to the fact that certain non-deterministic channel distortions, arising from nonlinear signal interactions with amplifier induced ASE noise, are not compensable by MC-NLC and set the upper bound on system performance. Nonetheless, according to Fig. 3.7a, the NLT is increased by 10 dB, with a related Q-factor improvement of ~8 dB and ~7.4 dB, when LC and SC-NLC are replaced by MC-NLC, respectively.

In Fig. 3.7b, we plot analytically calculated Q-factor curves, along with simulated data points, as a function of transmission distance for a PM-256QAM transmission system, employing SC-NLC and MC-NLC. The analytical curves are derived from [24–25], and include the effects of self phase modulation, cross phase modulation and four wave mixing. The results show that when intra-channel nonlinearities are compensated by SC-NLC, the analytical and numerical data shows excellent agreement— transmission system limited by the combined effect of XPM and FWM. However, for MC-NLC, where both intra- and inter-channel nonlinearities are compensated, it can be observed that the simulated performance does not even agree with performance bound imposed by XPM theory. This discrepancy is evident due to the fact that in the analytical models considered, non-deterministic channel nonlinearities were ignored as a first order approximation, consequently resulting in the model over-estimating the achievable Q-factor. In summary, even when deterministic XPM and FWM effects are fully compensated using MC-NLC, a substantial performance degradation is observed due to the non-deterministic interactions of signal and ASE noise—this phenomenon will be further explained in next section.

In Fig. 3.8 we extend the application of LC, SC-NLC and MC-NLC techniques across multiple QAM formats. In order to enable a fair comparison, the Q-factor after MC-DBP is fixed at ~9.8 dB, and corresponding Q-factor and NLTs are calculated for LC and SC-NLC systems for a given modulation format at a fixed transmission distance. The figure shows that, compared to LC, the benefit of SC-NLC reduces with increasing modulation order, whereas, the gain from MC-NLC increases with modulation complexity. This implies that the rate of return from employing NLC is much greater for short-reach applications, employing spectrally efficient modulations. These results provide an important insight into NLC methods, originally designed to improve transmission performance or maximum reach for niche ultra long-haul transmission applications; however, as shown here, NLC can enable substantial increase in capacity, especially in context of short distances. It is also worth mentioning that while MC-NLC is not practical for meshed network, as discussed earlier, it best suits the needs of short-reach transmission applications such as data center interconnect (DCI), where typically no intermediate network traffic add-drop is expected [26].

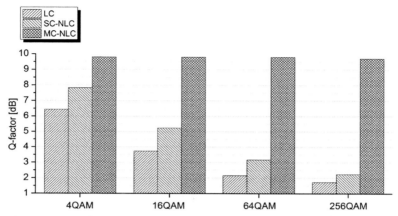

Figure 3.8 Performance scaling for WDM PM-mQAM, employing LC, SC-NLC, and MC-NLC. Performance is normalized to *Q*-factor of 9.8 dB after MC-NLC.

Finally, we discuss the application of LC, SC-NLC and MC-NLC across various optical market segments; typically categorized into four sectors, i.e., short reach, metro networks, regional networks and ultra long-haul networks. Figure 3.9 summarizes the findings of results presented earlier in this section by allocating the relevant technology to the appropriate market. It can be seen that while relatively low capacity modulation formats like 4QAM need not employ any form of NLC, highly spectrally efficient complex modulation formats needs one or the other form of NLC to enable application across different sectors.

Application	4QAM	16QAM	64QAM	256QAM
DCI*	LC	LC	LC	LC
Metro	LC	LC	LC	SC-NLC
Regional	LC	LC	SC-NLC	MC-NLC
Ultra Long-haul	LC	SC-NLC	MC-NLC	NA
Submarine	LC	MC-NLC	NA	NA

Figure 3.9 Summary of application of linear and nonlinear compensation in various market segments, including LC, SC-NLC and MC-NLC [11].

3.2.2   DBP Performance Limits

In this section, we aim to identify the source of non-deterministic performance limits—after digital back propagation, observed in previous section. Note that the existence of these bounds is not a new phenomenon for systems without NLC techniques. For instance, for dispersion-managed systems such effects were termed as Gordon–Haus [27] and Gordon–Mollenauer [28] effects, and as FWM [29], and in fact lead to widespread deployment of high local dispersion fiber infrastructures to minimize these distortions. While such solutions have been quite successful in counteracting the non-deterministic impairments for legacy systems, the rise of NLC-based coherent systems, allowing significantly higher transmission reaches, compared to traditional systems, has resulted in a substantial increase in the impact of such impairments.

Here, we consider an exemplary system, employing 100 Gb/s PM-4QAM and wideband NLC, and discuss the nature and extent of nonlinear interaction between signal and ASE. Figure 3.10a shows the system performance in terms of Q-factor as a function of launch power. As expected, the system is initially OSNR limited, before reaching the NLT, beyond which performance is limited by channel nonlinearities. While deterministic impairments are adequately compensated by NLC, it is the non-deterministic distortion that sets the performance bound, as also seen in Fig. 3.7. Note that in Fig. 3.10a, a flat response is seen from launch power of ~–3 dBm to ~10 dBm since no errors were detected in simulations due to limited number of bit counts. Nonetheless, the existence of an optimum launch power after wideband NLC suggests that the non-deterministic impairments not only limit the OSNR but in fact substantially degrade the system transmission performance beyond a given NLT. In order to confirm our hypothesis, we repeat the above transmission without any inline noise from amplifiers—equivalent ASE added at the receiver. It can be seen that the two cases in Fig. 3.10a overlap in OSNR-limited region; however as the launch power is increased, the Q-factor continues to improve with input launch power if inline amplifier noise is switched off, confirming that the signal-ASE is a distributed effect.

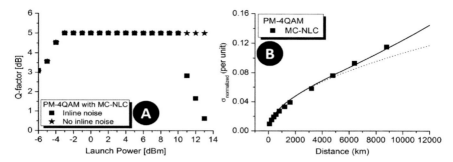

Figure 3.10 (a) *Q*-factor versus fiber launch power for single-channel 100 Gb/s PM-4QAM system employing SC-NLC. With distributed noise (squares), and without inline noise—equivalent noise added at receiver (stars). (b) Constellation variance as a function of transmission distance. Solid symbols: numerical data, solid line: signal-ASE FWM theory, dashed line: no inline nonlinearity [17].

The next question we address is the origin of such signal-ASE interactions, which is likely to be either XPM or FWM. Recently nonlinear interactions for OFDM signals have been analytically derived, where strong and weak phase-matched regions have been treated as separate contributors. Here we follow a similar approach, except that we replace one of the signal fields with noise component. This results in closed-form expressions for the parametric amplification of ASE [17], and may be given as follows:

$$K_{\text{Parametric}} = K_{\text{Signal}} \times K_{\text{ASE}}^2 (P_1 + P_2), \tag{3.2}$$

where, $K_{\text{Paramteric}}$ is the nonlinear noise power spectral density (PSD), K_{Signal} is the signal PSD, K_{ASE} is the linear noise PSD. P_1 and P_2 signify weakly and strongly phase-matched regimes, respectively [30]. For an *N*-span system, where each amplifier is assumed to contribute as an independent random variable with Gaussian statistics, the accumulated nonlinear noise is

$$K_{\text{Total}}^2 = N^2 K_{\text{ASE}}^2 + \sum_{M=1}^{N} K_{\text{Paramteric}}^2. \tag{3.3}$$

Figure 3.10b plots the evolution of the normalized constellation variance, for results in Fig. 3.10a. Also plotted is the variance calculated from Eq. 3.3. It can be seen that the numerical data show excellent agreement with analytical prediction of signal-ASE-based on FWM. On the other hand, Fig. 3.10b also shows the expected evolution of constellation variance in the absence of nonlinearity, corresponding to equivalent noise loading at the receiver. Comparing the data sets shows that in the studied system, the signal-ASE FWM-based nonlinear noise dominates after ~5,000 km.* One caveat that needs mentioning is that the DBP-based NLC approach also intrinsically adds distributed nonlinear noise. This phenomenon is intuitively depicted in Fig. 3.11, where the signal power is reversed by DBP as expected (Fig. 3.11a). On the contrary, while the noise contributions from first amplifier, all the way to the link end, are symmetrically counteracted by DBP; the noise terms added from intermediate amplifiers are only compensated up to their point of origin, beyond which they increase signal distortion by contributing artificially enhanced noise contributions. In summary, one should employ Eq. 3.3 to correctly consider the combined effects of forward transmission and digital back-propagation.

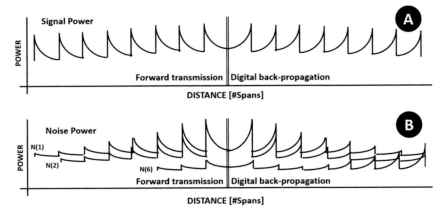

Figure 3.11 (a) Evolution of signal power profile as a function of transmission distance for a lumped EDFA-based system. (b) Evolution of noise component profile versus transmission distance for a lumped EDFA-based system [34].

---

*Note that the absolute values of the non-deterministic channel bound depends on interplay of several system parameters, e.g., polarization mode dispersion [31], regenerative transformations [32], channel memory [33], etc.

3.3   Phase Conjugation

In the previous section, we established that optical transmission performance is strongly limited by nonlinear fiber impairments for an LC-based system, whereas application of DBP substantially enhances the signal quality, and effectively the system reach and/or capacity. While DBP is a powerful algorithm to mitigate channel nonlinearities, for practical implementations, it necessitates substantial requirements on circuit size and power consumption [10, 35]. Previously, nonlinear fiber distortion has also been addressed using phase conjugation or spectral inversion. Different methods have been proposed to enable PC, including optical (periodically poled lithium niobate waveguides, etc.) [36], and electronic (based on intradyne detection) devices [37].

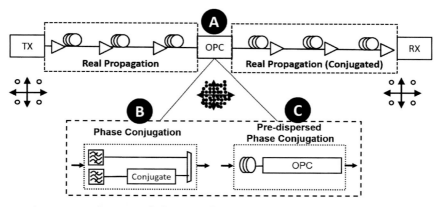

Figure 3.12 Conceptual diagram for application of phase conjugation in optical transmission systems. (a) 6-span transmission system, with PC placed at the center of the link. (b) Standard phase conjugation approach. (c) Pre-compensated phase-conjugated approach based on standard PC.

In its ideal form, phase conjugation is employed at the center of the link, and compensates fiber nonlinearities by conjugating the signal at the input of the PC device. Mathematically, this conjugation results in inverted signs of fiber dispersion and nonlinear parameters, a concept similar to that of DBP [38]. Figure 3.12a depicts a 6-span optical transmission system, where PC is employed after 3 spans. In one of the PC implementations, in the middle of the link, the signal is passed through a wavelength selective switch and is spectrally inverted, as shown in Fig. 3.12b. An improvement on the standard phase conjugation is shown in Fig. 3.12c, where the signal of interest is pre-dispersed before going through the PC stage [39]. In order to study the performance limits of the scheme, ideal spectral inversion is implemented by reversing the sign of the imaginary part of the signal. The signal is then re-multiplexed with its unprocessed neighbors. Similar to DBP, PC can also be employed in single-channel or multichannel configurations based on the properties of PC device, and availability of multiple co-propagating channels, as discussed in Section 3.2.

3.3.1 Pre-Dispersed PC

In order to enable optimum transmission performance, phase conjugation requires multiple criterions to be fulfilled: (1) mid-link placement, such that nonlinearities accumulated in first half of transmission link are symmetrically compensated in the second half, (2) symmetric link design, such that Raman-based amplification is used with a symmetric power profile in two halves of the link, (3) constant power profile, such that shortly spaced amplifiers lead to near constant power evolution on per span basis. In order to circumvent stringent requirements imposed by the conventional phase conjugation approach, pre-dispersed PC (PPC) has been proposed in literature, relaxing all of the above conditions by temporally aligning the two halves of the link using a small amount of pre-compensation before or after the typical PC stage. Without the loss of generality, in this work, we employ PPC before conventional PC, and review the operating principle of PPC.

Figure 3.13a shows a 6-span transmission example, where power evolution is shown as a function of transmission distance. Figure 3.13b shows the corresponding accumulated dispersion for a dispersion-unmanaged link as a function of reach, and shows the deployment of standard PC device exactly at the center of the link. The dark filled circles represent the fiber span effective length, representing the region of highest signal span power. It can be seen that after PC, the effective length regions are asymmetric, before and after PC, leading to a nonlinear compensation mismatch.

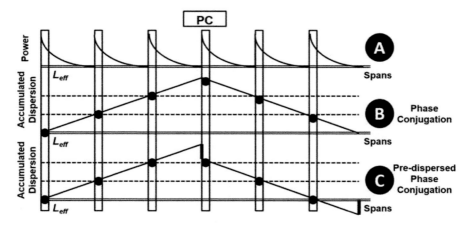

Figure 3.13 (a) Signal power evolution for a lumped EDFA-based system, (b) Accumulated dispersion as a function of transmission distance, employing standard phase conjugation, (c) Accumulated dispersion as a function of transmission distance, employing pre-compensated phase conjugation. Dark circles present effective length per span [12].

On the other hand, in case of PPC (Fig. 3.13c), a small amount of pre-compensation (using a dispersion compensating fiber or fiber Bragg grating) is added before the conventional PC stage, effectively aligning the high power regions before and after phase conjugation. PPC enables optimum performance for typical dispersion-unmanaged systems where fiber effective length (L_{eff}) is less than the amplifier spacing (L_{amp}), and the dispersion length (L_D) is sufficiently small ($L_D < L_{amp}$), thereby adding dispersion asymmetry. Note that the additional dispersion must be compensated at receiver, using either optical or electronic linear compensation methods.

3.3.2 Comparison of Single-Channel DBP and PC

In this section, we establish the performance gains from standard and pre-compensated phase-conjugated techniques and benchmark it against narrow band or single-channel digital back propagation. We study two transmission links with no inline dispersion compensation with either SSMF or NZDSF fiber. Furthermore, we employ ideal SC-NLC; however, similar results may be achieved using state-of-the-art algorithms with substantially reduced circuit complexity [40–41]. The transmission system comprised 28 Gbaud WDM channels, employing PM-mQAM formats, equally spaced at 50 GHz. In order to emulate next-generation network upgrade scenario, the target channel was set to be PM-64QAM, whereas neighboring channels were 100 Gb/s PM-4QAM.

Figure 3.14 depicts the performance after LC, PC, PPC, and SC-NLC. Generally, as the launch power is increased, different nonlinear compensation approaches show different NLTs and Q-gains—compared to LC only, confirming differences in their capabilities to mitigate channel nonlinearities. Figure 3.14a shows that, for SSMF, SC-NLC enables maximum system performance, followed by PPC, PC and LC. It is clear that the dispersion pre-compensation employed for PPC helps enable substantial performance improvements over the conventional PC approach. On the other hand, although PPC enables maximum Q-factor similar to SC-NLC, at higher launch powers it slightly diverges from its electronic counterpart (~0.3 dB) due to higher local dispersion of SSMF, leading to dispersion induced asymmetry on the two sides of PPC. However, this trend is reversed in case of low-dispersion LEAF fiber (Fig. 3.14b), where PPC is highly

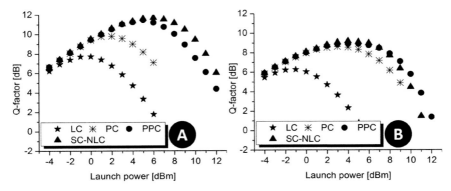

Figure 3.14 Q-factor versus fiber launch power for PM-64QAM signal with 100 Gb/s PM-4QAM co-propagating traffic, employing LC (stars), PC (asterisk), PPC (circles), and SC-NLC (up-triangle). (a) SSMF, and (b) LEAF [42].

symmetrical; however, DBP suffers from increased non-deterministic phase matching. Also note that the performance difference between PC and PPC is much narrower for LEAF fiber—compared to SSMF, since PPC corrects for higher dispersion inaccuracy in case of high local dispersion. Nonetheless, this analysis gives an interesting insight into the nature of trade-offs system designers need to consider for intra-channel fiber nonlinearity mitigation, i.e., SC-NLC prefers higher local dispersion (low L_D), minimizing non-deterministic distortion, whereas PPC favors lower link dispersion (high L_D) to avoid nonlinear asymmetric regions.

3.4   Commercial Applications and Perspective

In the final section of this chapter, we offer commercial perspective on applicability of NLC techniques in meshed networking scenarios. While it has been clear from results presented in this chapter, and from state-of-the-art literature, that NLC enables substantial gains, typically it is perceived otherwise. This is mainly due to two reasons: First, NLC gains are usually reported in worst-case configurations, leading to a pessimistic system view. Second, the associated complexity for low gains is considered too high. In this section, we focus on the real-world commercial use cases and review two key application scenarios, including homogeneous and heterogeneous networks, and dispersion-unmanaged and dispersion-managed link infrastructures.

Commercial networks typically follow a pay-as-you-grow model, where additional capacity is added on need basis, as opposed to over designed network on day one. Consequently, system engineers need to carefully balance legacy traffic with the latest technologies upgrading the network, as this leads to heterogeneous network configurations, which are not always optimum from specific system performance viewpoint but are the best available network solutions. This phenomenon has lately been seen in optical networks supporting legacy 10 and 40 Gb/s data rates, which not only continue to be upgraded with 100 Gb/s, but also with even higher data rate traffic of 200 Gb/s PM-16QAM, etc. In this context, it becomes apparent that a wide mix of data rates and modulation formats will give rise to diverse launch power and OSNR optima—typically higher order formats requiring higher power. It is thus important to evaluate the performance of NLC in such heterogeneous network architectures, where a higher data rate signal is introduced in a pre-existing network traffic topology. Figure 3.15a shows a conceptual diagram for this scenario, where mixed modulation formats may be launched at homogenous power or heterogeneous power. As mentioned earlier, the choice of launch power in itself is a system design parameter, and based on given topology, either self-favoring or altruistic launch power may be employed.

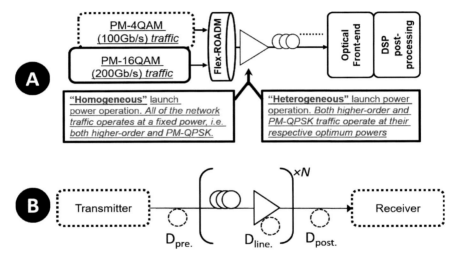

Figure 3.15 (a) Homogeneous and Heterogeneous launch power operation for mixed modulation format and mixed data rate network configurations. (b) Exemplary dispersion-managed optical transmission system employing transmitter, inline, and receiver-based dispersion compensation [10].

On the other hand, while the advent of coherent technologies have almost eliminated the need of inline dispersion compensation, most of the legacy links infrastructure are in fact dispersion managed, since they were primarily designed for 10 and 40 Gb/s traffic. This implies that while spectrally efficient modulation schemes favor transmission over NDM links, they will eventually end up on DM transmission paths simply due to lack of NDM links in a given network, or due to requirements enforced by their neighboring legacy traffic. Figure 3.15b shows a schematic diagram of such a link infrastructure, where dispersion management may be employed using transmitter dispersion pre-compensation, inline compensation, and receiver-based dispersion post-compensation, based on the system configuration and interplay of optical channel nonlinearities on a given path.

Figure 3.16a shows an example of heterogeneous traffic operation, where 200 Gb/s PM-16QAM is co-propagated with 100 Gb/s PM-4QAM, employing DBP-based SC-NLC. Here we consider both homogenous launch configuration—where all the traffic is operated at same launch power—and heterogeneous launch condition—where launch power PM-4QAM is fixed at 0 dBm. It can be seen that for the conventional homogenous configuration, SC-NLC enables ~1.3 dB performance gain, consistent with typically reported SC-NLC gain. In contrast, for heterogeneous systems SC-NLC gain of up to ~3.7 dB is observed, a significant increase from homogenous system. These results confirm that if upgrade traffic is allocated sufficient OSNR, SC-NLC may allow substantial performance gains. It is worth mentioning that the gain between 1.3 and 3.7 dB may depend on the overall system optimization based on performance of legacy traffic; however, even in case PM-16QAM traffic is operated at homogenous power of 0 dBm, SC-NLC gain of ~2 dB is obtained.

Figure 3.16b considers a heterogeneous system similar to Fig. 3.16a, and plots Q-factor as a function of the inline dispersion compensation ratio. As expected, performance degrades as the inline compensation ratio is increased; however, the SC-NLC gain increases with inline dispersion compensation ratio. This may be attributed to enhanced channel nonlinearities for highly phase-matched dispersion-managed links, where LC fails to mitigate any nonlinear impairments, whereas SC-NLC adequately compensates intra-channel nonlinearities. These results provides a powerful insight that while NLC algorithms were originally designed for dispersion-unmanaged links, the rate of return in dispersion-managed links is actually much higher than the NDM counterpart. Nonetheless, note that even the best-vase SC-NLC performance for a 95% DM link is ~3 dB worse than LC-based

NDM link, confirming that going forward, newly deployed links should be exclusively NDM.

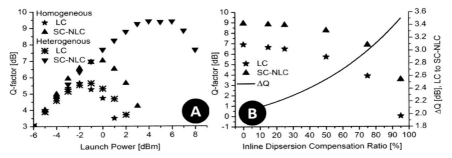

Figure 3.16 (a) Q-factor versus fiber launch power for 200 Gb/s PM-16QAM co-propagating with 100 Gb/s PM-4QAM. Homogeneous transmission: LC (stars) and SC-NLC (up-triangle), Heterogeneous transmission: LC (asterisk) and SC-NLC (down-triangle). (b) Left axis: Q-factor versus inline dispersion compensation ratio. LC (stars) and SC-NLC (up-triangle). Right axis: Delta Q-factor as a versus inline dispersion compensation ratio—solid line [43].

NLC	Research Effort Status	Commercial Effort Status
Performance Benchmarking	High	Medium
Algorithm Development	High	Medium
Hybrid Optoelec. NLC	Low	Low
Reduced Power Consumption	Low	Low
Reduced Circuit Size	Medium	Medium
NLC and FEC co-operation	Low	Low
Network Applications	Low	Medium

Figure 3.17 Various aspects and status of fiber nonlinearity compensation in academic and commercial realms.

Finally, in Fig. 3.17, we summarize the current status of NLC approach both from academic and industrial viewpoint. It can be seen that while focus has been relatively high on algorithm development and performance benchmarking, more attention needs to be given to simplifications, DSP multistage consolidation, and the application aspect of NLC techniques.

Acknowledgments

This work was partially funded by German Ministry of Education and Research (BMBF)—Project SpeeD under contract number 13N1374.[†]

References

1. Winzer, P. J. (2018). Scaling optical networking technologies for next generation SDM, *Optical Fiber Communication Conference*, paper Th1F.4.

2. Ellis, A. D., Zhao, J., Cotter, D. (2010), Approaching the non-linear shannon limit, *J. Lightwave Technol.*, 28(4), pp. 423–433.

3. Rafique, D. (2015). Putting flexible transmission rates to the test. Online: http://www.lightwaveonline.com/articles/2015/10/putting-flexible-transmission-rates-to-the-test.html.

4. ADVA Optical Networking (2016). FSP 3000 CloudConnect™. Online: http://www.advaoptical.com/en/products/scalable-optical-transport/fsp-3000-cloudconnect.aspx.

5. Infinera (2016). Infinera infinite capacity engine. Online: http://www.infinera.com/technology/engine/.

6. Rafique, D., et al. (2016). Multi-flex field trial over 762 km of G.652 SSMF using programmable modulation formats up to 64QAM, *Optical Fiber Communication Conference*, paper W4G.2.

7. Laperle, C., and O'Sullivan, M. (2014). Advances in high-speed DACs, ADCs, and DSP for optical coherent transceivers, *J. Lightwave Technol.*, 32(4), pp. 629–643.

8. Rahman, T., Rafique, D., Spinnler, B., Bohn, M., Napoli, A., Okonkwo, C. M., and de Waardt, H. (2016). 38.4Tb/s transmission of single-carrier serial line-rate 400 Gb/s PM-64QAM over 328 km for metro and data center interconnect applications, *Optical Fiber Communication Conference*, paper W3G.1.

9. Du, L. B., Rafique, D., Napoli, A., Spinnler, B., Ellis, A. D., Kuschnerov, M., and Lowery, A. J. (2014). Digital fiber nonlinearity compensation: Towards 1-Tb/s transport, *IEEE Signal Proc. Mag.*, 31(2), pp. 46–56.

10. Rafique, D. (2016). Fiber nonlinearity compensation: Commercial applications and complexity analysis, *J. Lightwave Technol.*, 34(2), pp. 544–553.

[†]Note that the majority of work presented in this chapter has been derived from material presented in [10–12, 17, 23, 42–43].

11. Rafique, D. (2012) *Electronic Signal Processing in Optical Communication: Analysis and Application of Nonlinear Transmission Limits*, Doctoral Thesis (Tyndall National Institute, University College Cork).

12. Rafique, D., and Ellis, A. D. (2012). Nonlinearity compensation via spectral inversion and digital back-propagation: A practical approach, *Optical Fiber Communication Conference*, paper OM3A.1.

13. Mateo, E., Zhu, L., and Li, G. (2008). Impact of XPM and FWM on the digital implementation of impairment compensation for WDM transmission using backward propagation, *Opt. Express*, 16, pp. 16124–16137.

14. Ip, E. (2010). Nonlinear compensation using backpropagation for polarization-multiplexed transmission, *J. Lightwave Technol.*, 28, pp. 939–951.

15. Du, L. B., and Lowery, A. J. (2010). Improved single channel backpropagation for intra-channel fiber nonlinearity compensation in long-haul optical communication systems, *Opt. Express*, 18, pp. 17075–17088.

16. Asif, R., Lin, C.-Yu, Holtmannspoetter, M., and Schmauss, B. (2010). Optimized digital backward propagation for phase modulated signals in mixed-optical fiber transmission link, *Opt. Express*, 18(22), pp. 22796–22807.

17. Rafique, D., and Ellis, A. D. (2011). The impact of signal-ASE four-wave mixing in coherent transmission systems, *Optical Fiber Communication Conference*, paper OThO.2.

18. Irukulapati, N. V., Wymeersch, H., Johannisson, P., and Agrell, E. (2014). Stochastic digital backpropagation, *IEEE Trans. Commun.*, 62(11), pp. 3956–3968.

19. Marsella, D., Secondini, M., and Forestieri, E. (2014). Maximum likelihood sequence detection for mitigating nonlinear effects, *J. Lightwave Technol.*, 32(5), pp. 908–916.

20. Lin, C.-Yu, Napoli, A., Spinnler, B., Sleiffer, V., Rafique, D., Kuschnerov, M., Bohn, M., and Schmauss, B. (2014). Adaptive digital back-propagation for optical communication systems, *Optical Fiber Communication Commun. Conference*, paper M3C.4.

21. Jiang, L., Yan, L., Chen, Z., Yi, A., Pan, Y., Pan, W., and Luo, B. (2016). Low Complexity and adaptive nonlinearity estimation module based on Godard's Error, *IEEE Photonics J.*, 8(1), pp. 7801007–916.

22. Chen, Z., et al. (2015). Low complexity and adaptive nonlinearity compensation, *Conference on Lasers ElectroOptics: Science and Innovations*, paper SM2M.6.

23. Rafique, D., Zhao, J., and Ellis, A. D. (2011). Performance improvement by fibre nonlinearity compensation in 112 Gb/s PM M-ary QAM, *Optical Fiber Communication Conference*, paper OWO.6.

24. Ellis, A. D., Zhao, J., and Cotter, D. (2010). Approaching the non-linear shannon limit, *J. Lightwave Technol.*, 28(4), pp. 423–433.

25. Poggiolini, A. P., Bosco, G., Carena, A., Curri, V., Jiang, Y., and Forghieri, F. (2014). The GN-model of fiber non-linear propagation and its applications, *J. Lightwave Technol.*, 28(4), pp. 694–721.

26. ADVA Optical Networking (2016). DCI Disaggregation: The next step is data functionality. Online: http://blog.advaoptical.com/dci-disaggregation-the-next-step-in-data-functionality.

27. Gordon, J. P., and Haus, H. A. (1986). Random walk of coherently amplified solitons in optical fiber transmission, *Opt. Lett.*, 11, pp. 665–667.

28. Gordon, J. P., and Mollenauer, L. F. (1990). Phase noise in photonic communication systems using linear amplifiers, *Opt. Lett.*, 15, pp. 1351–1353.

29. Hui, R., et al. (1997). Modulation instability and its impact in multispan optical amplified IMDD systems: Theory and experiments, *J. Lightwave Technol.*, 15, p. 1071.

30. Chen, X., and Shieh, W. (2010). Closed-form expressions for nonlinear transmission performance of densely spaced coherent optical OFDM systems, *Opt. Express*, 18(18), pp. 19039–19054.

31. Gao, G., Chen, X., and Shieh, W. (2012). Influence of PMD on fiber nonlinearity compensation using digital back propagation, *Opt. Express*, 20(13), pp. 14406–14418.

32. Sorokina, M., and Turitsyn, S. (2014). Nonlinear signal transformations: Path to capacity above the linear AWGN Shannon limit, *Photonics Society Summer Topical Meeting Series*, p. WD3.

33. Agrell, E. (2015). Conditions for a monotonic channel capacity, *IEEE Trans. Commun.*, 63(3), pp. 738–748.

34. Rafique, D., and Ellis, A. D. (2011). Impact of longitudinal power budget in coherent transmission systems employing digital back-propagation, *Opt. Express*, 19(26), pp. B40–B46.

35. Yamazaki, E., Sano, A., Kobayashi, T., Yoshida, E., and Miyamoto, Y. (2011). Mitigation of nonlinearities in optical transmission systems, *Optical Fiber Communication Conference*, paper OThF1.

36. Du, L. B., Morshed, M. M., and Lowery, A. J. (2012). Fiber nonlinearity compensation for OFDM super-channels using optical phase conjugation, *Opt. Express*, 20, pp. 14362–19928.

37. Mateo, E. F., Zhou, X., and Li, G. (2011). Electronic phase conjugation for nonlinearity compensation in fiber communication systems, *Optical Fiber Communication Conference*, paper JWA025.

38. Yariv, A., Fekete, D., and Pepper, D. M. (1979). Compensation for channel dispersion by nonlinear optical phase conjugation, *Opt. Lett.*, 4, pp. 52–54.

39. Minzioni, P., Christiani, I., Degiorgio, V., Marazzi, L., Martinelli, M., Langrock, C., and Fejer, M. M. (2006). Experimental demonstration of nonlinearity and dispersion compensation in an embedded link by optical phase conjugation, *IEEE Photonics Technol. Lett.*, 18, pp. 995–997.

40. Zhu, L., and Li, G. (2012). Nonlinearity compensation using dispersion-folded digital backward propagation, *Opt. Express*, 20(13), pp. 14362–14370.

41. Mussolin, M., Rafique, D., Mårtensson, J., Forzati, M., Fischer, J. K., Molle, L., Nölle, M., Schubert, C., and Ellis, A. D. (2011). Polarization multiplexed 224 Gb/s 16QAM transmission employing digital back-propagation, *European Conference on Optical Communications*, paper We.8.B.6.

42. Rafique, D., and Ellis, A. D. (2011). Nonlinearity compensation in multi-rate 28 Gbaud WDM systems employing optical and digital techniques under diverse link configurations, *Opt. Express*, 19(18), pp. 16919–16926.

43. Rafique, D., and Ellis, A. D. (2013). Intra-channel nonlinearity compensation for PM-16QAM traffic co-propagating with 28 Gbaud m-ary QAM neighbours, *Opt. Express*, 21(4), pp. 4174–4182.

Chapter 4

Phase-Conjugated Twin Waves and Phase-Conjugated Coding

Son Thai Le

Nokia Bell Labs, Crawford Hill, New Jersey, USA

son.thai_le@nokia-bell-labs.com

4.1   Introduction

As discussed in the Chapter 2, the nonlinear impairment due to Kerr effect limits the maximum signal power that could be launched into an optical fiber without degrading the effective signal-to-noise ratio (SNR) at the receiver [1–4]. As a result, fiber Kerr nonlinearity effect sets an upper bound on the achievable data rate in optical fiber systems with traditional linear transmission techniques [1–5].

In the past decade, extensive efforts have been made in attempting to surpass the Kerr nonlinearity limit through several nonlinearity compensation techniques. As shown in previous chapters, digital back propagation (DBP) (implemented at the receiver or the transmitter side) is a powerful nonlinearity

Optical Communication Systems: Limits and Possibilities
Edited by Andrew Ellis and Mariia Sorokina
Copyright © 2020 Jenny Stanford Publishing Pte. Ltd.
ISBN 978-981-4800-28-0 (Hardcover), 978-0-429-02780-2 (eBook)
www.jennystanford.com

compensation method, which effectively removes the signal × signal nonlinear distortion by solving the nonlinear Schrödinger equation digitally. However, accurate DBP requires a substantial increase in digital signal processing (DSP) complexity, proportional to the transmission distance [6]. Besides the enormous complexity, in wavelength division multiplexed (WDM) systems, the effectiveness of DBP is significantly reduced as the neighboring WDM channels are unknown to the compensator.

Digital and optical phase conjugations (OPC) are other well-known nonlinear compensation techniques that conjugate the signal phase after transmission in one segment of the link in order to achieve a net cancellation of the nonlinear phase shift using the nonlinearity generated in the second segment of the link [7–10]. However, OPC modifies the transmission link by employing an in-line phase conjugator (or several phase conjugators) and imposes significant symmetry conditions with respect to the phase conjugator, and thus, significantly reducing the flexibility in an optically routed network. The design and achievable performance of OPC will be discussed in more details in Chapter 9.

Recently, a breakthrough fiber nonlinearity compensation technique called phase-conjugated twin wave (PCTW) has been proposed by X. Liu et al. [11]. PCTW is a transponder-based technique that can be implemented with minimal additional optical hardware or DSP, providing a low-complexity and an effective solution in compensating fiber nonlinearity effect as it requires only an additional per symbol conjugate-and-add operation prior to symbol detection. In this scheme, the signal complex waveform and its phase-conjugate copy are simultaneously transmitted in x- and y-polarization states. If the dispersion map of the link is symmetrical, the nonlinear distortions on x-and y-polarizations are essentially anticorrelated [11–13], and thus, they can be subsequently mitigated at the receiver through coherent superposition of the two copies. However, PCTW halves the spectral efficiency (SE), meaning that the maximum achievable SE in a polarization division multiplexed system with quadrature phase shift keying (QPSK) modulation format and PCTW scheme is only ~2 bits/s/Hz, which is the same as those achieved in polarization division multiplexed (PDM) binary phase shift keying (BPSK) transmission [1, 2].

To address this problem, various modifications of PCTW and phase-conjugated coding techniques with and without spectral redundancy have been proposed for different transmission schemes, including single carrier, multi-carrier and transmissions over multimode and multicore fibers. In this chapter, we outline the general principle, benefit, application range and limitation of the PCTW technique. In addition, we discuss effective PCTW's modifications and phase-conjugated coding techniques to reduce the spectral redundancy and increase the flexibility of the original PCTW ideology.

4.2 General Principle

Under the common assumption that the nonlinear interaction length is much greater than the length scale of random polarization rotations, the propagation of an optical signal described by a PDM vector $E = (E_x, E_y)^T$ is governed by the Manakov equation (coupled nonlinear Schrödinger equation) as [14, 15]

$$\left[\frac{\partial}{\partial z} + \frac{\alpha(z) - g(z)}{2} + i\frac{\beta_2}{2}\frac{\partial^2}{\partial t^2} \right] E_{x,y}(z,t)$$

$$= i\frac{8}{9}\gamma(|E_x(z,t)|^2 + |E_y(z,t)|^2)E_{x,y}(z,t), \quad (4.1)$$

where i is the imaginary unit, and z, α, g, β_2, and γ, respectively, are the propagation distance, the loss coefficient, the gain coefficient, the second-order dispersion coefficient, and the fiber nonlinear Kerr coefficient along a transmission link.

Based on the well-known perturbation approach [11, 12], the first-order nonlinear distortion can be expressed in the frequency domain as

$$\delta E_{x,y}(L,\omega) = i\frac{8}{9}\gamma P_0 L_{\text{eff}} \int_{-\infty}^{+\infty}\int_{-\infty}^{+\infty} \eta(\omega_1,\omega_2)d\omega_1 d\omega_2$$

$$\cdot [E_{x,y}(\omega+\omega_1)E_{x,y}(\omega+\omega_2)E_{x,y}^*(\omega+\omega_1+\omega_2)$$

$$+ E_{y,x}(\omega+\omega_1)E_{x,y}(\omega+\omega_2)E_{y,x}^*(\omega+\omega_1+\omega_2)], \quad (4.2)$$

where $E_{x,y}(\omega) = \int_{-\infty}^{+\infty} E_{x,y}(0,t)\exp(-i\omega t)dt/\sqrt{2\pi}$, $L_{\text{eff}} = \int_0^L \exp(G(z))dz$ and $G(z) = \int_0^z [g(s) - \alpha(s)]ds$ are the effective length, distributed gain, respectively, and $\eta(\omega_1, \omega_2)$ is the nonlinear transfer function [12] defined as

$$\eta(\omega_1, \omega_2) = \frac{1}{L_{\text{eff}}} \int_0^L \exp[G(z) - i\omega_1\omega_2 C(z)]dz, \qquad (4.3)$$

where $C(z) = \int_0^z \beta_2(s)ds$ is the cumulative dispersion along the link.

When symmetric dispersion and power evolution maps are applied such that $C(z) = -C(L - z)$, $G(z) = G(L - z)$, then $\eta(\omega_1, \omega_2)_s$ become real-valued [11, 12], i.e.,

$$\eta^*(\omega_1, \omega_2) = \eta(\omega_1, \omega_2). \qquad (4.4)$$

In practice, a symmetric dispersion map can be effectively obtained by a mean of pre-EDC at the transmitter. On the other hand, a nearly symmetric power condition can be achieved using links with low loss excursion in the transmission path, e.g., through fiber spans with low loss or Raman amplification [16–18].

4.2.1 Phase-Conjugated Twin Waves

The basic idea of PCTW transmission scheme is presented in Fig. 4.1. In this scheme, the signals on x- and y-polarizations at the transmitter satisfy the relations

$$E_y(0,t) = E_x^*(0,t), \; E_y(\omega) = E_x^*(-\omega).$$

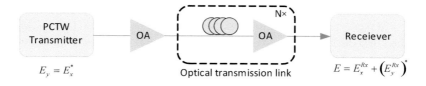

Figure 4.1 PCTW-based transmission for fiber nonlinearity mitigation.

As shown in the previous section, under the assumption of symmetric dispersion map and symmetric power evolution along the link $\eta(\omega_1, \omega_2)$ is real-valued. In this case we have

$$\delta E_y(L,\omega) = i\frac{8}{9}\gamma P_0 L_{\text{eff}} \int\limits_{-\infty}^{+\infty}\int\limits_{-\infty}^{+\infty} \eta(\omega_1,\omega_2)d\omega_1 d\omega_2$$

$$\cdot[E_y(\omega+\omega_1)E_y(\omega+\omega_2)E_y^*(\omega+\omega_1+\omega_2)$$

$$+ E_x(\omega+\omega_1)E_y(\omega+\omega_2)E_x^*(\omega+\omega_1+\omega_2)]$$

$$\approx i\frac{8}{9}\gamma P_0 L_{\text{eff}} \int\limits_{-\infty}^{+\infty}\int\limits_{-\infty}^{+\infty} \eta^*(\omega_1,\omega_2)d\omega_1 d\omega_2$$

$$\cdot[E_x^*(-\omega-\omega_1)E_x^*(-\omega-\omega_2)E_x(-\omega-\omega_1-\omega_2)$$

$$+ E_y^*(-\omega-\omega_1)E_x^*(-\omega-\omega_2)E_y(-\omega-\omega_1-\omega_2)]$$

$$\approx -[\delta E_y(L,-\omega)]^* \qquad (4.5)$$

By taking the inverse Fourier transform, we have

$$\delta E_y(L,t) = -[\delta E_x(L,t)]^*. \qquad (4.6)$$

This relation indicates that the nonlinear distortions experienced by the PCTWs are anticorrelated. As a result, the first-order nonlinear distortions can be cancelled and the original signal $E(0, t)$ can be restored using coherent superposition of the received PCTWs as

$$E_x(L,t) + E_y(L,t)^* = 2E(0,t). \qquad (4.7)$$

The benefit of coherent superposition in PCTW-based transmission is twofold. First of all, it remarkably cancels the nonlinear signal distortion including those resulting from the interaction between Kerr nonlinearity and dispersion. Second, as exploited in many fields of physic and engineering, the coherent superposition halves the variance of the linear noise resulting from the ASE of optical amplifiers, and thus, further improving the overall system performance.

The effectiveness of the PCTW technique is preliminarily demonstrated in Fig. 4.2 for single-channel 20 Gbaud PCTW-based transmission. As indicated in Fig. 4.2, the received signal quality is significantly increased after coherent superposition of the two phase-conjugated twin waves at the receiver. The Q-factor is increased from ~12 to ~21 dB showing an improvement of 9 dB. This clearly indicates the high performance of the PCTW technique in cancelling the first-order nonlinear distortion through coherent superposition of PCTWs at the receiver. The performance benefit and limitation of the PCTW technique will be discussed in more details in Section 4.3.

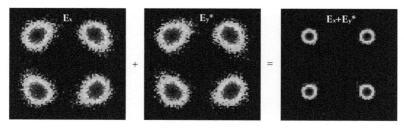

Figure 4.2 Simulated constellations in *x*- and *y*-polarizations and the final constellation after coherent superposition of the received PCTW waves in a 20 Gbaud single-channel PCTW-based transmission. The transmission link was 50 × 80 km of SSMF amplified by EDFAs with 5 dB of noise figure; the launch power was 2 dBm.

4.2.2 Nonlinear Noise Squeezing

If the transmit signal field is real-valued in the time domain, such as PAM signals, i.e., $E_{x,y}(t) = E^*_{x,y}(t)$. In the frequency domain, we have

$$E_{x,y}(\omega) = E^*_{x,y}(-\omega).$$

(4.8)

Using Eq. (4.2), we have

$$\delta E_{x,y}(L,\omega) \approx i\frac{8}{9}\gamma P_0 L_{\text{eff}} \int_{-\infty}^{+\infty}\int_{-\infty}^{+\infty} \eta^*(\omega_1,\omega_2)d\omega_1 d\omega_2$$

$$\cdot [E^*_{x,y}(-\omega-\omega_1)E^*_{x,y}(-\omega-\omega_2)E_{x,y}(-\omega-\omega_1-\omega_2)$$

$$+ E^*_{y,x}(-\omega-\omega_1)E^*_{x,y}(-\omega-\omega_2)E_{y,x}(-\omega-\omega_1-\omega_2)]$$

$$\approx -[\delta E_{x,y}(L,-\omega)]^*.$$

(4.9)

Using the conjugation property of the Fourier transform again, we have

$$\delta E_{x,y}(L,t) = -[\delta E_{x,y}(L,t)]^*,$$

(4.10)

which means that the nonlinear distortions $\delta E_{x,y}(L,t)$ are imaginary to first order. As the same time, as the original $E_{x,y}(L,t)$ is real-valued, the nonlinear distortions $\delta E_{x,y}(L,t)$ are thus, "squeezed" along the direction which is orthogonal to the decision line. As a consequence, the nonlinear distortion $\delta E_{x,y}(L,t)$ does not lead to decision errors to the first order. It can also be shown that the conclusion is still valid if there is an arbitrary phase offset between the x- and y-polarization components. This means that the original signals only need to be real-valued after a constant phase rotation. It is also interesting to see that the nonlinear noise squeezing (NLNS) effect occurs even in the presence of cross-polarization nonlinear interactions.

To conclude, in polarization multiplexed real-valued signals, if symmetric dispersion map and symmetric power evolution conditions are satisfied, nonlinear distortions to the first order have no impact on the transmission performance after coherent superposition of PCTWs.

4.2.3 Connection between NLNS and PCTW

In general, a complex signal can be considered as the multiplexing of two independent, orthogonal real-valued signals. As a result, there exists a unitary transformation matrix U that transforms a PCTW vector to a vector of two real-valued signals as

$$\begin{pmatrix} B_1(0,t) \\ B_2(0,t) \end{pmatrix} = U \begin{pmatrix} E_x(0,t) \\ E_y(0,t) \end{pmatrix} = U \begin{pmatrix} E_x(0,t) \\ E_x^*(0,t) \end{pmatrix} = \sqrt{2} \begin{pmatrix} \mathrm{Re}(E_x(0,t)) \\ \mathrm{Im}(E_x(0,t)) \end{pmatrix}, \quad (4.11)$$

where $B_1(0,t)$ and $B_2(0,t)$ are two independent real-valued signals and orthogonally polarized and U is the uniform transformation matrix defined as

$$U = \frac{1}{\sqrt{2}} \begin{pmatrix} 1 & 1 \\ -i & i \end{pmatrix}. \tag{4.12}$$

After nonlinear transmission over a distance L, we have for the symmetric dispersion map case

$$\begin{pmatrix} E_x(L,t) \\ E_y(L,t) \end{pmatrix} \approx \begin{pmatrix} E(0,t) + \delta E_x(L,t) \\ E^*(0,t) - \delta E_x(L,t)^* \end{pmatrix}. \tag{4.13}$$

By calculating $B_1(L,t)$ and $B_2(L,t)$ we have

$$\begin{pmatrix} B_1(L,t) \\ B_2(L,t) \end{pmatrix} = \frac{1}{\sqrt{2}} \begin{pmatrix} 1 & 1 \\ -i & i \end{pmatrix} \begin{pmatrix} E_x(L,t) \\ E_y(L,t) \end{pmatrix}$$

$$\approx \begin{pmatrix} B_1(0,t) \\ B_2(0,t) \end{pmatrix} + i\sqrt{2} \begin{pmatrix} \mathrm{Im}(\delta E_x(L,t)) \\ -\mathrm{Re}(\delta E_x(L,t)) \end{pmatrix}, \qquad (4.14)$$

indicating that the nonlinear distortions of $B_1(L,t)$ and $B_2(L,t)$ are imaginary to the first order, or "squeezed" along the direction orthogonal to the decision line. As a result, the nonlinear cancellation effect in PCTW scheme can also be considered as the NLNS effect after using a unitary transformation matrix.

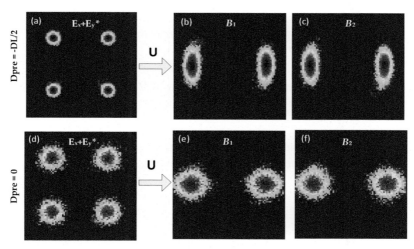

Figure 4.3 Illustration of the relationship between the PCTW and the NLNS effects. (a) and (d) simulated QPSK signal constellations after coherent superposition of the PCTWs; (b, c) and (e, f) recovered BPSK signal constellations (with x-and y-polarization components overlaid); (a–c) with pre-dispersion compensation (Dpre = –DL/2); (d–f) without dispersion pre-compensation (Dpre = 0). The transmission link was 50 × 80 km of SSMF amplified by EDFAs with 5 dB of noise figure; the launch power was 2 dBm.

The relationship between the PCTW and the NLNS effects is illustrated in Fig. 4.3 for QPSK PCTW-based transmission. With 50% pre-EDC (for creating a symmetric dispersion map), the nonlinear noises of the decomposed BPSK constellations

($B_1(L,t)$ and $B_2(L,t)$) are effectively squeezed along the imaginary direction, which is orthogonal to the decision line. On the other hand, without pre-EDC, NLNS effect is not observed when the nonlinear noises of $B_1(L,t)$ and $B_2(L,t)$ are "rounded" rather than being "squeezed" along the imaginary axis. In addition, it is clear that the signal's quality is significantly improved when optimum pre-EDC is applied. This result highlights the importance of symmetric dispersion map in PCTW-based transmissions.

4.2.4 Generalized Phase-Conjugated Twin Waves (G-PCTW)

In general, rather than "twinning" a signal with its complex conjugate copy on an orthogonal polarization state, it is also possible to twin a signal with its complex conjugate in any other orthogonal dimension [12] as well. More specifically, instead of using the two orthogonal polarizations of a single optical carrier, the phase-conjugated twin can also propagate along an orthogonal spatial path or it can occupy two different time segments or (with certain limitations) also two different carrier frequencies [19]. As suggested by X. Liu in [12], the concept of PCTW can also be extended such that the twin waves are vector waves, i.e., PDM signals with independent complex-valued fields in x- and y-polarizations. For example, one may twin two general vector waves across two orthogonal physical dimensions, the vector $\boldsymbol{E_1} = (E_x, E_y)^T$ in one dimension and $\boldsymbol{E_2} = \boldsymbol{E_1}^*$ in another dimension. It can be readily shown that [12]

$$\delta E_2(L,t) = -[\delta E_1(L,t)]^*, \tag{4.15}$$

which also indicates that the nonlinear distortions experienced by the vector PCTW are anti-correlated to first order, which leads to the first-order cancellation of nonlinear distortions of the original optical vector field using coherent superposition of the two received vectors PCTWs

$$E_1(L,t) + E_2(L,t)^* = 2E_1(0,t) \tag{4.16}$$

In summary, the phase-conjugated twin of a complex-valued signal can be transmitted on a second orthogonal dimension.

This effectively provides nonlinear compensation with the cost of halving the system throughput. However, it should be noted that the nonlinear compensation provided by the PCTW technique is up to first order only. In addition, so far, we neglected the effects of PMD, polarization-dependent loss (PDL), and third-order dispersion, all of which may reduce the effectiveness of NLNS and PCTWs in mitigating fiber nonlinear impairments.

4.3 Benefit and Limitation of PCTW

4.3.1 SNR and Capacity Gain in PCTW-Based Transmissions

To estimate the performance gain offered by PCTW-scheme, it is useful to take into consideration the well-known calculation of SNR at the receiver of coherent optical communication systems based on the Gaussian noise (GN) model as [20–22]

$$\text{SNR} = \frac{P_{\text{S}}}{N P_{\text{ASE}} + \eta N P_{\text{S}}^3}, \tag{4.17}$$

where P_{S} and P_{ASE} are the signal power and in-band ASE noise power, respectively, η is the nonlinear coefficient which can be defined for optical link with erbium-doped fiber amplifiers (EDFA) as [22]

$$\eta = \frac{\gamma^2}{\pi \alpha |\beta_2|} \log \left(\frac{2\pi^2 |\beta_2| B^2}{\alpha} \right), \tag{4.18}$$

where α, β_2, and γ are the fiber's loss, second-order dispersion and nonlinear coefficients, respectively, and B is the signal's bandwidth.

In PCTW-based transmission, the first-order nonlinear distortion is cancelled and the ASE noise power is reduced by a factor of two as a consequence of coherent superposition. As a result, the SNR in PCTW-based transmission can be calculated as

$$\text{SNR}_{\text{PCTW}} = \frac{P_{\text{S}}}{N P_{\text{ASE}} / 2 + \eta_{\text{PCTW}} N P_{\text{S}}^3}, \tag{4.19}$$

where η_{PCTW} is the reduced nonlinear coefficient as a result of coherent superposition. In general, η_{PCTW} depends on the link power evolution profile and signal's bandwidth.

From Eqs. (4.17) and (4.19) the optimum SNR in the conventional and PCTW-based systems can be calculated as

$$\text{SNR}^{\text{opt}} = \frac{1}{3\sqrt[3]{\dfrac{\eta N^3 P_{\text{ASE}}^2}{4}}}, \quad \text{SNR}_{\text{PCTW}}^{\text{opt}} = \frac{1}{3\sqrt[3]{\dfrac{\eta_{\text{PCTW}} N^3 P_{\text{ASE}}^2}{16}}}. \quad (4.20)$$

As a result, the SNR gain in dB can be expressed as

$$\text{SNR}_G = 2dB + \frac{1}{3}\left[\frac{\eta}{\eta_{\text{PCTW}}}\right][dB]. \quad (4.21)$$

For convenience, we define the nonlinear reduction coefficient in PCTW-based transmission as

$$R_{\text{PCTW}} = \left[\frac{\eta}{\eta_{\text{PCTW}}}\right][dB]. \quad (4.22)$$

The relationship between overall SNR gain and nonlinear reduction coefficient is presented in Fig. 4.4. If the nonlinear distortions on x- and y-polarizations are uncorrelated, a nonlinear reduction coefficient of 3 dB is obtained. This also leads to 3 dB gain in overall system performance. However, as PCTW provides nonlinear cancellation instead of nonlinear de-correlation, the nonlinear reduction coefficient can be higher than 3 dB. From Eq. (4.22) and Fig. 4.4, it is clear that in order to increase the overall SNR by every 1 dB the nonlinear distortion has to be decreased by 3 dB.

For a quick estimation, we assume that the residual nonlinear impairment in uncompensated optical communication can be approximated by a Gaussian noise [20] (please also see Chapter 2 for more detail). In this case, the optical channel impaired by the fiber nonlinearity and ASE noise can be approximated as an AWGN channel. Based on the well-known Shannon theory, the capacity in conventional and PCTW-based systems can be estimated as [1, 2]

$$C = 2\log_2(1 + \text{SNR}^{\text{opt}}) \tag{4.23}$$

$$C_{\text{PCTW}} = \log_2(1 + \text{SNR}^{\text{opt}}_{\text{PCTW}}). \tag{4.24}$$

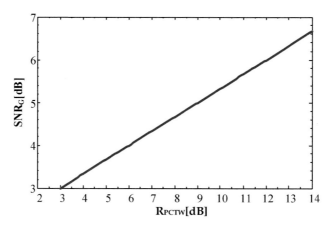

Figure 4.4 SNR gain as a function of nonlinear reduction coefficient in PCTW-based transmission.

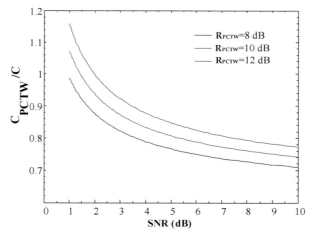

Figure 4.5 Overall capacity gain provided by the PCTW technique as a function of the SNR for different values of nonlinear reduction coefficient (8 dB, 10 dB, 12 dB).

We should note that in the PCTW scheme, one polarization transmits the phase-conjugated copy and thus, does not carrying any useful data.

The overall capacity gain offered by PCTW is presented in Fig. 4.5, as a function of the received SNR. It is clear that the system total capacity can be increased with the PCTW technique only in the region of low SNR and high nonlinear reduction coefficient. Such conditions usually are not satisfied as PCTW provides nonlinear compensation to the first order only putting a limitation on the achievable nonlinear reduction coefficient, and a very low SNR (<3 dB) is not of practical interest. As a result, the PCTW technique should be considered as a technique for improving the system performance, such as distance reach and signal's quality rather than improving the total capacity.

4.3.2   Benefit and Application Range of PCTW

The benefit of the PCTW technique in highly dispersive fiber link is demonstrated in Fig. 4.6 for single-channel 56 Gbaud transmission. For simplicity, we consider here only uncoded transmission systems. The system performance was estimated using the well-known error-vector-magnitude and then subsequently converted to the SNR. In Fig. 4.6a, it is clear that the SNR is increased by ≈4.8 dB after coherent superposition, indicating the effectiveness of PCTW. The received constellations at the optimum launch power before and after superposition of PCTW are shown in Fig. 4.6b,c.

To analyze in more details the benefit of PCTW, Fig. 4.7 plots the SNR gains offered by coherent superposition of PCTW as functions of the launch power in 56 Gbaud transmissions, with and without 50% pre-EDC and ASE noise. Without 50% pre-EDC, the SNR improvement increases from 3 dB to ~4 dB when the signal launch power is increased from –6 dB to 5 dBm. It is noted that an improvement that is bigger than 3 dB indicates some nonlinearity cancellation offered by the PCTW rather than nonlinearity de-correlation observed in scrambled coherent superposition techniques [23]. On the other hand, if 50% pre-EDC is applied, a more significant SNR gain, up to 8 dB can be achieved. This indicates that the symmetric dispersion map condition should be met in order to achieve the optimum benefit of the PCTW technique. If only the nonlinear effect is considered (by ignoring the ASE noise) then the SNR gain offered by coherent superposition decreases with the launch power, as shown in Fig. 4.7.

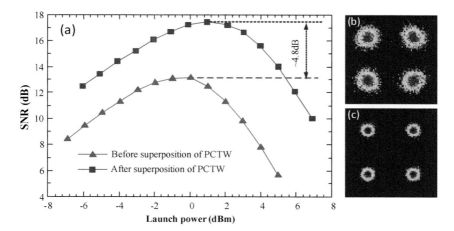

Figure 4.6 Demonstration of nonlinearity cancellation in 56 Gbaud PCTW-based transmission. (a) Estimated SNRs before and after superposition of PCTW. (b) Received constellation at optimum launch power before superposition of PCTW. (c) Received constellation at optimum launch power after superposition of PCTW.

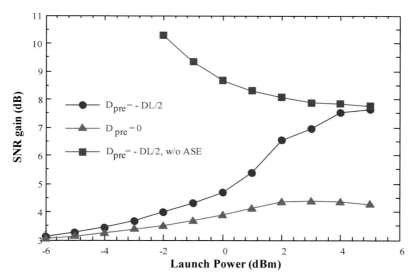

Figure 4.7 Demonstration of nonlinearity cancellation in 56 Gbaud PCTW-based transmission–SNR gain after superposition of PCTW with and without 50% pre-EDC.

As discussed above, the nonlinear benefits of PCTW comes at the expense of spectral efficiency or the overall system capacity. Therefore, it is important to discuss the key tradeoff and application range of the PCTW technique in more details.

For dispersion-unmanaged transmission, the SNR improvement (equivalent to Q^2 factor improvement) brings reach increase of approximately the same amount (in dB). Based on the results presented above, we can assume that the PCTW approach reduces the nonlinear and linear noise variances by about 8.5 and 3 dB, respectively. Using Eq. (4.21) we have an improvement brought by the PCTW approach is expected to be ~4.8 dB, indicating a ~3-fold increase in the transmission distance. This has been experimentally verified in [12] for transmission systems at 15 Gbaud. Figure 4.8 shows the Q^2 factors of PDM-BPSK, PDM-QPSK, PDM-QPSK PCTW, PDM-8QAM PCTW and PDM 16QAM PCTW after transmission over a fiber link with EDFA-only amplification. As indicated in Fig. 4.8, at the same Q^2 factor of 12.9 dB the reach of 30 Gb/s PDM QPSK PCTW was 6400 km, which is ~3 times as long as that of 60 Gb/s PDM-QPSK, which was ~2100 km. This is in good agreement with the prediction by Eq. (4.21). At the same data rate of 30 Gb/s the reach of PDM-BPSK was ~4000 km, which is ~2 dB less than that of PDM-QPSK PCTW. This is a clear benefit of PCTW in systems with low spectral efficiency, i.e., with BPSK.

The PCTW technique can also be used to improve the nonlinear transmission performance of higher order modulation formats such as 8-QAM and 16-QAM or to use higher order modulation formats combined with PCTW to achieve the same single-path capacity as that of PDM signals with lower order formats. In particular, PDM-16QAM PCTW can be compared with PDM-QPSK as both systems offer 4 raw bits per symbol. However, as indicated in Fig. 4.8, the reach of 10 Gb/s PDM-16QAM PCTW at a Q^2 factor of 8 dB (~3200 km), is two times less than the reach of 60 Gb/s PDM-QPSK (6400 km). This is due to the fact that the sensitivity of 16QAM is much lower as compared to PDM-QPSK (~9 dB lower for the same symbol rate taking into account typically higher implementation penalties for higher order constellations).

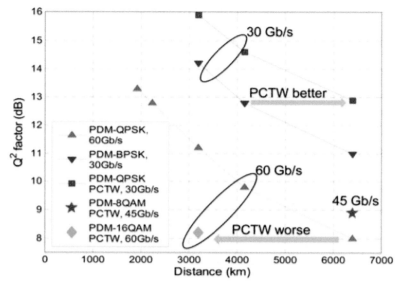

Figure 4.8 Experimental performance comparison of various signal formats (PDM-BPSK, PDM-QPSK, PDM-QPSK PCTW, PDM-8QAM PCTW and PDM 16QAM PCTW, all modulated at 15 Gbaud) after transmission over a fiber link consisting of 80 km SSMF spans with EDFA-only amplification. Reproduced from [12], with permission.

Figure 4.9 shows transmission system options with the same capacity of 4 raw bits per symbol and the same long-haul reach, assuming that the transmission distance requirement can be just met by PCTW (Fig. 4.9). As the transmission distance of PDM-QPSK PCTW is three times longer than PDM-QPSK, the conventional PDM-QPSK systems would need to be regenerated two times along the link, thus requiring three coherent transceivers in total. In order to achieve the same capacity in comparison with PDM-QPSK, two parallel PCTW-based PDM QPSK links are required. As a result, with the PCTW technique, only two optical transceivers are needed instead of three, representing 33% transceiver savings of the PCTW system compared to the PDM system. For systems, where the cost of optical transceivers is the dominating portion of the total cost, the use of PCTW can thus, provide substantial cost savings. For transmission links, such as transoceanic optical links, where signal regeneration is not allowed, the performance improvement of PCTW is even more valuable [12].

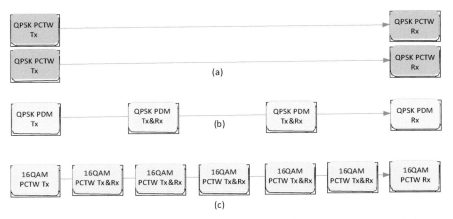

Figure 4.9 Transmission system options with the same capacity of 4 (raw) bits per symbol and the same reach. (TX: transmitter; RX: receiver) For fixed capacity and long-haul reach, PDM-QPSK PCTW requires the lowest number of transponders. (a) Transmission system based on two PCTW paths. (b) Transmission system based on one PDM path. (c) Transmission system based on one higher-order PCTW path.

When PDM-16QAM PCTW is used for achieving the same capacity of 4 raw bits per symbol, 6 optical transceivers are required. This is two times higher than that of the conventional PDM-QPSK transmission. This clearly indicates that the PCTW technique is not beneficial in high-spectral-efficiency transmissions.

Regarding the DSP complexity of the PCTW scheme, the additional DSP needed for coherent superposition is negligible, leaving the main complexity difference to conventional DSP schemes with the requirement for split CD compensation between transmitter and receiver. We can assume that split CD compensation between the transmitter and receiver doubles the complexity of CD compensation in conventional systems. This is, however, still over one order of magnitude lower than that of DBP [24, 25]. In fact, the PCTW approach relies on the physical link itself (rather than on extensive DSP) to generate anticorrelated nonlinear distortions so that the distortion cancellation becomes simple.

4.4 Phase-Conjugated Pilot (PCP)

4.4.1 Principle

As discussed above, one important shortcoming of the PCTW technique is that it sacrifices half the transmission capacity. To partially solve this problem, a novel nonlinearity compensation technique for coherent optical orthogonal frequency division multiplexing (CO-OFDM) systems based on the transmission of phase-conjugated pilots (PCPs), without scarifying 50% of the transmission capacity has been proposed in [26].

In this scheme, a portion of the OFDM subcarriers (up to 50%) is transmitted as phase-conjugates of other subcarriers. The PCPs are then used at the receiver to estimate the nonlinear distortion of their respective original subcarriers. The estimated distortion can also be used to compensate the nonlinear impairments in other subcarriers close to the PCP, thanks to the narrow OFDM subcarrier spacing (tens of MHz), which enhances the correlation between nonlinear phase shifts of neighboring subcarriers.

The concept of inserting PCP is illustrated in Fig. 4.10. Suppose the information symbol carried by the k-th subcarrier is $S_k = A_k \cdot \exp(i \cdot \phi_k)$ where A_k and ϕ_k are the amplitude and the phase of this information symbol, then the phase-conjugated symbol can be transmitted in the h-th subcarrier, $S_h = S_k^* = A_k \cdot \exp(\phi i \cdot \phi_k)$, where $(.)^*$ represents complex conjugation. To simplify the exposition, we assume that during propagation over a fiber link nonlinear phase shifts, represented by θ_k and θ_h, are added to these subcarriers. The received information symbols on the k-th and h-th subcarriers are $R_k = A_{r,k} \cdot \exp(i \cdot \phi_k + i \cdot \theta_k)$ and $R_h = A_{r,h} \cdot \exp(\phi i \cdot \phi_k + i \cdot \theta_h)$, respectively. If the frequency spacing between the k-th and h-th subcarriers is small enough, the nonlinear phase shifts will be highly correlated, $\theta_k \approx \theta_h$ providing the opportunity of cancelling the nonlinear phase shift on the k-th subcarrier by averaging the received information symbol of a subcarrier and the subcarrier that carries its phase conjugate (after a second conjugation):

$$\overline{R}_k = (R_k + R_h^*)/2 \approx A_{r,k} \cdot \cos(\theta_k) \cdot \exp(j\phi_k) \qquad (4.25)$$

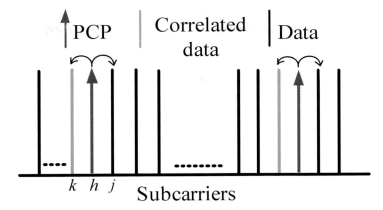

Figure 4.10 Inserting phase-conjugated pilots for fiber nonlinearity compensation.

Note that the nonlinear phase shift on the original subcarrier k can be estimated as

$$\theta_k = \arg(R_k \cdot R_h)/2. \tag{4.26}$$

Ideally, a data-carrying subcarrier and its PCP should be closely spaced in frequency (adjacent) to maximize the level of correlation of the nonlinear phase shifts between these subcarriers. For those data-carrying subcarriers which do not have PCPs, the nonlinear phase shift of the j-th subcarrier can be estimated and compensated as

$$\theta_j = \arg\left(\sum_{k,h}\eta_{jkh} \cdot R_k \cdot R_h\right)/2 \tag{4.27}$$

$$\overline{R_j} = R_j \cdot e^{-i \cdot \theta_j}, \tag{4.28}$$

where η_{jkh} is the FWM efficiency coefficient. Here, η_{jkh} is approximated as 1 if j is the closest subcarrier to k or h and 0 otherwise.

By applying this fiber compensation technique, the fiber nonlinearity phase shifts on data subcarriers in an OFDM system can be compensated without conjugating all pairs of subcarriers. In this system configuration, several data-carrying subcarriers are placed between conjugate pairs. The nonlinear phase shifts for all of these subcarriers are similar as long as the frequency spacing is small. These nonlinear distortions can be compensated using the estimated nonlinear distortion on the closest pair of subcarrier data and phase-conjugated pilot. As a result, one phase-conjugated pilot can be used to compensate the nonlinear distortions on several subcarriers and the overhead due to phase-conjugated pilots is relaxed and can be designed according to the requirement of a specific application.

While the most accurate nonlinear compensation will be achieved by weighting and summing the nonlinear distortion estimated from all of the phase conjugate pairs, the nonlinear distortion on subcarriers that are not part of phase conjugate pairs can be estimated in various ways. The first approximation is to simply use the estimated nonlinear distortion from the nearest phase conjugate pair. The second approximation is to use a linear interpolation of the estimated nonlinear distortions from the two closest phase conjugate pairs (two points).

In common with PCTW, the performance of a system based on PCP can be further improved with 50% electrical dispersion pre-compensation (pre-EDC), which is applied to create a dispersion-symmetry along the transmission link. This dispersion map enhances the similarity between nonlinear distortions on subcarrier data and its phase conjugate, thus further improving the effectiveness of the PCP nonlinearity compensation scheme.

4.4.2 Performance Benefit of PCP

As a proof of concept, a simulation of the PCP scheme for a single-channel 112 Gb/s PDM QPSK CO-OFDM transmission was conducted. The OFDM useful duration was 50 ns (20 MHz subcarrier spacing), IFFT size of 2048 where 1400 subcarriers were used for data transmission. Firstly, to demonstrate the effectiveness of the nonlinear noise cancellation scheme based on the coherent superposition of the PCP pairs, the ASE noise was turned off.

The simulation results are shown in Fig. 4.11 for a 1200 km optical link with 5 dBm of the launch power. After coherent superposition, a dramatic reduction (~7 dB) of the SNR was observed, indicating that the nonlinear distortion on a data-carrying subcarrier and those of its PC are highly anticorrelated, especially if the frequency spacing is small.

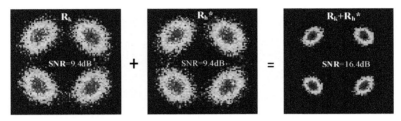

Figure 4.11 Nonlinearity cancellation based on the coherent superposition of PCP pairs. The transmission distance is 1200 km, launch power is 5 dBm, ASE noise is not considered, and 50% pre-EDC is adopted.

Figure 4.12 plots the reduction in the signal variance (σ^2), which is equivalent to the SNR improvement, when coherent superposition is applied for PCP pairs in systems with and without 50% pre-EDC. The difference between the open and solid symbols illustrates the impact of ASE noise. When the launch power is small, the dominant limiting factor in the system is the ASE noise. As a result, the coherent superposition of the PCP pairs using 50% pilots offers around ~3 dB reduction of the signal variance, as expected from the linear effects of coherent superposition of two copies of the same signal [23]. However, with increasing launch powers a larger reduction in σ^2 eventually occurs, indicating the maximum effectiveness of the proposed nonlinear noise cancellation scheme. Note that in this regime, there is little impact from the addition of ASE noise suggesting that the system is limited by compensation accuracy rather than the fundamental parametric noise amplification process. When 50% pre-EDC is performed, the reduction of σ^2 is further enhanced, reaching around 7 dB at a high level of the launch power. When considering only the nonlinear noise distortion, an even higher reduction of ~8 dB can be observed, which is comparable to the original PCTW technique.

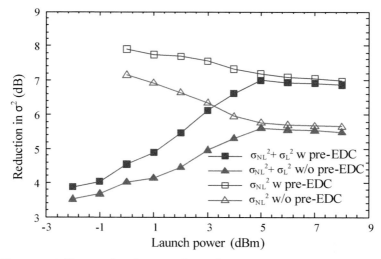

Figure 4.12 Measured reduction of signal variance from both nonlinear distortion (σ_{NL}^2) and linear noise (σ_L^2) as a function of the launch power, in PCP systems with 50% overhead with and without 50% pre-EDC. The transmission distance is 1200 km, and ASE noise is included.

The effective SNR of 112 Gb/s OFDM systems with 50% PCPs and without any PCPs are compared in Fig. 4.13. Note that, the spectral efficiency is reduced by a factor of 2 when 50% of the subcarriers are transmitted as PCPs. In Fig. 4.13, by combining the pre-EDC and PCP techniques, a reduction of around 4.5 dB in the signal variance can be achieved at a transmission distance of 3200 km at the cost of 50% overhead. The nonlinear threshold is also increased by 9 dB with PCP compensation. This result clearly indicates that a substantial fraction of the nonlinear distortion can be mitigated by coherently adding the phase-conjugated pilot and its correlated data subcarrier. As a result of this improvement, a longer transmission distance can be achieved. In Fig. 4.13 the effective SNR of a system with 50% PCP after 6400 km of transmission distance is also plotted for comparison purpose. As can be seen, this system still offers around 1.5 dB advantage in performance in comparison with OFDM system without PCP after 3200 km of transmission distance. This comparison indicates that the product of spectral efficiency and transmission distance can be significantly increased with PCP techniques and is consistent with results obtained for PC-TW.

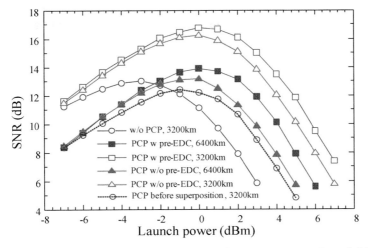

Figure 4.13 Nonlinear cancellation using coherent superposition of PCP pairs in long-haul 56 Gb/s (net data rate) CO-OFDM transmission. The transmission distances are 3200 km (open symbols) and 6400 km (closed symbols).

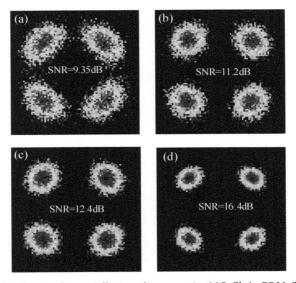

Figure 4.14 Received constellation diagrams in 112 Gb/s PDM CO-OFDM systems without (with the same bandwidth, before CS) (a) and with PCPs for fiber nonlinearity compensation (b: 12.5%, c: 25%, d: 50% overhead). The transmission distance is 1200 km, the launch power is 5 dBm.

This implementation offers excellent performance but it requires 50% overhead. The required overhead can be reduced by using the estimated nonlinear distortion on one pair of subcarrier data and its PCP to compensate the nonlinear distortions on other subcarriers. Specifically, one PCP can be used to compensate the nonlinear distortion of 2, 3, 4, or more data subcarriers at the cost of 33%, 25%, 20%, or smaller overhead, respectively. As an example, the received constellation diagrams in 112 Gb/s PDM CO-OFDM systems with PCPs with various overheads are shown in Fig. 4.14. In Fig. 4.15, the received constellation diagrams of systems with and without PCPs for fiber nonlinearity compensation are shown for different values of PCP overhead with the launch power deliberately set in the highly nonlinear region (+5 dBm). The trade-off between overhead due to PCPs and performance can be clearly observed. A better performance comes with the cost of larger overhead due to the transmission of additional PCPs.

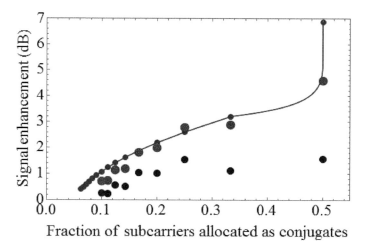

Figure 4.15 Signal enhancement of a 3200 km PDM CO-OFDM system at the optimum launch power as a function of the fraction of subcarriers allocated as phase conjugate pilots, showing measured reduction in signal variance (red symbols), net gain after subtraction of overhead (purple) and predicted signal to noise ratio gain (blue symbols).

The reduction in the signal variance (in dB) at the optimum launch power (difference of the minimum achievable σ^2 in systems without and with PCPs) and the net benefit in dB (after extracting the spectral efficiency reduction due to the PCPs) as a function of the overhead due to PCPs are shown in Fig. 4.15. With 50%, 33%, and 20% overhead the achievable reduction in σ^2 are 4.6, 3.2 and 2.1 dB, respectively, or approximately 0.1 dB per 1% of overhead. This is confirmed by a theoretical estimation (shown in blue). To obtain this estimation, the normalized difference in nonlinear distortion from FWM was computed for all possible subcarrier triplets whose nonlinear distortions fall on either the carrier or its conjugate. An excellent fit is observed over a wide range of configurations with the exception of the case where 50% of the subcarriers are phase conjugate pilots. This error is due to a reduced benefit of coherent superposition when the noise fields are no longer statistically independent, due to their parametric amplification by the signal.

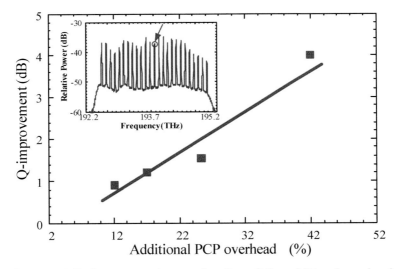

Figure 4.16 Performance gain as a function of the additional overhead due to PCPs for the center channel in 25×20 Gb/s WDM CO-OFDM transmission, after 3200 km of distance. Without PCPs, an overhead of ~8% was required for CPE compensation.

The effectiveness of PCP is experimentally confirmed in Fig. 4.16 for a 25 × 20 Gb/s WDM CO-OFDM transmission. Without PCP, 8% OFDM subcarrier were allocated for phase noise estimation [27, 28]. As a result, the additional overhead for nonlinear compensations were 12%, 17%, 25%, and 42% when

20%, 25%, 33%, and 50% of OFDM subcarriers, respectively, are transmitted as PCPs. As shown in Fig. 4.16 the achieved performance gains were 0.9, 1.2, 1.5, and 4 dB, respectively. This result clearly shows the flexibility of the proposed PCP fiber nonlinearity compensation technique, allowing the number of PCP to be chosen to meet the performance requirement.

To conclude, phase-conjugated pilot is an effective and flexible nonlinear compensation technique, which can be applied in both single-polarization and PMD systems, in both single-channel and WDM systems. With this technique, the fiber nonlinearity impairments due to the Kerr effect in OFDM systems can be effectively compensated without the complexity of DBP or 50% loss in capacity of PC-TW. The technique proposed here can be effectively implemented in both single-polarization and PDM systems, in both single-channel and WDM systems. In other words, nonlinearity compensation using PCPs offers a simple, easy implementation applicable to any optical links where the level of nonlinear compensation may be readily tuned by selecting an appropriate number of PCPs.

4.5 Phase-Conjugated Subcarrier Coding (PCSC)

4.5.1 Principle of PCSC

As discussed in the previous section, the PCP technique allows nonlinearity mitigation to be achieved in a flexible way, where the overhead can be adjusted according to the required performance gain. However, it is still desirable to remove the overhead completely to maximize the system spectral efficiency. To address this issue, a phase-conjugated subcarrier coding (PCSC) scheme has been proposed in [29, 30] for CO-OFDM transmission by adopting the concept of dual PCTW [31] to encoding and processing neighboring OFDM subcarriers simultaneously.

In the PCSC scheme (Fig. 4.17), each pair of neighboring OFDM subcarriers (with the indexes of $2k - 1$ and $2k$, where k is an integer) after symbol mapping are encoded before being fed into the IFFT block to generate the time-domain signal as

$$\begin{cases} S_1(2k-1) = S(2k-1) + S(2k) \\ S_1(2k) = S^*(2k-1) - S^*(2k), \end{cases} \tag{4.29}$$

where (.)* stands for the complex conjugation operation.

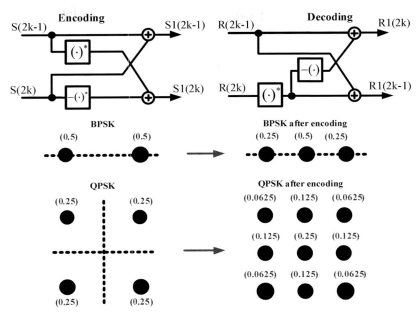

Figure 4.17 Phase-conjugated subcarrier coding scheme for CO-OFDM transmission. The numbers (in red) are the probabilities of symbols in the constellation set.

At the receiver, before symbol demapping, the received information symbols in this subcarrier pair are decoded as

$$\begin{cases} R_1(2k-1) = R(2k-1) + R^*(2k) \\ R_1(2k) = R(2k-1) - R^*(2k). \end{cases} \quad (4.30)$$

It should be noted that the PCSC can be considered as a one-by-one mapping scheme that does not require any overhead. The only requirement of PCSC is that the number of OFDM subcarriers is even. The PCSC scheme modifies both the constellation set and probabilities of constellation points. As shown in Fig. 4.17, if the input modulation format is BPSK with equal probability (0.5, 0.5) for each constellation point (−1, 1), the output constellation set will be a 3 ASK (−2, 0, 2) in which the symbol "0" occurs twice as often as the two other information symbols (−2, 2). This indicates that 50% of BPSK OFDM subcarriers will be turned off after encoding. Similarity, if the input modulation format is QPSK, after encoding, the output constellation set will be a 9QAM with unequal probabilities (Fig. 4.17), which can potentially reduce the nonlinear distortions on OFDM subcarriers due to the unequal power distribution across the OFDM band [32].

The sensitivities of OFDM systems with and without PCSC scheme in the additive white Gaussian noise channel are compared in Fig. 4.18, for different modulation formats, namely BPSK, QPSK, and 8QAM. It can be seen that independently of the modulation format used, PCSC gives no performance gain or penalty (the same sensitivity) in linear transmission channels. This result indicates that PCSC is ineffective for CO-OFDM systems if the distortions on neighboring subcarriers are Gaussian distributed and uncorrelated. However, if the OFDM subcarrier frequency spacing is small (tens of MHz) we can expect that the nonlinear phase shifts on neighboring subcarriers will be highly correlated. Thus, potential performance gain can be achieved by encoding and processing neighboring subcarriers simultaneously at the transmitter and receiver. In order to enhance the similarity of nonlinear distortions on neighboring OFDM subcarriers, pre-EDC is applied in this work to create a dispersion-symmetry along the transmission link.

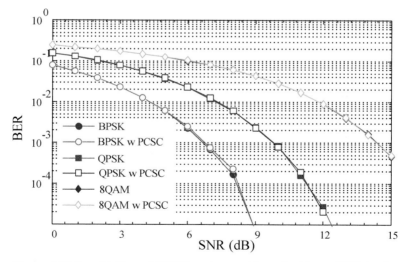

Figure 4.18 Sensitivities of OFDM systems with and without PCSC in the linear channel with white Gaussian noise.

4.5.2   Performance of PCSC

As a proof of concept, a simulation of the PCSC scheme in a single-channel 80 Gbaud PDM CO-OFDM transmission system was firstly conducted with BPSK and QPSK modulation formats. The IFFT size was 2048, where 1000 subcarriers were modulated with data while zeros occupying the remainder. The OFDM useful symbol duration was 12 ns and a cyclic prefix of 0.4 ns was added for polarization mode dispersion (PMD) compensation. The net bit-rate (after extracting 7% FEC) is 150 and 300 Gb/s when BPSK and QPSK are adopted.

Performances of the 150 Gb/s PDM CO-OFDM systems with and without the PCSC scheme (with and without pre-EDC) are compared in Fig. 4.19. In this figure, the performance of the PCTW technique with QPSK modulation format providing the same SE (~2 bits/s/Hz) is also presented. As PCTW halves the SE, despite the effective nonlinear noise cancellation effect, PCTW with QPSK gives only around 0.5 dB advantage over the traditional BPSK PDM CO-OFDM transmission scheme. On the other hand, when the PCSC coding scheme combined with pre-EDC is applied, a performance improvement of 1.5 dB can be achieved without reducing the SE, confirming the effectiveness of the PCSC scheme.

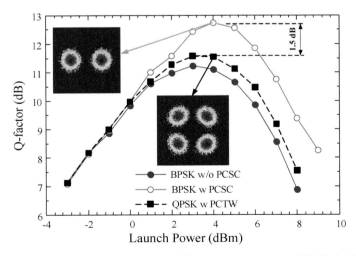

Figure 4.19 Q-factor as a function of the launch power in 150 Gb/s PDM CO-OFDM system with and without PCSC. The transmission distance is 8000 km.

Interestingly, the NLNS effect was also observed (Figs. 4.20, and 4.21) in a similar manner as in single carrier system with real-valued signal and the symmetrical dispersion map. Without PCSC, the real and imaginary parts of each constellation point have the same distribution. However, with PCSC and the optimized pre-EDC, the PDF of the real part of each constellation point is significantly narrowed. This nonlinear noise squeezing effect significantly reduces the BER in a transmission system using BPSK modulation format.

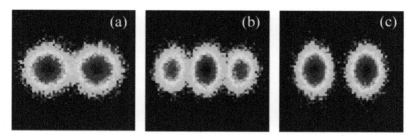

Figure 4.20 Constellation diagrams on *x*-polarization, 8 dBm of the launch power: (a) Without PCSC. (b, c) With PCSC, before and after decoding.

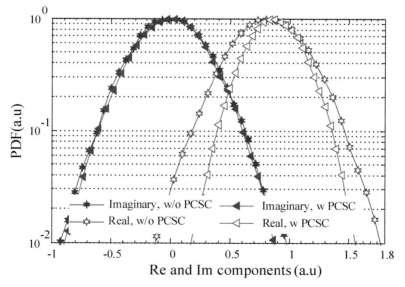

Figure 4.21 PDF of real and imaginary components for the "1" symbol in systems with and without the PCSC, the launch power was 7 dBm.

When the PCSC scheme combined with the optimized pre-EDC is applied for 300 Gb/s QPSK PDM CO-OFDM system, a performance improvement of around 0.7 dB is achieved, as shown in Fig. 4.22. This result clearly indicates that the proposed PCSC scheme also effectively mitigates the nonlinear distortions on OFDM subcarriers when QPSK modulation format is adopted. However, as QPSK cannot take the advantage of the nonlinear noise squeezing effect, the performance improvement in this case is only a half of those achieved with BPSK modulation format. In addition, without 50% pre-EDC, PCSC does not provide a significant improvement in the system's performance. This result confirms the benefit of pre-EDC in applying the proposed coding scheme (for both BPSK and QPSK modulation formats).

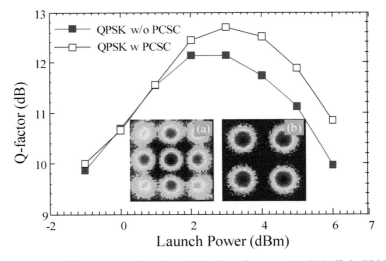

Figure 4.22 Q-factor as a function of the launch power in 300 Gb/s PDM CO-OFDM system with and without PCSC and constellation diagrams (before (a) and after (b) decoding) at 4 dBm, after 3200 km of transmission distance.

The PCSC technique can also be applied effectively in WDM CO-OFDM transmission. The experimental setup for 25 × 10 Gbaud WDM CO-OFDM transmission is shown in Fig. 4.23. The OFDM signals (400 symbols each of 20.48 ns length, 2% cyclic prefix) encoded with BPSK and QPSK modulation formats were generated offline in MATLAB using an IFFT size of 512, where 210 subcarriers

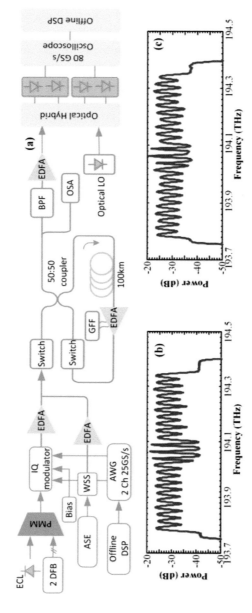

Figure 4.23 (a) Schematic of experimental setup of WDM CO-OFDM transmission with PCPs for fiber nonlinearity compensation. ECL, external cavity laser; PMM, polarization maintaining multiplexer; WSS, Wavelength Selective Switch; DFB, distributed feedback laser; BPF, band-pass filter (optical); GFF, gain flatten filter; OSA, optical spectrum analyser; LO, local oscillator. (b) Optical spectrum after the transmitter. (c) Optical spectrum after 2400 km of transmission distance.

were filled with data and the remainder zeros giving a line rate of 10 and 20 Gb/s (9.1 and 18.2 Gb/s after cyclic prefix and FEC overhead are removed) for BPSK and QPSK modulation formats, respectively. The DSP at the receiver included synchronization, x- and y-polarizations combination using the maxima-ratio combining method, frequency offset compensation, chromatic dispersion compensation using an overlapped frequency domain equalizer with overlap-and-save method, channel estimation and equalization with the assistance of initial training sequence (2 training symbols every 100 symbols), phase noise compensation with the help of 8 pilot subcarriers, and symbol detection.

The Q-factor as a function of the launch power in BPSK transmissions with and without PCSC and pre-EDC is plotted in Fig. 4.24 for a transmission distance of 6000 km. The constellation diagrams at the optimum launch power for both cases are also included. In Fig. 4.24, a performance improvement of around 1.5 dB is observed, which is equivalent with the simulation result plotted in Fig. 4.19 for single-channel transmission. This result indicates that PCSC with pre-EDC is also effective in compensating the nonlinear distortions due to cross phase modulation as long as the OFDM frequency spacing is small. This result confirms that PCSC is effective in both single and WDM transmission configurations.

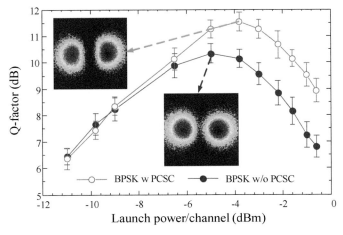

Figure 4.24 Q-factor as a function of the launch power/channel for the center channel in BPSK WDM CO-OFDM systems with and without PCSC. The transmission distance is 6000 km.

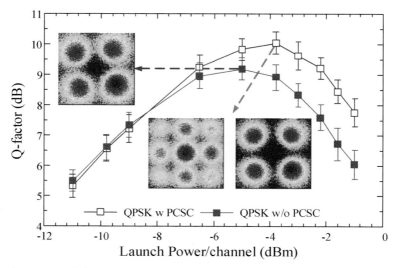

Figure 4.25 Q-factor as a function of the launch power/channel for the center channel in QPSK WDM CO-OFDM systems with and without PCSC. The transmission distance is 4000 km.

To conclude, the fiber nonlinearity impairments in CO-OFDM transmission can be mitigated by processing neighboring subcarriers simultaneously using the PCSC scheme. This coding scheme is very simple and can be effectively combined with pre-EDC to achieve a performance improvement of up to 1.5 dB. In addition, it can be effectively applied in both single-polarization and PDM systems, in both single-channel and WDM systems without suffering from carrier uncertainty problem

4.6 Other Variants of PCTW

4.6.1 Temporally Multiplexed PCTW (TM-PCTW)

As indicated before, it is also possible to "twin" a signal with its phase-conjugated copy in any other orthogonal dimension rather than polarization state. In particular, one may "twin" a signal and its phase-conjugated copy in different time segments as

$$E((2k+2)\cdot T) = E^*((2k+1)\cdot T),$$
(4.31)

where k is an integer and T is the symbol period. At the receiver, coherent superposition can be performed as

$$E((2k+1)\cdot T) = (E((2k+1)\cdot T) + E^*((2k+2)\cdot T))/2. \qquad (4.32)$$

This scheme can be referred to as temporally multiplexed PCTW [33] or conjugate data repetition [34]. A theoretical study in [33] has shown that the nonlinear distortions on neighboring pulses are also anticorrelated to the first order in TM-PCTW transmission. As a result, coherent superposition of PCTWs can also effectively mitigate the nonlinearity effect of fiber links and thus enhance the transmission performance and system reach. In a similar manner to the original PCTW scheme, symmetric dispersion map is mandatory in order to achieve the best performance.

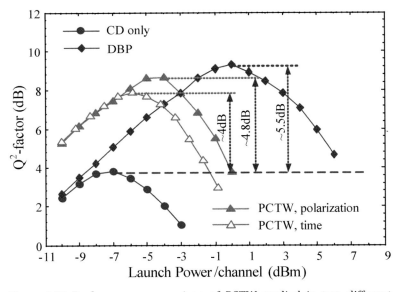

Figure 4.26 Performance comparison of PCTW applied in two different polarization states and time segments with DBP for a 7×15 Gbaud 16QAM Nyquist-spaced PDM WDM transmission over 40×80 km standard single mode fiber link with ideal Raman amplification.

The performance of the TM-PCTW technique is compared with the conventional PCTW technique and DBP in Fig. 4.26 for a 7 × 15 Gbaud 16QAM Nyquist-spaced PDM WDM transmission over 40 × 80 km of SSMF. For PCTW techniques, optimized pre- and post-dispersion was applied digitally to create a symmetrical dispersion map along the link. In Fig. 4.26, PCTW applied in two different time segments provides a performance improvement of 4 dB, meaning that 1 dB net gain is achieved due to the nonlinearity suppression. This result clearly indicates that nonlinear impairments can also be mitigated by applying PCTW in two different time segments. However, PCTW applied in two polarization states is a more effective compensation scheme giving a performance gain of 4.8 dB. In comparison to PCTW, DBP outperforms by 0.7 dB, providing a performance gain of 5.5 dB in line with the expected improvement in SNR. This slight difference may be explained by the fact that ideal DBP compensates all the inter signal nonlinearity while PCTW only compensates the nonlinear distortions to the first-order.

4.6.2 Modified PCTW (M-PCTW)

It has been shown recently [35, 36] that the SE of the PCTW technique can be increased by modulating one of the conjugated signals by an additional bits sequence. This scheme is referred to as modified PCTW (M-PCTW). The principle of M-PCTW scheme is demonstrated in Fig. 4.27, where E_a is a constant field with unit amplitude. It has been shown in [35] that the coherent superposition can be effectively applied provided that E_a is accurately detected:

$$E_r(t) = E_x(t)) + (E_y(t)/E_a(t))*$$

(4.33)

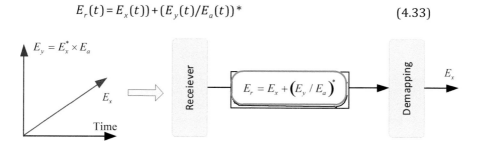

Figure 4.27 Principle of the modified PCTW.

Usually, detecting E_a requires a maximum likelihood approach, where all possible value of E_a is tested and the most likely is chosen. However, one major drawback of this scheme is that an error from E_a directly leads to an error from E_x and it has been shown that E_a is the dominant source of error [35]. To solve this issue, the authors of [35] proposed to apply a FEC with high overhead to protect the E_a bit sequence. Using this technique, the SE of PCTW-based scheme can be increased by 25% without compromising the system's performance.

4.6.3 PCTW for Multimode and Multi-Core Fibers

In general, the concept of PCTW can be extended to any orthogonal dimensions. In particular, conjugated signals can be simultaneously transmitted in two spatial modes [37] and/or two cores in a single fiber [38, 39]. As shown in [37], there are several options for applying the PCTW technique for few-mode fibers, such as PDM-wise and mode-wise PCTW. The best performance was achieved in the strong mode coupling regime due to the high correlation of nonlinear distortions on spatial modes, enhancing the effectiveness of the coherent superposition. In addition, strong mode coupling regime has been shown to be advantageous in suppressing the nonlinear distortion [40]. For weakly mode coupling regime in few mode fibers, PCTW simply acts as scrambled coherent superposition, which increases the SNR proportionally to the number of spatial modes [23]. In this case, PCTW simply enhances the linear performance rather than nonlinear performance. Similar result can be expected for the case of multicore fiber, where the best performance is likely to be achieved in the high nonlinear crosstalk regime. However, the optimum design of PCTW-based systems for multimode and multi-core fibers is still an open subject for further investigation.

4.7 Conclusion

Phase-conjugated twin wave is an interesting, low-complexity technique for compensating the nonlinearity impairment in fiber-optical communication link. In general, PCTWs can be transmitted effectively in any orthogonal dimensions such as two polarization states, time segments, frequencies, spatial modes and fiber cores. The original concept of PCTW can offer ~ 4.8 dB gain in the received SNR, which can be transferred to 3 times longer in the distance reach. As a result, PCTW is an effective technique to increase the reach of long-haul optical transmission and to reduce the system cost due to in-line regeneration. However, one major drawback of the PCTW technique is that it halves the system capacity or spectral efficiency. Therefore, the PCTW technique is not suitable for high-spectral-efficiency transmission. Several modification schemes have been proposed to mitigate the spectral efficiency loss of the original PCTW scheme, including PCP, PCSC, dual-PCTW, and M-PCTW. All of these schemes retain the advantage of low complexity and can be flexibly considered for different system configurations.

References

1. R. Essiambre, G. Kramer, P. J. Winzer, G. J. Foschini, and B. Goebel, Capacity limits of optical fiber networks, *J. Lightwave Technol.,* 28, pp. 662–701, 2010.

2. A. D. Ellis, Z. Jian, and D. Cotter, Approaching the Non-Linear Shannon Limit, *J. Lightwave Technol.,* 28, pp. 423–433, 2010.

3. A. Mecozzi and R.-J. Essiambre, Nonlinear Shannon limit in pseudolinear systems, *IEEE/OSA J. Lightwave Techn.,* 30(12), pp. 2011–2024, June 2012.

4. M. Secondini, E. Forestieri, and G. Prati, Achievable information rate in nonlinear WDM fiber-optic systems with arbitrary modulation formats and dispersion maps, *IEEE/OSA J. Lightwave Techn.,* 31(23), pp. 3839–3852, Dec. 2013.

5. S. T. Le, J. E. Prilepsky, and S. K. Turitsyn, Nonlinear inverse synthesis for high spectral efficiency transmission in optical fibers, *Opt. Express,* 22, pp. 26720–26741, 2014.

6. D. Rafique, M. Mussolin, M. Forzati, J. Mårtensson, M. N. Chugtai, and A. D. Ellis, Compensation of intra-channel nonlinear fibre impairments using simplified digital back-propagation algorithm, *Opt. Express,* 19(10), pp. 9453–9460, 2011.

7. Pepper and A. Yariv, Compensation for phase distortions in nonlinear media by phase conjugation, *Opt. Lett.,* 5, pp. 59–60, 1980/02/01 1980.

8. W. Pieper, C. Kurtzke, R. Schnabel, D. Breuer, R. Ludwig, K. Petermann, and H. Weber, Nonlinearity-insensitive standard-fibre transmission based on optical-phase conjugation in a semiconductor-laser ampli- fier, *Electronics Lett.*, 30(9), pp. 724–726, 1994.

9. K. Solis-Trapala, M. Pelusi, H. N. Tan, T. Inoue, and S. Namiki, Transmission optimized impairment mitigation by 12 stage phase conjugation of WDM 24×48 Gb/s DP-QPSK signals, *Proceedings Optical Fiber Communications Conference and Exhibition (OFC)*, p. Th3C.2, 2015.

10. W. A. D. Ellis, M. E. McCarthy, M. A. Z. Al-Khateeb, and S. Sygletos, Capacity limits of systems employing multiple optical phase conjugators, *Opt. Express*, 23(16), pp. 20381–20393, 2015.

11. X. Liu, R. A. Chraplyvy, P. J. Winzer, W. R. Tkach, and S. Chandrasekhar, Phase-conjugated twin waves for communication beyond the Kerr nonlinearity limit, *Nat. Photon*, 7, pp. 560–568, 2013.

12. X. Liu, S. Chandrasekhar, P. J. Winzer, R. W. Tkach, and A. R. Chraplyvy, Fiber-nonlinearity-tolerant superchannel transmission via nonlinear noise squeezing and generalized phase-conjugated twin waves, *J. Lightwave Technol.*, 32(4), pp. 766–775, Feb. 15, 2014.

13. X. Liu, H. Hu, S. Chandrasekhar, R. M. Jopson, A. H. Gnauck, M. Dinu, C. Xie, and P. J. Winzer, Generation of 1.024-Tb/s Nyquist-WDM phase- conjugated twin vector waves by a polarization-insensitive optical parametric amplifier for fiber-nonlinearity-tolerant transmission, *Opt. Express*, 22(6), pp. 6478–6485, 2014.

14. P. K. A. Wai, C. R. Menyuk, and H. H. Chen, Stability of solitons in randomly varying birefringent fibers, *Opt. Lett.*, 16, 1231–1233, 1991.

15. D. Marcuse, C. R. Menyuk, and P. K. A. Wai, Application of the Manakov-PMD equation to studies of signal propagation in optical fibers with randomly varying birefringence, *J. Lightwave Technol.*, 15(9), pp. 1735–1746, Sep. 1997.

16. X. Wei, Power-weighted dispersion distribution function for characterizing nonlinear properties of long-haul optical transmission links, *Opt. Lett.*, 31, pp. 2544–2546, 2006.

17. J. D. Ania-Castañón, Quasi-lossless transmission using second-order Raman amplification and fibre Bragg gratings, *Opt. Express*, 12(19), pp. 4372–4377, 2004.

18. I. Phillips, M. Tan, M. F. Stephens, M. McCarthy, E. Giacoumidis, S. Sygletos, et al., Exceeding the nonlinear-Shannon limit using Raman laser based amplification and optical phase conjugation,

Optical Fiber Communication Conference, San Francisco, California, 2014.

19. Y. Tian, Y.-K. Huang, S. Zhang, P. R. Prucnal, and T. Wang, Demonstration of digital phase-sensitive boosting to extend signal reach for long-haul WDM systems using optical phase-conjugated copy, *Opt. Express*, 21, 5099–5106 2013.

20. P. Poggiolini, G. Bosco, A. Carena, V. Curri, Y. Jiang, and F. Forghieri, The GN-model of fiber non-linear propagation and its applications, *J. Lightwave Technol.*, 32(4), pp. 694–721, 2014.

21. G. Bosco, P. Poggiolini, A. Carena, V. Curri, and F. Forghieri, Analytical results on channel capacity in uncompensated optical links with coherent detection: erratum, *Opt. Express*, 20(17), pp. 19610–19611, Aug. 2012.

22. X. Chen and W. Shieh, Closed-form expressions for nonlinear transmission performance of densely spaced coherent optical OFDM systems, *Opt. Express*, 18(18, pp. 19039–19054, Aug. 2010.

23. X. Liu, S. Chandrasekhar, P. J. Winzer, A. R. Chraplyvy, R. W. Tkach, B. Zhu, T. F. Taunay, M. Fishteyn, and D. J. DiGiovanni, Scrambled coherent superposition for enhanced optical fiber communication in the nonlinear transmission regime, *Opt. Express*, 20, 19088–19095, 2012.

24. E. Ip and J. M. Kahn, Compensation of Dispersion and nonlinear impairments using digital backpropagation, *J. Lightwave Technol.*, 26, 3416–3425, 2008.

25. O. V. Sinkin, R. Holzlöhner, J. Zweck, and C. R. Menyuk, Optimization of the split-step Fourier method in modeling optical-fiber communications systems, *J. Lightwave Technol.*, 21(1), pp. 61–68, 2003.

26. S. T. Le, M. E. McCarthy, N. Mac Suibhne, A. D. Ellis, and S. K. Turitsyn, Phase-conjugated pilots for fibre nonlinearity compensation in CO-OFDM Transmission, *J. Lightwave Technol.*, 33, 1308–1314, 2015.

27. S. T. Le, T. Kanesan, M. McCarthy, E. Giacoumidis, I. Phillips, M. F. Stephens, et al., Experimental demonstration of data-dependent pilot-aided phase noise estimation for CO-OFDM, *Optical Fiber Communication Conference*, San Francisco, California, 2014, p. Tu3G.4.

28. S. T. Le, T. Kanesan, E. Giacoumidis, N. J. Doran, and A. D. Ellis, Quasi-pilot aided phase noise estimation for coherent optical OFDM systems, *Photonics Technol. Lett. IEEE,* 26, pp. 504–507, 2014.

29. S. T. Le, E. Giacoumidis, N. Doran, A. D. Ellis, and S. K. Turitsyn, Phase-conjugated subcarrier coding for fibre nonlinearity mitigation in CO-OFDM transmission, presented at the ECOC, Cannes, France, paper We.3.3.2, 2014.

30. S. T. Le, M. E. McCarthy, N. M. Suibhne, M. A. Z. Al-Khateeb, E. Giacoumidis, N. Doran, et al., Demonstration of phase-conjugated subcarrier coding for fiber nonlinearity compensation in CO-OFDM transmission, *J. Lightwave Technol.*, 33, pp. 2206–2212, 2015.

31. T. Yoshida, T. Sugihara, K. Ishida, and T. Mizuochi, Spectrally-efficient dual phase-conjugate twin waves with orthogonally multiplexed quadrature pulse-shaped signals, *Optical Fiber Communication Conference*, San Francisco, California, 2014, p. M3C.6.

32. S. T. Le, K. Blow, and S. Turitsyn, Power pre-emphasis for suppression of FWM in coherent optical OFDM transmission, *Opt. Express,* 22, pp. 7238–7248, 2014.

33. S. T. Le, M. E. McCarthy, S. K. Turitsyn, I. Phillips, D. Lavery, T. Xu, P. Bayvel, A. D. Ellis optical and digital phase conjugation techniques for fiber nonlinearity compensation, *Opto-Electronics and Communications Conference (OECC), 2015*, Shanghai, 2015.

34. H. Eliasson, P. Johannisson, M. Karlsson, and P. A. Andrekson, Mitigation of nonlinearities using conjugate data repetition, *Opt. Express*, 23, 2392–2402, 2015.

35. Y. Yu and J. Zhao, Modified phase-conjugate twin wave schemes for fiber nonlinearity mitigation, *Opt. Express*, 23, 30399–30413, 2015.

36. Y. Yu, W. Wang, P. D. Townsend and J. Zhao, Modified phase-conjugate twin wave schemes for spectral efficiency enhancement, *Optical Communication (ECOC), 2015 European Conference on*, Valencia, 2015, pp. 1–3.

37. J. S. Tavares, L. M. Pessoa, and H. M. Salgado, Phase conjugated twin waves based transmission in few modes fibers, 2015 17th International Conference on Transparent Optical Networks (ICTON), Budapest, 2015.

38. R. Asif, F. Ye, and T. Morioka, Phase conjugated twin-waves in 8×21×224 Gbit/s DP-16QAM multi-core fiber transmission, *Asia Communications and Photonics Conference 2014*, OSA Technical Digest (online) (Optical Society of America, 2014), paper AF1F.5.

39. J. K. Hmood, K. A. Noordin, S. W. Harun, Effectiveness of phase-conjugated twin waves on fiber nonlinearity in spatially multiplexed

all-optical OFDM system, *Opt. Fiber Technol.*, 30, Pages 147–152, July 2016.

40. F. Ferreira, N. Mac Suibhne, C. Sánchez, S. Sygletos, and A. D. Ellis, Advantages of strong mode coupling for suppression of nonlinear distortion in few-mode fibers, *Optical Fiber Communication Conference*, OSA Technical Digest (online) (Optical Society of America, 2016), paper Tu2E.3.

Chapter 5

Information-Theoretic Concepts for Fiber Optic Communications

Mariia Sorokina[a] and Metodi P. Yankov[b]

[a]*Aston Institute of Photonic Technologies,*
Aston University, A4 7ET, UK
[b]*Department of Photonics Engineering,*
Technical University of Denmark, 2800 Kgs. Lyngby, Denmark

m.sorokina@aston.ac.uk, meya@fotonik.dtu.dk

In this chapter, we briefly review fundamental information-theoretic concepts related to fiber optic communications, channel modeling, and calculations of channel capacity. We discuss different fiber-optic channel models and the challenges related to capacity calculations.

5.1 Communication Channel

A communication system consists of

1. information source—a message, a string of symbols chosen from an alphabet;

Optical Communication Systems: Limits and Possibilities
Edited by Andrew Ellis and Mariia Sorokina
Copyright © 2020 Jenny Stanford Publishing Pte. Ltd.
ISBN 978-981-4800-28-0 (Hardcover), 978-0-429-02780-2 (eBook)
www.jennystanford.com

2. transmitter, which samples, compresses, and encodes a message to produce a signal suitable for transmission;
3. channel—a medium through which the signal is transmitted; a variety of channels and mathematical models include description of the signal distortion by noise and degradation due to nonlinearity, fading, and other effects that induce errors at the receiver;
4. receiver, which reconstructs the message from the received signal;
5. destination, the person or device for whom the message is intended.

A digital communication system can be represented by the basic building blocks shown in Fig. 5.1. To correct transmission errors, one has to add redundancy, so the transmitted data (a sequence of uncoded bits) is transformed to coded bits $A_{[k]}$ by a channel encoder (see Fig. 5.1) according to the chosen forward error correction (FEC) coding scheme, e.g., Reed–Solomon, or low- density parity check codes (see Chapter X for more details). To differentiate between a sequence and a set we denote them as $A_{[k]}$ and A_k correspondingly. A constellation mapper assigns the bits to the chosen modulation format: $A_{[k]} \rightarrow X_{[k]}$. Bit-to-symbol mapping is usually performed by channel coding techniques such as Gray coding (adjacent constellation symbols have a maximum Hamming distance of 1) or differential coding. The symbols $A_{[k]}$ are real or complex numbers depending on the modulation format. There are a variety of different modulation formats, where information is encoded in phase (phase-shift keying [PSK]) or amplitude (amplitude-shift keying [ASK]), a combination of amplitude and phase modulation (APSK), or quadrature amplitude modulation (QAM).

Further, the discrete-time signal $X_{[k]}$ is transformed by a pulse shaper to continuous time signal: $X(t)$, which is then transmitted through the media (for example, launched into the fiber channel).

The continuous-time signal $Y(t)$ at the input of the receiver is converted to the electrical baseband domain, analogue-to-digital converted, and then the obtained digital signal is resampled. Afterwards, digital signal processing algorithms are applied to compensate for the transmission distortions (e.g., clock recovery algorithm, chromatic dispersion, polarization mode dispersion in the optical fiber case, etc.). Then the signal is resampled to one sample per symbol to obtain discrete-time signal representation $Y_{[k]}$ (this might also include maximum likelihood or maximum a posteriori probability estimation to remove degradations caused by signal transmission). Further, we denote deterministic outcomes of a channel input $X_{[k]}$ and output $Y_{[k]}$ by lower case x_k and x_k correspondingly, and their probability of occurrence by, e.g., $P_X(X_{[k]} = x_k) = p(x_k)$. We also notate probability density and probability mass functions with lower and upper case correspondingly.

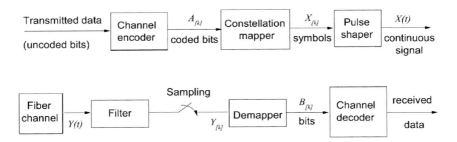

Figure 5.1 Basic components of a communication system.

Afterwards, $Y_{[k]}$ is converted into bits $B_{[k]}$ by a de-mapper and then a decoder removes redundancy added due to FEC and corrects the errors; thus, the received data are obtained. To characterize the system performance, the bit error rate (BER) or symbol error rate (SER) is usually used. The BER is defined as the number of bit errors (incorrectly received bits $B_{[k]}$ for given input bits $A_{[k]}$) per second or probability to receive incorrect bit $B_{[k]}$ for a given bit $A_{[k]}$. Similarly, SER is defined as the number of symbol errors (incorrectly received symbols $Y_{[k]}$ for given input symbols $X_{[k]}$) per second or probability to receive incorrect symbol $Y_{[k]}$ for a given symbol $X_{[k]}$.

5.2 Fiber Optic Communications

A typical communication system is presented in Fig. 5.2a. At the transmitter, the message is modulated to a discrete time set of constellation symbols (here, 64-QAM plotted in Fig. 5.2b and after pulse shaping it is mapped to a continuous time signal, see Fig. 5.2c, which is subsequently launched into an optical fiber link. The propagation of the continuous-time signal $U(z, t)$ in the optical fiber, see Fig. 5.2d, is governed by the well-known nonlinear Schrödinger equation (NLSE) (we follow here notations and assumptions used in Chapter 1, Eqs. 1.15–1.17 and refer to them for details):

$$\frac{\partial U(z,t)}{\partial z} = [\hat{L} + \hat{N}]U(z,t) \tag{5.1}$$

$$\hat{L} = -\frac{\alpha}{2} + j\frac{\beta_2}{2}\frac{\partial^2}{\partial t^2} + \frac{\beta_3}{6}\frac{\partial^3}{\partial t^3} \tag{5.2}$$

$$\hat{N} = -j\gamma(1 - f_r)|U(z,t)|^2 - j\gamma f_r \int h_r(\tau)|U(z,t-\tau)|^2 \delta\tau \tag{5.3}$$

The deterministic distortions, see Fig. 5.2e, are introduced mainly by the fiber loss α, second-order dispersion parameter β_2, and by Kerr nonlinearity characterized by the coefficient γ. The signal attenuation in fiber caused by the losses is compensated by optical signal amplification leading to an in-line noise due to the amplified spontaneous emission effect. The stochastic distortions arising from signal amplification can be described by including a term representing zero-mean additive white Gaussian noise (AWGN) $\eta(t, z)$. An accurate channel model needs to convey these effects. It can be continuous-time model (operating with time-dependent signals, e.g., the received signals $U_{out}(t)$, see Fig. 5.2f) or discrete-time (operating with discrete set of symbols, e.g., the output symbols $\{Y_{[k]}\}$).

In the simplest case of linear transmission, the noise can be included through a simple model of a linear continuous-time channel: $U_{out}(t) = U_{in}(t) + \eta(t)$, when the output signal $U_{out}(t)$ is dependent on the input signal $U_{in}(t)$ (in the same time slot) and the noise term $\eta(t)$ represents the accumulated noise at the receiver. The corresponding discrete-time model can be written as: $Y_{[k]} = X_{[k]} + \eta_{[k]}$, assuming the output signal was sampled, resulting in a discrete set of symbols.

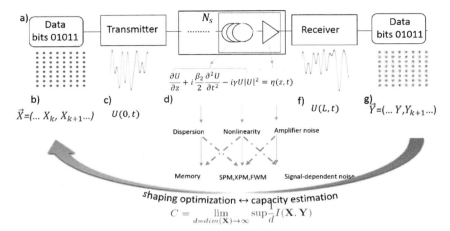

Figure 5.2 A schematic diagram of fiber-optic communication system. (a) The fundamental building blocks of a communication system where data are encoded to a discrete set of symbols $X_{[k]}$ (panel (b) and transformed into continuous-time form $U(0,t)$ (panel c) to be transmitted via the fiber channel. During transmission, the signal is governed by the NLSE (panel d), which results in distortions: dispersion, nonlinearity, noise. At the receiver, the distorted signal $U(L,t)$ (panel f) is converted to discrete-time output symbols $X_{[k]}$ (g). The shaping (the process of encoding bits to symbols) needs to be optimized for each set of transmission system parameters to receive the maximum achievable transmission rate–channel capacity C.

However, in a highly nonlinear regime, signal-noise interference becomes a dominant effect (see Chapters 1–2), therefore accurate modeling of nonlinear interaction of signal with noise during propagation is important.

Moreover, a discrete-time channel model is crucial for an information-theory–based analysis as it enables estimation and optimization of the mutual information functional $I(X, Y)$ for deriving the optimum probability density function (pdf) $p_X(X)$ or probability mass function $p_X(X)$ (probabilistic shaping discussed in Section 1.6) and for calculating the maximum reliable transmission rate, which is the channel capacity $C = \sup_{p_X \cdot (X)} I(X, Y)$. Note, that here we used supremum rather than maximum, this is because the definition $C = \max_{p_X(X)} I(X, Y)$ derived by C. Shannon in 1948 [1] holds only for memoryless channels (i.e. the output symbol depends only on the input symbol in the same time slot and is independent of the neighboring symbols, see example of a linear channel above). In case of channels with memory (e.g., fiber channel, which has memory due to dispersion caused signal broadening and nonlinear interference between neigboring symbols) one needs to use the generalization $C = \sup_{P_X(X)} I(X, Y)$ [2] (see discussion of convergence and further generalization in [3]).

The transition from continuous-time modeling, given by Eq. (5.1), to discrete-time modeling for fiber channel is not straightforward, since it requires expansion over a complete orthogonal set of basis functions $\{f_k(t)\}$. This is equivalent to matched filter demodulation at the receiver for generating observable discrete-time variables $\{Y_{[k]}\}$, see Fig. 5.2g. At the transmitter, signal expansion over the carrier pulses is considered, that is $U(0, t) = \sum_{k=-8}^{\infty} X_{[k]} f(t - kT)$, where $X_{[k]}$ are the complex modulated symbols, $f(t)$ is the time-varying pulse waveform, and T is the symbol period. At the receiver, the signal undergoes matched filtering, dispersion compensation and equalization.

Channel modeling is an important stage of optimization and capacity estimations as will be discussed in the next section, where requirements for accurate information-theoretic model will be also discussed. The overview of the derived fiber-optic channel models will be given in Section 5.4.

5.3   Shannon Capacity and Mutual Information

In 1948, Claude Shannon developed the concept of quantifying the information and estimating the efficiency of information transmission. Representing a discrete information source as a Markov process, Shannon used the entropy as a fundamental information measure. Shannon exploited the properties of Boltzmann entropy in statistical mechanics to develop the information entropy, commonly referred to as the Shannon entropy.

Shannon [1] defined the entropy of a discrete source with probability mass function $(P_1, ..., P_K)$ as

$$H = \Sigma_{i = 1 \ldots K} P_i \log_2 P_i, \qquad\qquad (5.4)$$

where the sum is over the support of the distribution. Thus, exploiting the properties of information, Shannon introduced the basic notion of information theory—the information entropy. Shannon was a pioneer in borrowing the statistical methods and notions for the analysis of information. The new approach enabled Shannon to develop two fundamental theorems in information theory, by which he defined the compression and transmission limits.

As one transmits information, one can expect noise distortions, which induce errors at the receiver. Since noise is a stochastic process, the corresponding impairments, being nondeterministic effects, cannot be compensated for and information losses will be inevitable. Therefore, one has to introduce redundancy to ensure successful information transmission. Consequently, to quantify the information transmitted via noisy channel—here two statistical processes (the source and the noise) affect the channel output—one can use mutual information $I(X, Y)$.

The mutual information measures the maximum amount of information that can be conveyed through the noisy channel with a given input distribution.

Since mutual information represents the achievable rate for a specific input distribution, one can optimize the latter to achieve the maximum of mutual information, which is defined as Shannon capacity—maximum achievable error-free transmission rate. As was discussed above, for efficient communication over a noisy channel, one has to introduce redundancy. Shannon showed that any stochastic channel can be used for reliable communication at nonzero rate (when the input and output are correlated) and defined the analytical expression for the optimal transmission rate. Shannon introduced the concept of the channel, as a mathematical model where noise corruption to the signal is described, and showed that the transmitter has to add redundancy. Shannon further defined the concept of capacity [1] for a discrete memoryless channel as:

$$C = \max_{P_X(X)} I(X, Y), \tag{5.5}$$

where optimization is over input probability $P_X(X)$. Capacity here is measured in bits/symbol.

Let us consider a channel, where the constellation mapper outputs (see Fig. 5.2) $X_{[k]}$ are discrete symbols from a discrete alphabet X and the outputs of the demapper $Y_{[k]}$ are discrete symbols of the alphabet Y. If the channel is memoryless (the output symbol $Y_{[k]}$ sampled at time k is dependent only by the input $X_{[k]}$ at the same time slot k), then the channel is described by the set of conditional probabilities: $P_{Y|X}(Y|X)$, with the mutual information given by

$$I(X; Y) = \Sigma_{x \in X,\ y \in Y}\, P_X(x)\, P_{Y|X}(y|x) \log_2 \frac{P_{Y|X}(y|x)}{P_X(x)} \qquad (5.6)$$

In case of continuous alphabet the sums are changed by integrals. The statistical properties of the channel are given by the conditional probability: $P_{Y|X}(Y|X)$. In fiber-optic systems, the channel memory is crucial as dispersion induces pulse broadening, while nonlinearity induces complex nonlinear interactions. Thus, to take into account memory effects one can use the above expression for mutual information with probabilities operating with d-dimensional input and output vectors becoming multivariate probabilities $P_{Y|X}(Y|X)$, while the capacity definition becomes [2]

$$C = \lim_{d \to \infty} \frac{1}{d} \sup_{P_X(X)} I(X, Y), \qquad (5.7)$$

Here the input and output vectors of length d are given by: $X = (X_1, ..., X_d)$ and similarly for $Y = (Y_1, ..., Y_d)$. To optimize the mutual information over the input distribution, one must derive the multivariate conditional distribution that takes into account the memory effects defined by the channel model. The channel model needs to incorporate nonlinear effects in the system: signal-signal and signal-noise beating as a result of the intersymbol interference.

5.4 Information-Theoretic Channel Modeling

A number of channel models have been proposed for the information-theoretic description of fiber-optic systems. These are generalized in Table 5.1. We classify them using the following features: whether the model is discrete-time (DT) or continuous time (CT), infinite or finite (I/F) memory, and if the capacity or low bounds have been derived. One can identify the following class of infinite-memory Gaussian noise models based on the assumption of Gaussian noise interference [4–8]. A number of models are related to expansion in Volterra series [9, 10] including multiple order terms [11]. However, the Volterra approach allows receiving only continuous time-form. The finite memory models (e.g., [12]) have been obtained only more recently.

Table 5.1 Review of channel models

Model	Time/ frequency	Memory	Capacity
Signal-signal			
Infinite memory Gaussian noise [4–8]	DT	I	Lower bound
Volterra series based model [9–11]	CT	F	Lower bound
Finite memory Gaussian noise [12]	DT	F	Lower bound
Nonlinear interference noise (NLIN) [13, 14]	DT	F	Lower bound
NLIN with Gaussian approximation [15]	DT	I	Lower bound
NLIN for phase noise [16]	DT	I	Lower bound
Nonlinear phase noise – 1 [17–18]	DT	no	Calculated
Signal-noise			
Nonlinear phase noise – 2 [18]	DT	no	Lower bound
Nonlinear signal-noise [19]	DT	I	Lower bound
Multivariate channel model [20]	DT	F	Lower bound

A discrete-time finite-memory channel model was derived in [13–15] taking into account signal-signal interactions. In order to estimate the capacity, the model was simplified using Gaussian noise statistics [16], thus losing the exact knowledge of the inter-symbol interference and leading to a concave capacity lower bound that vanishes at high powers. As it has been shown in [12], such approximations inevitably lead to underestimated capacity bounds.

However, as deterministic distortions (signal-signal beating) can be in principle ideally compensated, the impact of stochastic distortions (signal-noise beating) poses a fundamental limitation on communication rates. The impact of signal-signal [17] and signal noise [18] effects has been studied in memoryless channels, where it has been shown for the first time that the capacity of such channels is an increasing function of power even at the highly nonlinear regime. Finally, in [19] the conditional pdf for signal-dependent noise under continuous-time approximation was derived. Recently, a discrete-time finite-memory channel model taking into account signal-signal and signal-noise interference was obtained and monotonically increasing lower bound on its capacity was demonstrated [20].

From Table 5.1, it is clear that channel capacity has been obtained only for a few channel models. Below we discuss the procedure for calculating lower bounds on capacity numerically.

5.5 Numerical Calculations of Lower Bounds on Shannon Capacity

One can calculate lower bounds on Shannon capacity numerically for a discrete memoryless channel for a given $P_{Y|X}(Y|X)$ with fixed input and output alphabets by using the properties of Shannon capacity:

(1) $C \geq 0$;
(2) $C \geq \log_2 |\mathcal{X}|$, where $|\mathcal{X}|$ is the size of the input alphabet;
(3) $C \geq \log_2 |\mathcal{Y}|$, where $|\mathcal{Y}|$ is the size of the output alphabet;
(4) $I(X, Y)$ is a continuous function of $P_X(X)$;
(5) $I(X, Y)$ is a concave function of $P_X(X)$.

Hence, from the last property follows that the capacity has a local maximum, whereas from the properties 2 and 3 it follows that the maximum is finite for the finite alphabet. To calculate capacity lower bounds numerically, one can use different methods, such as the following:

- Constrained maximization using the Kuhn–Tucker conditions [21, 22], generalizes the method of Lagrange multipliers for inequality constraints.
- The gradient search algorithm of Frank–Wolfe [23] is an iterative first-order optimization algorithm for constrained convex optimization, also known as the conditional gradient method, reduced gradient algorithm and the convex combination algorithm. It is based on a linear approximation of the objective function in each iteration and moves slightly towards a minimum of this linear function (taken over the same domain).
- The Blahut–Arimoto algorithm [24, 25] is an iterative method that maximizes the capacity of arbitrary finite input/output alphabet sources. The optimization is performed recursively. The algorithm was modified for faster convergence using a natural-gradient-based or accelerated Blahut–Arimoto algorithm [26]. A number of algorithms were proposed for discrete alphabet [27] or continuous: using computation of a sequence of finite sums [28] or particle-based Blahut–Arimoto algorithm [29]. In the latter, the particles x_k are moved to increase the relative entropy while keeping the output probability fixed.
- For discrete-input and discrete-output channels with memory an efficient technique for calculation of transmission rate was proposed [30–32], later generalized for continuous channels in [33].
- Most recently, a new hybrid iterative method for computing the channel capacity of both discrete and continuous input, continuous output channels was proposed [34], which is a modification of the Blahut–Arimoto algorithm with additional adaptive iterative tuning of the discrete approximations of the input signal. The method can be particularly advantageous for highly nonlinear memoryless channels.

For fiber optic channels, which have memory, the methodology becomes even more complex and very few numerical results are known. A number of works have been dedicated to estimating upper [36] and lower bounds on capacity (see e.g., [15, 36, 37]). However, such bounds do not provide the input distribution $p(X)$, which would enable to achieve high transmission rates in practice. From analysis of achievable rates where Gaussian distribution was used [38–41], it is clear that there is a need for novel types of input distributions $p(X)$. Most recently, a number of works have been dedicated to the search of new probability functions [12, 20, 42, 43], which usually have different high order standardized moments compared to Gaussian distributions. The overview of progress on probabilistic shaping will be given in the following section.

5.6 Probabilistic Shaping

The pdf of a discrete variable X can be defined as

$$f_X(x) = \sum_{x_i \in X} P_i \delta(x - x_i),$$

where $\mathcal{X}$ is the signaling alphabet, δ is the Dirac delta function and P_i is the weight of the i-th point in the alphabet. The PMF of $\mathcal{X}P_X(X)$ is the set of the probabilities of occurrence P_i. When optimizing the transmission strategy, also referred to as constellation shaping, the pdf is typically optimized either in terms of the positions of the points x_i in the signaling space, known as geometric shaping, or their weights, known as probabilistic shaping.

In practice, the digital nature of data transmission requires that the signaling alphabet be of finite size. In such cases, the true capacity cannot be reached and instead, the term constellation constrained capacity (CCC) can be adopted. As the size of the signaling alphabet $|\mathcal{X}| \to \infty$, geometric shaping can also be performed implicitly by probabilistic shaping by setting the weights of certain points to 0. In such cases, the CCC approaches the true capacity. For standard linear channels, it can be proven that the CCC is arbitrarily close to the true capacity in the limit of large constellations [44, 45].

Probabilistic shaping is typically easier to perform than geometric shaping because the PMF appears in the mutual information (MI) functional (see Eq. (5.5)). For standard communication channels, e.g., binary channels, discrete, memoryless channels (DMCs) and Gaussian channels, the constellation constrained capacity can be efficiently found by performing convex optimization of MI, since under certain conditions, such as fixed channel likelihood (also known as the transition probability for DMC) $P(y|x)$ [46], the MI is concave in the input PMF. For DMCs, the optimal input PMF and the corresponding capacity can be found with the Blahut–Arimoto (BA) algorithm [24, 25]. The algorithm can be extended to linear additive white Gaussian noise (AWGN) channels [27] and also linear multiple-input multiple-output (MIMO) channels [47]. For such channels, it can also be proven that the MI is strictly concave in the PMF [48], resulting in a unique optimal solution for the PMF. In the limit of large signal-to-noise ratio (SNR), this solution was found to be the Maxwell-Boltzmann (MB) distribution over a sufficiently large signaling set, also known as quantized Gaussian [44, 45].[a] An example of an AWGN channel output for standard uniform 256QAM transmission and transmission with an optimized MB distribution for an SNR = 25 dB is given in Fig. 5.3. The optimized PMF exhibits larger mass towards the origin of the coordinate system and for a fixed power constraint, the result is an increased Euclidean distance. It is worth noting that for large SNR and when the signaling set is insufficiently small, the conditional entropy $H(X|Y)$ vanishes (the symbol error rate in such cases is 0, which means that as much information can be transferred reliably as the signaling allows). The optimal input PMF in such cases is the uniform as it maximizes the entropy $H(X)$ and thus the MI.

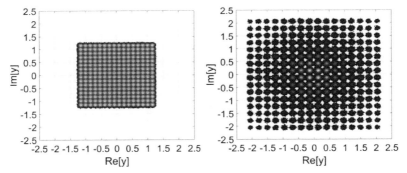

Figure 5.3 Received constellations for an AWGN channel with SNR=25 dB. (Left) Uniform PMF. (Right) Maxwell–Boltzmann PMF.

---

[a]When the signaling variable is continuous, the optimal input distribution is the Gaussian. In the discrete case, the quantized Gaussian is currently proven optimal in some boundary conditions, e.g., the above-mentioned SNR $\rightarrow$ 0.

5.6.1   Optimization for the Optical Fiber Channel

Two major obstacles exist to optimizing the input PMF and estimating the CCC on the optical fiber channel. First, the channel likelihood is not known in closed and tractable form. Moreover, the channel likelihood function is dependent on the input, which does not allow for a strict convex optimization as in the trivial Gaussian channel case. The dependency stems from the fact that the refractive index of silica, and thus the propagation velocity changes with the intensity of the optical field, which is known as the Kerr effect [49]. This effect results in data dependent nonlinear phase shift, which in combination with the amplification noise makes the likelihood function difficult to estimate. Suitable likelihood functions of tractable forms for the highly nonlinear region of transmission which are receiver friendly are currently unavailable, and receivers usually resort to the standard Gaussian function [50].

The second obstacle is related to the channel memory, which due to the interactions of dispersion and nonlinearities can easily exceed hundreds of symbols in metro-networks. For memoryless channels, it can be easily verified that memoryless inputs are optimal.[b] For such inputs, the PMF factorizes as $P_{X_1^K}(x_1^K) = \prod_{k=1}^{K} P_X(x_k)$, where x_1^K is the sequence from time 1 to K. Such cases are easy to treat, as the optimization is one-dimensional, and the MI can be estimated as a sum of the MIs on each dimension (in this case, the dimensionality describes time). For the fiber channel, the factorization assumption is often too strong. The memory that stems from the interplay of noise, nonlinearities and dispersion is in a typical optical system treated as noise and thus cannot be compensated effectively. Since the power of the nonlinear interference is proportional to the launch power, the achievable information rates (AIRs) of typical optical systems, operating under the memoryless assumption and the Gaussian likelihood function exhibit an optimal launch power which

---

[b]The multidimensional entropy $H(X_1^K)$ is maximized when the dimensions, in this case time, are independent. Then for memoryless channels, the conditional entropy $H(X_1^K|Y_1^K)$ is unchanged, and the MI is thus maximized when $H(X_1^K)$ is maximized.

maximizes the AIR, and degrading performance when the power is increased. An example of this behavior is given in Fig. 5.4, where the MI is estimated with the assumptions above for a given point-to-point fiber communication system. The MI is estimated in bits/symbol. In this case, the signaling alphabet is 256-(QAM), which limits the MI to the maximum value of the entropy $H(X)$ $\leq \log_2 256 = 8$, achieved with a uniform input PMF. The system parameters are given in Table 5.2. An erbium doped fiber amplified (EDFA) system is considered. The channel is simulated with the split-step Fourier method (SSFM) with a step size of 8 km and dual polarization is employed (polarization mode dispersion is neglected for simplicity). The dashed curves correspond to the AIR with a uniform input PMF, and the solid curves to a probabilistically optimized input PMF with a modified BA algorithm for the fiber channel [43]. The optimization is also performed under the memoryless assumption and with the Gaussian function used as channel likelihood.

Table 5.2 Transmission parameters for estimating the MI

Number of channels	5
Baudrate	28 GBaud
Channel spacing	30 GHz
Pulse-shape	Raised cosine
Roll-off factor	0.01
EDFA noise figure	5 dB
Fiber loss	0.2 dB/km
Dispersion	17 ps/(nm km)
Nonlinear coefficient	1.3 1/(W km)
Span length	80 km
Number of spans	10

In the case of a linear approximation of the channel with nonlinear coefficient $\gamma = 0$, the channel is well-modeled as an AWGN. For sufficiently small SNR (launch power, respectively), the AWGN channel capacity of $\log_2(1 + \text{SNR})$ found by Shannon is approached by the CCC. When the SNR is increased, the cardinality of the alphabet is insufficient, and the CCC approaches the above-mentioned maximum entropy. The shaping gain in this case is upper bounded by 1.53 dB in the limit of SNR $\to \infty$ and $| \mathcal{X} | \to \infty$ [51], and is $\approx$1.2 dB in the case of 256QAM at Pin $\approx$ –0.8 dBm per channel.

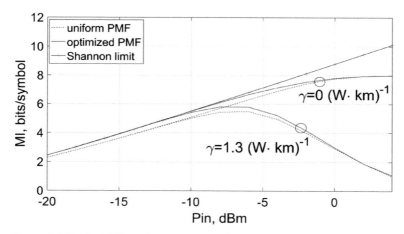

Figure 5.4 Typical MI performance as a function of the launch power of the optical fiber. After the optimal launch power, the performance degrades due to the significant nonlinear noise.

When the model is more realistic, the nonlinearities reduce the effective SNR on the channel. The effective SNR may be defined as SNR = $\mathbb{E}_k [|x_k|^2]/\mathbb{E}_k [|x_k - y_k|^2]$, where $\mathbb{E}_k[\cdot]$ denotes expectation over the time k. Reduced effective SNR results in reduced MI after the optimal launch power of –6 dBm, at which the effects of linear and nonlinear noise are balanced. The shaping gain can be defined in this case to be $\approx$0.4 bits/symbol at the optimal launch power. As pointed out in [52], shaping can have a negative effect on the effective SNR due to the enhanced nonlinear noise, however, positive shaping gain can still be achieved in the weakly nonlinear region. Similar shaping gains can be achieved with optimized MB distributions and various other optimization strategies, which rely on the above-mentioned memoryless, Gaussian assumptions [53].

Several recent works attempt at lifting these heavy constraints. Using multi-dimensional spheres, the authors in [54] demonstrate that the shaping gain in fibers can potentially exceed the ultimate *linear* Gaussian channel shaping gain of 1.53 dB mentioned above. In [55], an approximate channel model is adopted, which allows for the memory to be estimated and taken into account when estimating the rates. This results in an MI that is nondecreasing with power. A perturbation model is derived for the fiber in [20], which allows for the signal-noise interactions to be taken into account in the likelihood function. The continuous input alphabet is optimized to a *ripple* distribution, which allows for increasing the rate with launch power in the considered model. Finite state machines emitting QAM symbols are probabilistically optimized in [55], allowing for increased optimal launch power by suppressing sequences, which are associated with a high nonlinear noise power.

While the above mentioned works are promising for unlocking the highly-nonlinear region of the fibers, current transceivers are generally limited in their complexity to memoryless operation, and the gains from probabilistic shaping in currently deployed networks are limited by the channel memory.

5.6.2 Probabilistic Shaping of Binary Data

Achieving an optimized PMF with binary user data, which are also encoded for error protection, is not trivial. A straightforward technique for achieving a nonuniform distribution from i.i.d. bits is to employ source decoding [56, 57]. Most of the available methods for source decoding are variable length as they are data dependent. Source coding/decoding is typically performed after error-correction, where the bit sequences are error-free. However, in the case of shaping, the shaped data must appear on the channel, and will thus be erroneous at the receiver side. A single error in variable length decoders results in an error-propagation, making such strategies infeasible for coded data.

Some of the early works on probabilistic shaping employ a trellis at the transmitter side to encode the transmitted symbol sequence to one of lower energy, which allows for increased Eucledian distance and shaping gains of up to ≈1 dB [58]. Later works employ a look-up table to achieve a similar outcome—emit the low-energy symbols of the constellation more often and thus achieve shaping gain [59]. Shell mapping was also used for this purpose in the context of optical fibers [60].

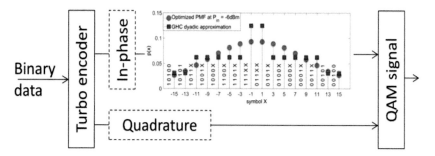

Figure 5.5 Block diagram for many-to-one mapping transmitter, together with an optimized PMF and its corresponding mapping function.

The concept of many-to-one mapping was employed [61–64] to achieve a nonuniform PMF of the output symbols. The basic idea is to assign multiple different bit sequences to the same constellation symbol, making it appear more often on average. Typically, those will be the low-energy symbols. This type of mapping results in PMFs of dyadic form, for which the probability of each constellation symbol can be expressed as $P_i = 2^{-l_i}$, where l_i is an integer. The optimal PMF found by, e.g., optimizing a MB distribution [56] or the BA algorithm [24, 25] can then be quantized to a dyadic form and many-to-one mapping can be used to achieve this PMF. The block diagram of such a transmitter is given in Fig. 5.5. The binary data are encoded by a FEC code, in this case–a turbo code, then the many-to-one mapping function given on the right is applied. In this example, this is a mapping function for 16PAM in each in-phase and quadrature component, or 256QAM in total. The optimized PMF and the effect of the dyadic quantization can also be seen. The dyadic quantization in this case is performed with the geometric Huffman code (GHC), which can be shown to be optimal [57]. Each symbol can take multiple bit labels, making some symbols more likely to be produced when the input binary data are i.i.d. For example symbol "3" in Fig. 5.5 takes all bit strings of length 5, which begin with the prefix "0101" (2 different bit strings in total), making its probability 2^{-4}. Symbol "15" only takes one bit string, and its probability is therefore 2^{-5}. The data rate of this scheme is easily controlled by puncturing of the turbo encoder.

The many-to-one mapping results in ambiguities, which are resolved by the error-correcting code. Iterative de-mapping and decoding is sometime required in order to resolve all the ambiguities are achieve shaping gain, which may become problematic for high-speed communications where decoding complexity and latency are key aspects. The dyadic quantization also results in a shaping loss, which can be significant especially for small constellations. Many-to-one mapping strategies are employed in the context of optical communications in, e.g., [40, 43].

A fixed-to-fixed length distribution matching (DM) method was proposed in [65, 66], which asymptotically achieves the desired PMF without quantizing it. The distribution matcher can then be employed on the systematic bits of the error-correcting code, which allows for the bits to be protected and thus circumvents the issue of error-propagation. A block diagram of such a transmitter is given in Fig. 5.6 for the in-phase component. The parity bit sequence, which has a uniform probability distribution is used to select the sign of the symbol (or the quadrant when the in-phase and quadrature components are combined), and the rate-matched systematic data are used to select the amplitude of each in-phase and quadrature components. In this case, standard Gray codes can be applied as mapping function. The DM in this scheme also directly controls the data rate for a fixed rate of the FEC, which allows for flexible-rate hardware implementation at constant decoder complexity.

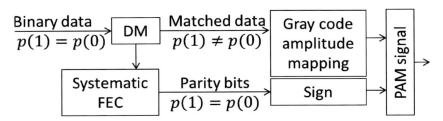

Figure 5.6 A transmitter block diagram for a distribution matched probabilistic amplitude shaping scheme.

The DM can introduce latency, especially when the PMF must be closely approached. A parallelized structure is proposed in [67], which circumvents these issues. The ability to combine with arbitrary error-correcting codes, together with the simplicity for implementation has made the DM method very popular in the optical communications community in recent years, particularly after its several experimental demonstrations, e.g., [41, 68] including at transoceanic distances [69].

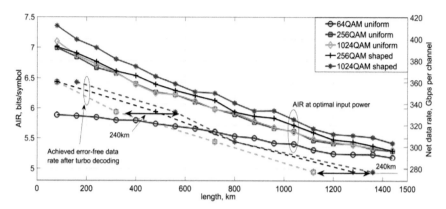

Figure 5.7 Maximum AIRs for the many-to-one mapping system [43] (solid lines) and error-free operating points (dashed lines) after turbo decoding.

Probabilistic shaping can typically be applied in order to either increase the maximum data rate by about 10% (similar to the gain of Fig. 5.4), or alternatively, translate this gain to increased maximum transmission reach for a fixed data rate by a few hundreds of kilometers [41, 43, 60]. The AIR performance of, for instance, the many-to-one mapping scheme from [43] is given in Fig. 5.7 as a function of the transmission distance at the optimal launch power. The transmission parameters are similar to those of Table 5.2, however, the transceiver imperfections in this case are taken into account (for details, see [43]). We see that 64QAM suffers from the limited number of points, and is therefore inferior to the other constellations. The best performing overall constellation is a dyadic 1024QAM with a distance increase of about 200 km for fixed data rate w.r.t. standard 1024QAM. The 256QAM is also able to achieve shaping gains of around 100 km. In Fig. 5.7, the maximum distances, where error-free transmission is achieved at different rates net data rates (after turbo decoding) are also shown (dashed lines). We see that shaping provides a steady increase of between 0.3 and 0.5 bits/symbol, or between 8.4 and 14 Gbps per channel. This translates to between 80 km (short distance) and 240 km (long distance) of maximum reach increase at the same data rate. Similar gains are achieved with, e.g., the distribution matching scheme[c] [41].

Achieving larger shaping gains would require lifting the above-mentioned assumptions on the channel memory and/or the likelihood function.

5.7 Concluding Remarks

In this chapter, we have described the information-theoretic approach to channel modeling and recalled related fundamental concepts. We discussed the complexities associated with information theory friendly channel modeling and gave an overview of different channel models.

Further, an overview of the current progress in optimization of the transmission strategy was given. We formalized the concept of constellation shaping and reviewed some of the popular methods for shaping, particularly for the optical fiber channel.

References

1. C. E. Shannon, Mathematical theory of communication, *Bell Syst. Tech. J.*, 27, 379, 623 (1948).

2. R. L. Dobrushin, General formulation of Shannon's main theorem in information theory, *Amer. Math. Sot. Trans.*, 33, pp. 323–438, AMS, Providence, RI (1963).

3. S. Verdu and T. S. Han, A general formula for channel capacity, *IEEE Trans. Inf. Theory*, 40(4), 1147–1157 (1994).

4. Splett, C. Kurtzke, and K. Petermann, Ultimate transmission capacity of amplified optical fiber communication systems taking into account fiber nonlinearities, *Technical Digest of European Conference on Optical Communication paper* MoC2.4. (1993).

5. P. P. Mitra and J. B. Stark, Nonlinear limits to the information capacity of optical fibre communications, *Nature*, 411, 1027–1030 (2001).

6. R.-J. Essiambre, G. J. Foschini, G. Kramer, and P. J. Winzer, Capacity limits of information transport in fiber-optic networks, *Phys. Rev. Lett.*, 101, 163901 (2008).

---

[c]The performance overall is improved with the system from [41] due to the lack of PMF quantization and the superior FEC, however, the shaping gains are similar.

7. P. Poggiolini, A. Carena, V. Curri, G. Bosco, and F. Forghieri, Analytical modeling of non-linear propagation in uncompensated optical transmission links, *IEEE Photon. Technol. Lett.*, 23(11), 742–744 (2011).

8. P. Johannisson and M. Karlsson, Perturbation analysis of nonlinear propagation in a strongly dispersive optical communication system, *J. Lightwave Technol.*, 31(8), 1273–1282 (2013).

9. K. Peddanarappagari and M. Brandt-Pearce, Volterra series transfer function of single-mode fibers, *J. Lightwave Technol.*, 15(12), 2232–2241 (1997).

10. M. Schetzen, *The Volterra and Wiener Theories of Nonlinear Systems* (Krieger, 2006).

11. A. Amari, P. Ciblat, and Y. Jaouen, Fifth-order Volterra series based nonlinear equalizer for long-haul high data rate optical fiber communications, *Asilomar Conference ACSSC* (2014).

12. E. Agrell, A. Alvarado, G. Durisi, and M. Karlsson, Capacity of a nonlinear optical channel with finite memory, *J. Lightwave Technol.*, 32(16), 2862–2876 (2014).

13. R. Dar, M. Feder, A. Mecozzi, and M. Shtaif, Properties of nonlinear noise in long, dispersion-uncompensated fiber links, *Opt. Express*, 21(22), 25685–25699 (2013).

14. R. Dar, M. Feder, A. Mecozzi, and M. Shtaif, Inter-channel nonlinear interference noise in WDM systems: Modeling and mitigation, *J. Lightwave Technol.*, 33(5), 1044–1053 (2015).

15. R. Dar, M. Shtaif, and M. Feder, New bounds on the capacity of the nonlinear fiber-optic channel, *Opt. Lett.*, 39, 398–401 (2014).

16. M. Shtaif, R. Dar, A. Mecozzi, and M. Feder, Nonlinear interference noise in WDM systems and approaches for its cancelation, *Optical Communication (ECOC 2014), 39th European Conference and Exhibition on Optical Communications*, paper We1.3.1.

17. E. E. Narimanov and P. Mitra, The channel capacity of a fiber optics communication system: Perturbation theory, *J. Lightwave Technol.*, 20(3), 530–537 (2002).

18. K. S. Turitsyn, S. A. Derevyanko, I. V. Yurkevich, and S. K. Turitsyn, Information capacity of optical fiber channels with zero average dispersion, *Phys. Rev. Lett.*, 91, 203901 (2003).

19. M. Secondini, E. Forestieri, and C. R. Menyuk, A combined regular-logarithmic perturbation method for signal-noise interaction in amplified optical systems, *J. Lightwave Technol.*, 27(16), 3358–3369 (2009).

20. M. Sorokina, S. Sygletos, and S. Turitsyn, Ripple distribution for nonlinear fiber-optic channels, *Opt. Express*, 25, 2228–2238 (2017).

21. W. Karush, Minima of functions of several variables with inequalities as side constraints. M.Sc. Dissertation Deptartment of Mathematics, University of Chicago, Chicago, Illinois (1939).

22. H. W. Kuhn and A. W. Tucker, Nonlinear programming. *Proceedings of 2nd Berkeley Symposium*. Berkeley: University of California Press. 481–492 (1951).

23. M. Frank and P. Wolfe, An algorithm for quadratic programming, Naval Research Logistics Quarterly 3: 95. doi:10.1002/nav.3800030109 (1956).

24. S. Arimoto, An algorithm for computing the capacity of arbitrary discrete memoryless channels, *IEEE T. Inform Theory*, 18, 14–20 (1972).

25. R. Blahut, Computation of channel capacity and rate-distortion functions, *IEEE T. Inform Theory*, 18, 460–473 (1972).

26. G. Matz and P. Duhamel, Information geometric formulation and interpretation of accelerated Blahut–Arimoto-type algorithms, *Proceedings 2004 IEEE Information Theory* Workshop, San Antonio, TX, USA, October 24–29 (2004).

27. N. Varnica, X. Ma, and A. Kavcic, Capacity of power constrained memoryless AWGN channels with fixed input constellations, *Proceedings IEEE Global Telecommunications Conference (GLOBECOM)*, Taipei, Taiwan, China, November 2002, pp. 1339–1343.

28. C.-I. Chang and L. D. Davisson, On calculating the capacity of an infinite-input finite (infinite)-output channel, *IEEE Trans. Inf. Theory*, 34, 1004–1010 (1988).

29. J. Dauwels, Numerical computation of the capacity of continuous memoryless channels, *Proceedings of the Symposium on Information Theory in the Benelux, Brussels, Belgium*, May 2005, pp. 221–228.

30. D. Arnold and H.-A. Loeliger, On the information rate of binary-input channels with memory, *Proceedings 2001 IEEE International Conference on Communications*, Helsinki, Finland, June 2001, pp. 2692–2695.

31. V. Sharma and S. K. Singh, Entropy and channel capacity in the regenerative setup with applications to Markov channels, *Proceedings 2001 IEEE International Symposium on Information Theory*, Washington, DC, USA, June 24–29, 2001, p. 283.

32. H. D. Pfister, J. B. Soriaga, and P. H. Siegel, On the achievable information rates of finite-state ISI channels, *Proceedings 2001 IEEE Globecom*, San Antonio, TX, pp. 2992–2996, November 25–29 (2001).

33. J. Dauwels and H.-A. Loeliger, Computation of information rates by particle methods, *IEEE Trans. Inf. Theory*, 54, 406–409 (2008).

34. E. G. Shapiro, D. A. Shapiro, and S. K. Turitsyn, Method for computing the optimal signal distribution and channel capacity, *Opt. Express*, 23, 15119–15133 (2015).

35. G. Kramer, M. I. Yousefi, and F. R. Kschischang, Upper bound on the capacity of a cascade of nonlinear and noisy channels, IEEE Information Theory Workshop (2015).

36. E. Agrell, M. Karlsson, A. R. Chraplyvy, D. J. Rochardson, P. M. Krummrich, P. Winzer, K. Roberts, J. K. Fisher, S. J. Savory, B. J. Eggleton, M. Secondini, F. R. Kschischang, A. Lord, J. Prat, I. Tomkos, J. E. Bowers, S. Srinivasan, M. B. Pearce, and N. Gisin, Roadmap of optical communications, *J. Opt.*, 18(6), 063002 (2016).

37. M. Secondini and E. Forestieri, Scope and limitations of the nonlinear Shannon limit, online version of 24.10.2016 http:// ieeexplore.ieee.org/document/7637002/.

38. I. B. Djordjevic, H. G. Batshon, L. Xu, and T. Wang, Coded polarization-multiplexed iterative polar modulation (PM-IPM) for beyond 400 Gb/s serial optical transmission, *Proceedings Optical Fiber Communication Conference*, Los Angeles, CA, March 2010, p. OMK2.

39. T. Fehenberger, A. Alvarado Segovia, G. Bocherer, and N. Hanik, Sensitivity gains by mismatched probabilistic shaping for optical communication systems, *IEEE Photon. Technol. Lett.*, 28(7) 786–789 (2016).

40. C. Pan and F. R. Kschischang, Probabilistic 16-QAM shaping in WDM systems, *J. Lightwave Technol.*, 34(18), 4285–4292 (2016).

41. F. Buchali, G. Bocherer, W. Idler, L. Schmalen, P. Schulte, and F. Steiner, Rate adaptation and reach increase by probabilistically shaped 64-QAM: An experimental demonstration, *J. Lightwave Technol.*, 34(7), 1599–1609 (2016).

42. T. Fehenberger, A. Alvarado, G. Böcherer, N. Hanik, On probabilistic shaping of quadrature amplitude modulation for the nonlinear fiber channel, *J. Lightwave Technol.*, 34(21), 5063–5073 (2016).

43. M. P. Yankov, F. Da Ros, E. P. da Silva, S. Forchhammer, K. J. Larsen, L. K. Oxenløwe, M. Galili, and D. Zibar, Constellation shaping for WDM systems using 256QAM/1024QAM with probabilistic optimization, *J. Lightwave Technol.*, 34(22), 5146–5156 (2016).

44. S. Verdu and Y. Wu, Functional properties of minimum mean-square error and mutual information, *IEEE Trans. Information Theory*, 58(3), pp. 1289–1301 (2012).

45. M. P. Yankov, S. Forchhammer, K. J. Larsen, and L. P. B. Christensen, Approximating the constellation constrained capacity of the MIMO channel with discrete input, *IEEE International Conference on Communications*, London (2015).

46. T. M. Cover and J. A. Thomas, *Elements of Information Theory*, Hoboken, New Jersey: John Wiley & Sons, Inc. (2006).

47. J. Bellorado, S. Ghassemzadeh, and A. Kavcic, Approaching the capacity of the MIMO Rayleigh flat-fading channel with QAM constellations, independent across antennas and dimensions, *IEEE Trans. Commun.*, 5(6), pp. 1322–1332 (2006).

48. I. C. Abou-Faycal, M. D. Trott, and S. Shamai, The capacity of discrete-time memoryless Rayleigh fading channels, *IEEE Trans. Information Theory*, 47(4), pp. 1290–1301 (2001).

49. R.-J. Essiambre, G. Kramer, P. J. Winzer, G. J. Foschini, and B. Goebel, Capacity limits of optical fiber networks, *IEEE/OSA J. Lightwave Technol.*, 28(4), pp. 662–701 (2010).

50. T. A. Eriksson, T. Fehenberger, P. A. Andrekson, M. Karlsson, N. Hanik, and E. Agrell, Impact of 4D channel distribution on the achievable rates in coherent optical communication experiments, *IEEE/OSA J. Lightwave Technol.*, 34(9), pp. 2256–2266 (2016).

51. R. F. H. Fischer, *Precoding and Signal Shaping for Digital Transmission*, New York: John Wiley & Sons, Inc. (2002).

52. J. Renner, T. Fehenberger, M. P. Yankov, F. Da Ros, S. Forchhammer, G. Bocherer, and N. Hanik, Experimental comparison of probabilistic shaping methods for unrepeated fiber transmission, 2017. [Online]. Available: https://arxiv.org/abs/1705.01367.

53. R. Dar, M. Feder, A. Mecozzi, and M. Shtaif, On shaping gain in the nonlinear fiber-optic channel, *IEEE International Symposium on Information Theory*, Honolulu, Hawaii (2014).

54. E. Agrell, A. Alvarado, G. Durisi, and M. Kralsson, Capacity of a nonlinear optical channel with finite memory, *IEEE/OSA J. Lightwave Technol.*, 32(16), pp. 2862–2876 (2015).

55. M. P. Yankov, K. J. Larsen, and S. Forchhammer, Temporal probabilistic shaping for mitigation of nonlinearities in optical fiber systems, *IEEE/OSA J. Lightwave Technol.*, 35(10), pp. 1803–1810 (2017).

56. F. R. Kschischang and S. Pasupathy, Optimal nonuniform signaling for Gaussian channels, *IEEE Trans. Information Theory*, 39(3), pp. 913–929 (1993).

57. G. Bocherer and R. Mathar, Matching dyadic distributions to channels, *Data Compression Conference*, Snowbird, USA (2011).

58. G. D. Forney Jr, Trellis shaping, *IEEE Trans. Information Theory*, 38(2), pp. 281–300 (1992).

59. M. C. Valenti and X. Xiang, Constellation shaping for Bit-interleaved LDPC coded APSK, *IEEE Trans. Commun.*, 60(10), pp. 2960–2970 (2012).

60. L. Beygi, E. Agrell, J. M. Kahn, and M. Karlsson, Rate-adaptive coded modulation for fiber-optic communications, *IEEE/OSA J. Lightwave Technol.*, 32(2), pp. 333–343 (2014).

61. D. Raphaeli and A. Gurevitz, Constellation shaping for pragmatic turbo-coded modulation with high spectral efficiency, *IEEE Trans. Commun.*, 52(3), pp. 341–345 (2004).

62. M. P. Yankov, S. Forchhammer, K. J. Larsen, and L. P. B. Christensen, Rate-adaptive constellation shaping for near-capacity achieving turbo coded BICM, *IEEE International Conference on Communications*, Sydney, Australia (2014).

63. F. Schreckenbach, Approaching AWGN channel capacity using non-unique symbol mappings, *7th Australian Communications Theory Workshop*, Perth, Australia (2006).

64. T. Liu, C. Lin, and I. B. Djordjevic, Advanced GF(3^2) nonbinary LDPC coded modulation with non-uniform 9-QAM outperforming star 8-QAM, *Opt. Express*, 24(13), pp. 13866–13874 (2016).

65. P. Schulte and G. Böcherer, Constant composition distribution matching, *IEEE Trans. Information Theory*, 62(1), pp. 430–434 (2016).

66. G. Böcherer, F. Steiner, and P. Schulte, Bandwidth efficient and rate-matched low-density parity-check coded modulation, *IEEE Trans. Commun.*, 63(12), pp. 4651–4665 (2015).

67. G. Böcherer, F. Steiner, and P. Schulte, High throughput probabilistic shaping with product distribution matching, 2017. [Online]. Available: https://arxiv.org/abs/1701.07371.

68. W. Idler, F. Buchali, L. Schamel, E. Lach, R. Braun, G. Böcherer, P. Schulte, and F. Steiner, Field trial of a 1 Tb/s super-channel network using probabilistically shaped constellations, *IEEE/OSA J. Lightwave Technol.*, 35(8), pp. 1399–1406 (2017).

69. A. Ghazisaeidi, I. F. de Jauregui Ruiz, R. Rios-Muller, L. Schmalen, P. Tran, P. Brindel, A. C. Mesequer, Q. Hu, F. Buchali, G. Charlet, and J. Renaudier, 65Tb/s transoceanic transmission using probabilistically shaped PDM-64QAM, *European Conference on Optical Communications*, Dusseldorf, Germany (2016).

Chapter 6

Advanced Coding for Fiber-Optics Communications Systems

Ivan B. Djordjevic

Department of Electrical and Computer Engineering,
University of Arizona, 1230 E. Speedway Blvd, Tucson, Arizona 85721, USA

ivan@email.arizona.edu

This chapter is devoted to advanced channel coding and coded modulation techniques for fiber-optics communication systems. Two classes of codes on graphs suitable for fiber-optics communications together with corresponding decoding algorithms are described, namely, turbo product and LDPC codes. Given that LDPC codes are channel capacity achieving, they are described with more details. Decoding algorithms for both binary and nonbinary LDPC codes are described including their FPGA implementation. To deal with time-varying optical channel conditions, various rate adaptive schemes are described. The second half of the chapter is related to coded modulation for fiber optics communications. After providing the coded modulation fundamentals, we describe

Optical Communication Systems: Limits and Possibilities
Edited by Andrew Ellis and Mariia Sorokina
Copyright © 2020 Jenny Stanford Publishing Pte. Ltd.
ISBN 978-981-4800-28-0 (Hardcover), 978-0-429-02780-2 (eBook)
www.jennystanford.com

multilevel coded modulation, bit-interleaved coded modulation (BICM), and various hybrid multidimensional coded modulation schemes suitable for ultra-high-speed optical transmission. To demonstrate high potentials of advanced coded modulation, we describe an experiment in which the concepts described in this chapter are used as key enabling technologies to improve the spectral efficiency and/or extend the transmission distance.

6.1 Introduction

As a response to these never-ending demands for higher data rates and distance-independent connectivity, 100 Gb/s Ethernet (100 GbE) has been already standardized, and the effort has moved to 400 GbE/1 TbE (see [1] and references therein). It has become evident that soft-decision forward error correction (FEC) represents one of the key enabling technologies for the next generation of optical transport networks [1–23].

In this chapter, we describe different codes on graphs of interest for optical communications including turbo-product and low-density parity-check (LDPC) codes. Given channel capacity achieving properties of LDPC codes, both binary and nonbinary LDPC decoding processes have been described with more details including their FPGA implementations. Some attractive classes of LDPC codes, recently proposed, include large-girth quasi-cyclic LDPC codes [3], staircase LDPC codes [6, 7], spatially coupled codes [8–10], convolutional [11] (and references therein) LDPC codes, and nonbinary irregular quasi-cyclic (QC) LDPC codes derived from pairwise block designs (PBDs) introduced in [12]. We further describe how to perform the code rate adaptation, based on optical channel conditions by shortening, partial reconfiguration of the parity-check matrix of a template regular QC-LDPC code, puncturing, and re-encoding approach introduced in [13].

The spectral efficiency of fiber-optics communications can be improved by considering coding and modulation jointly in so-called coded modulation fashion [14–21]. After providing the coded modulation fundamentals, the following coded modulation techniques suitable for fiber-optics communications are described:

multilevel coded modulation, bit-interleaved coded modulation (BICM), and various hybrid multidimensional coded modulation schemes [5, 15].

6.2 Turbo-Product Codes

A turbo-product code (TPC), also known as block turbo code (BTC) [4, 24–28], represents an $(n_1 n_2, k_1 k_2, d_1 d_2)$ code, wherein n_i, k_i and d_i (i = 1,2) denote the codeword length, dimension, and minimum distance, respectively, of the i-th component code (i = 1,2). A TPC codeword, shown in Fig. 6.1, forms an $n_1 \times n_2$ array in which each row represents a codeword from an (n_1, k_1, d_1) code C_1, while each column represents a codeword from an (n_2, k_2, d_2) code C_2. The code rate of TPC is given by $R = R_1 R_2 = (k_1 k_2)/(n_1 n_2)$, where R_i is the code rate of the i-th (i = 1,2) component code, namely $R_i = k_i/n_i$. The corresponding overhead (OH) of each component code is defined as $OH_i = (1/R_i - 1) \cdot 100\%$. The TPC can be considered as a particular instance of serial concatenated block codes. In this interpretation, rows belong to the outer code and columns to the inner code, and the interleaver is a deterministic column-row interleaver.

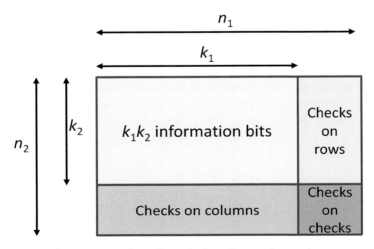

Figure 6.1 The structure of a codeword of a turbo-product code.

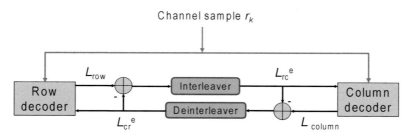

Figure 6.2 Decoder configuration of a turbo-product code.

The corresponding TPC decoder is shown in Fig. 6.2. In soft-decision decoding, the extrinsic reliabilities are iterated between soft-input-soft-output (SISO) decoders for C_1 and C_2. The extrinsic information in iterative decoders is obtained from corresponding companion decoder. The key idea behind extrinsic information is to provide to the companion decoder only with soft information not already available to it. The product codes were proposed by Elias [27], but the term "turbo" is used when two SISO decoders exchange the extrinsic information [28]. As already mentioned above, the minimum distance of a product code is the product of minimum distances of component codes, $d = d_1 d_2$. The constituent codes are typically extended BCH codes, because with extended BCH codes we can increase the minimum distance by $d_1 + d_2 + 1$ compared to the nominal BCH codes. Given the high complexity maximum likelihood (ML) decoding is typically not used in BTC decoding, but simple *Chase II-like decoding algorithms* have been used instead [1, 24, 26, 28]. One such algorithm, which is independent of the channel model [4], is described below. Let v_i be i-th bit in a codeword $\mathbf{v} = [v_1 \dots v_n]$, and y_i be the corresponding received sample in a received vector of samples $\mathbf{y} = [y_1 \dots y_n]$. The role of decoder is to iteratively estimate the *a posteriori* probabilities (APPs) $\Pr(v_i|y_i)$. The knowledge of APPs allows for optimum decisions on the bits v_i by MAP rule

$$\frac{P(v_i = 0 \mid \mathbf{y})}{P(v_i = 1 \mid \mathbf{y})} \overset{0}{\underset{1}{\gtrless}} 1, \tag{6.1}$$

or, more conveniently we can write

$$\hat{v}_i = \frac{1 - \text{sign}\left[L(v_i)\right]}{2}, \quad \text{sign}\left[x\right] = \begin{cases} +1, & x \geq 0 \\ -1, & x < 0 \end{cases}, \tag{6.2}$$

where $L(v_i)$ is the log *a posteriori* probability (log-APP), commonly referred to as the log-likelihood ratio (LLR), defined as

$$L(v_i) = \log\left[\frac{P(v_i = 0 \mid \boldsymbol{y})}{P(v_i = 1 \mid \boldsymbol{y})}\right]. \tag{6.3}$$

The component SISO decoders for C_1 and C_2 calculate $L(v_i)$ and exchange extrinsic information as explained above. The initial/channel bit LLRs, to be used in SISO decoding including Chase II algorithm, can be calculated by

$$L_{ch}(v_i) = \log\frac{P(v_i = 0 \mid y_i)}{P(v_i = 1 \mid y_i)}, \tag{6.4}$$

where the conditional probability $P(v_i \mid y_i)$ is determined by

$$P(v_i \mid y_i) = \frac{P(y_i \mid v_i)P(v_i)}{P(y_i \mid v_i = 0)P(v_i = 0) + P(y_i \mid v_i = 1)P(v_i = 1)}, \tag{6.5}$$

In the presence of fiber nonlinearities, the conditional $P(y_i \mid \cdot)$ can be evaluated by estimation of histograms. On the other hand, the channel LLRs for AWGN channel can be calculated as follows:

$$L_{ch}(v_i) = 2\frac{y_i}{\sigma^2}, \text{ for binary input AWGN}, \tag{6.6}$$

where σ^2 is the variance of the Gaussian distribution of the AWGN. It has been shown in experimental study for beyond 100 Gb/s transmission [29, 30] that in the links without in-line chromatic dispersion (CD) compensation, the distribution of samples upon compensation of CD and nonlinear phase compensation is still Gaussian-like, which justifies the use of the Gaussian assumption (6.6) in an amplified spontaneous emission (ASE) noise dominated scenario.

The constituent *SISO decoding* algorithms are based on modified Chase II decoding algorithm [1, 4]:

(1) Determine p least reliable positions starting from either (6.4) or (6.6). Generate 2^p test patterns to be added to the hard-decision word obtained after (6.2).

(2) Determine the i-th ($i = 1,..., 2^p$) perturbed sequence by adding (modulo-2) the test pattern to the hard-decision word (on least reliable positions).

(3) Perform the algebraic or hard decoding to create the list of candidate codewords. Simple syndrome decoding is suitable for high-speed implementation.

(4) Calculate the k-th candidate codeword c_k LLRs by

$$L[v_k = (v_k(1)\cdots v_k(n))] = \sum_{i=1}^{n} \log\left[\frac{e^{(1-v_k(i))L(v_k(i))}}{1+e^{L(v_k(i))}}\right],\qquad(6.7)$$

where $v_k(i)$ denotes the i-th bit in the k-th candidate codeword and $L(\cdot)$ is a corresponding LLR.

(5) Calculate the extrinsic bit reliabilities, denoted as $L_e(\cdot)$, for the next decoding stage using

$$L_e(v_i) = L'(v_i) - L(v_i),\qquad(6.8)$$

where

$$L'(v_i) = \log\left[\frac{\sum_{v_k(i)=0} L(v_k)}{\sum_{v_k(i)=1} L(v_k)}\right]\qquad(6.9)$$

In (6.9), summation in numerator (denominator) is performed over all candidate codewords having 0 (1) at position i.

(6) Set $L(v_i) = L'(v_i)$ and move to step 1.

The following "max-star" operator can be applied to (6.7) and (6.9) in a recursive fashion as follows:

$$\max{}^*(x,y) \triangleq \log(e^x + e^y) = \max(x,y) + \log[1+e^{-|x-y|}]\qquad(6.10)$$

Given this description of SISO constituent decoding algorithm, the TPC decoder, shown in Fig. 6.2, operates as follows.

Let $L^{e}_{rc,j}$ denote the extrinsic information to be passed from row to column decoder, and let $L^{e}_{cr,j}$ denote the extrinsic information to be passed in opposite direction. Then, assuming that the column-decoder operates first, the *TPC decoder* performs the following steps:

(0) *Initialization*: $L^{e}_{cr,i} = L^{e}_{rc,i} = 0$ for all i.

(1) *Column decoder*: Run the SISO decoding algorithm described above with the following inputs $L(v_j) + L^{e}_{rc,j}$ to obtain $\{L_{column}(v_i)\}$ and $\{L^{e}_{cr,i}\}$, as shown in Fig. 6.2. The extrinsic information is calculated by (6.8). Pass the extrinsic information $\{L^{e}_{cr,j}\}$ to companion row decoder.

(2) *Row decoder*: Run the SISO decoding algorithm with the following inputs $L(v_i) + L^{e}_{cr,i}$ to obtain $\{L_{row}(v_i)\}$ and $\{L^{e}_{rc,i}\}$. Pass the extrinsic information $\{L^{e}_{cr,i}\}$ to companion column decoder.

(3) *Bit decisions*: Repeat the steps 1 to 2 until a valid codeword is generated or a predetermined number of iterations has been reached. Make the decisions on bits by sign $[L_{row}(u_k)]$.

Unlike [24, 26, 28], because not any approximation is used in calculation of the extrinsic reliabilities, there is no need to introduce the scaling factors and/or the correction factors.

6.3 LDPC Codes

LDPC codes were invented by Robert Gallager (from MIT) in 1960, in his PhD dissertation [31] but received no attention from the coding community until THE 1990s [32].

6.3.1 LDPC Codes Fundamentals and Large-Girth Code Design

LDPC codes belong to the class of linear block codes, and as such they can be described as a k-dimensional subspace C of the n-dimensional vector space of n-tuples, F^{n}_{2}, over the binary field F_2. For this k-dimensional subspace, we can find the basis $B = \{g_0, g_1, ..., g_{k-1}\}$ that spans C so that every codeword v can be written as a linear combination of basis vectors $v = m_0 g_0 +$

$m_1 g_1 + \dots + m_{k-1} g_{k-1}$ for message vector $m = (m_0, m_1, \dots, m_{k-1})$; or in compact form we can write $v = mG$, where G is so-called generator matrix with the i-th row being g_i. The $(n-k)$-dimensional null space $C^{\perp}$ of G comprises all vectors x from F_2^n such that $xG^{\mathrm{T}} = 0$ and it is spanned by the basis $B^{\perp} = \{h_0, h_1, \dots, h_{n-k-1}\}$. Therefore, for each codeword v from C, $vh_i^{\mathrm{T}} = 0$ for every i; or in compact form we can write $vH^{\mathrm{T}} = 0$, where H is so-called parity-check matrix whose i-th row is h_i.

An LDPC code can now be defined as an (n, k) linear block code whose parity-check matrix H has a low density of 1's. A *regular* (w_c, w_r) *LDPC code* is a linear block code whose H matrix contains exactly w_c 1's in each column and exactly $w_r = w_c n/(n-k)$ 1's in each row, where $w_c \ll n-k$. The code rate of the regular LDPC code is determined by $R = k/n = 1 - w_c/w_r$. The graphical representation of LDPC codes, known as bipartite (Tanner) graph representation, is helpful in efficient description of LDPC decoding algorithms. A *bipartite (Tanner) graph* is a graph whose nodes may be separated into two classes (*variable* and *check* nodes), and where *undirected edges* may only connect two nodes not residing in the same class. The Tanner graph of a code is drawn according to the following rule: check (function) node c is connected to variable (bit) node v whenever element h_{cv} in a parity-check matrix H is a 1. In an $m \times n$ parity-check matrix, there are $m = n - k$ check nodes and n variable nodes. As an illustration, the Tanner graph of a regular (3,4) LDPC(36,13) code is shown in Fig. 6.3. The blue-circles represent variable (bit) nodes and green-squares represent the parity-check (function) nodes. In principle, the Tanner graph can be defined for any linear block code. For instance, the parity-check matrix for Hamming (15,11) code is given by

$$H = \begin{bmatrix} 1 & 0 & 0 & 0 & 1 & 0 & 0 & 1 & 1 & 0 & 1 & 0 & 1 & 1 & 1 \\ 0 & 1 & 0 & 0 & 1 & 1 & 0 & 1 & 0 & 1 & 1 & 1 & 1 & 0 & 0 \\ 0 & 0 & 1 & 0 & 0 & 1 & 1 & 0 & 1 & 0 & 1 & 1 & 1 & 1 & 0 \\ 0 & 0 & 0 & 1 & 0 & 0 & 1 & 1 & 0 & 1 & 0 & 1 & 1 & 1 & 1 \end{bmatrix}.$$

For any valid codeword $v = [v_0 \, v_1 \dots v_{n-1}]$ ($n = 15$), the parity-check checks used to decode the codeword can be represented as follows:

- Parity-check equation (c_0): $v_0 + v_4 + v_7 + v_8 + v_{10} + v_{12} + v_{13} + v_{14} = 0 \pmod 2$
- Parity-check equation (c_1): $v_1 + v_4 + v_5 + v_7 + v_9 + v_{10} + v_{11} + v_{12} = 0 \pmod 2$
- Parity-check equation (c_2): $v_2 + v_5 + v_6 + v_8 + v_{10} + v_{11} + v_{12} + v_{13} = 0 \pmod 2$
- Parity-check equation (c_3): $v_3 + v_6 + v_7 + v_9 + v_{11} + v_{12} + v_{13} + v_{14} = 0 \pmod 2$

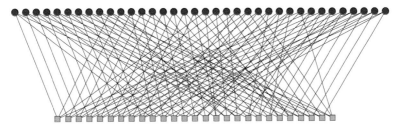

Figure 6.3 Tanner graph of a regular (3,4) LDPC (36,13) code of girth-8.

Clearly, the parity-check equation (c_i) corresponds to the i-th row, and variable nodes v_i correspond to the columns in the parity-check matrix. The nonzero elements in the i-th row determine which bits are involved in corresponding parity-check equation. Since the column-weight is not fixed, this code is an example of the irregular code. The corresponding Tanner graph is shown in Fig. 6.4. Parity-check equations defined above are denoted by green-squares (c_0)–(c_4). The variable nodes v_0, v_4, v_7, v_8, v_{10}, v_{12}, v_{13}, and v_{14} are involved in the parity-check equation (c_0) and therefore, the edges are established between these variable nodes and the check-node (c_0). Other edges in Tanner graph are established by following the same strategy. A closed path in a bipartite graph comprising l edges that closes back on itself is called a *cycle* of length l. The shortest cycle in the bipartite graph is called the *girth*. The shortest cycle in the Tanner graph of the Hamming (15,11) code is of length 4 and it is denoted by red-colored edges in Fig. 6.4, indicating that the girth of this code is 4. The girth influences the minimum distance of LDPC codes, correlates the extrinsic LLRs, and therefore affects the decoding performance. The use of large (high) girth LDPC codes is preferable because the large girth increases the minimum distance and prevents early correlations in the extrinsic information during decoding process. To improve the iterative decoding performance, we have to avoid cycles of length 4 and 6. As an illustration, the girth of LDPC code shown in Fig. 6.3 is 8.

The code description can also be done by the degree distribution polynomials $\mu(x)$ and $\rho(x)$, for the variable-node (v-node) and the check-node (c-node), respectively

$$\mu(x) = \sum_{d=1}^{d_v} \mu_d x^{d-1}, \qquad \rho(x) = \sum_{d=1}^{d_c} \rho_d x^{d-1}, \tag{6.11}$$

where μ_d and ρ_d denote the fraction of the edges that are connected to degree-d v-nodes and c-nodes, respectively, and d_v and d_c denote the maximum v-node and c-node degrees, respectively.

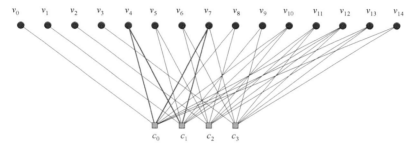

Figure 6.4 Tanner graph of Hamming (15,11) code.

The most obvious way to design LDPC codes is to construct a low-density parity-check matrix with prescribed properties. Some important LDPC designs, among others, include: Gallager codes (semi-random construction) [31], MacKay codes (semi-random construction) [32], combinatorial design-based LDPC codes [33] (see [34] for combinatorial designs), finite-geometry-based LDPC codes [35, 36], array [also known as quasi-cyclic (QC)] LDPC codes [37, 38], to mention a few. The generator matrix of a QC-LDPC code can be represented as an array of circulant sub-matrices of the same size B indicating that QC-LDPC codes can be encoded in linear time using simple shift-register-based architectures [39]. A QC-LDPC code can be defined as an LDPC code for which every sectional cyclic shift to the right (or left) for $l \in [0, B-1]$ places of a codeword $\boldsymbol{v} = [\boldsymbol{v}_0\ \boldsymbol{v}_1\ ...\ \boldsymbol{v}_{B-1}]$ (each section $\boldsymbol{v}_i$ contains B elements) results in another codeword.

Based on Tanner's bound for the minimum distance of an LDPC code [40]

$$d \geq \begin{cases} 1 + \dfrac{w_c}{w_c - 2}((w_c - 1)^{\lfloor (g-2)/4 \rfloor} - 1), \ g/2 = 2m + 1 \\[3mm] 1 + \dfrac{w_c}{w_c - 2}((w_c - 1)^{\lfloor (g-2)/4 \rfloor} - 1) + (w_c - 1)^{\lfloor (g-2)/4 \rfloor}, \ g/2 = 2m \end{cases} \qquad (6.12)$$

(where g and w_c denote the girth of the code graph and the column weight, respectively, and where d stands for the minimum distance of the code), it follows that large girth leads to an exponential increase in the minimum distance, provided that the column weight is at least 3. ($\lfloor \rfloor$ denotes the largest integer less than or equal to the enclosed quantity.) For example, the minimum distance of girth-10 codes with column weight $r = 3$ is at least 10. The parity-check matrix of regular QC LDPC codes [1–5] can be represented by

$$H = \begin{bmatrix} I & I & I & \cdots & I \\ I & P^{S[1]} & P^{S[2]} & \cdots & P^{S[c-1]} \\ I & P^{2S[1]} & P^{2S[2]} & \cdots & P^{2S[c-1]} \\ \cdots & \cdots & \cdots & \cdots & \cdots \\ I & P^{(r-1)S[1]} & P^{(r-1)S[2]} & \cdots & P^{(r-1)S[c-1]} \end{bmatrix}, \qquad (6.13)$$

where I is $B \times B$ (B is a prime number) identity matrix, P is $B \times B$ permutation matrix given by $P = (p_{ij})_{B \times B}$, $p_{i,i+1} = p_{B,1} = 1$ (zero otherwise), and where r and c represent the number of block-rows and block-columns in (6.13), respectively. The set of integers S are to be carefully chosen from the set $\{0, 1, ..., B-1\}$ so that the cycles of short length, in the corresponding Tanner (bipartite) graph representation of (6.13), are avoided. We have to avoid the cycles of length $2k$ ($k = 3$ or 4) defined by the following equation.

$$S[i_1]j_1 + S[i_2]j_2 + \cdots + S[i_k]j_k = S[i_1]j_2 + S[i_2]j_3 + \cdots + S[i_k]j_1 \bmod B, \qquad (6.14)$$

where the closed path is defined by (i_1, j_1), (i_1, j_2), (i_2, j_2), (i_2, j_3), ..., (i_k, j_k), (i_k, j_1) with the pair of indices denoting row-column indices of permutation-blocks in (6.13) such that $l_m \neq l_{m+1}$, $l_k \neq l_1$ ($m = 1, 2, .., k; l \in \{i, j\}$). Therefore, we have to identify the sequence of integers $S[i] \in \{0, 1, ..., B-1\}$ ($i = 0, 1, ..., r-1; r < B$) not satisfying Eq. (6.14), which can be done either by computer search or in a combinatorial fashion. We add an integer at a time from the set $\{0, 1, ..., B-1\}$ (not used before) to the initial set S and check if Eq. (6.14) is satisfied. If Eq. (6.13) is satisfied, we remove that integer from the set S and continue our search with another integer from set $\{0, 1, ..., B-1\}$ until we exploit all the elements from $\{0, 1, ..., B-1\}$. The code rate of these QC codes, R, is lower-bounded by

$$R \geq \frac{|S|B - rB}{|S|B} = 1 - r/|S|,\qquad (6.15)$$

and the codeword length is $|S|B$, where $|S|$ denotes the cardinality of set S. For a given code rate R_0, the number of elements from S to be used is $\lfloor r/(1 - R_0) \rfloor$. With this algorithm, LDPC codes of arbitrary rate can be designed.

As an illustration, by setting $B = 2311$, the set of integers to be used in (6.13) is obtained as $S = \{1, 2, 7, 14, 30, 51, 78, 104, 129, 212, 223, 318, 427, 600, 808\}$. The corresponding LDPC code has rate $R_0 = 1-3/15 = 0.8$, column weight 3, girth-10 and length $|S|B = 15\cdot2311 = 34665$. In the example above, the initial set of integers was $S = \{1, 2, 7\}$, and the set of rows to be used in (6.13) is $\{1,3,6\}$. The use of a different initial set will result in a different set from that obtained above.

6.3.2 Decoding of Binary LDPC Codes

The sum-product algorithm (SPA) is an iterative LDPC decoding algorithm in which extrinsic probabilities are iterated forward and back between variable and check nodes of the Tanner graph representation of a parity-check matrix [1–5, 41]. To facilitate the explanation of the various versions of the SPA, we use $N(v)$ [$N(c)$] to denote the neighborhood of v-node v (c-node c), and introduce the following notations:

- $N(c) = \{v\text{-nodes connected to } c\text{-node } c\}$

- $N(c)\backslash\{v\}$ = {v-nodes connected to c-node c except v-node v}
- $N(v)$ = {c-nodes connected to v-node v}
- $N(v)\backslash\{c\}$ = {c-nodes connected to v-node v except c-node c}
- $P_v = \Pr(v = 1|y)$
- $E(v)$: the event that the check equations involving variable node v are satisfied
- $M(c')\backslash\{c\}$ = {messages from all c'-nodes except node c}
- $q_{vc}(b) = \Pr(v = b\,|\,E(v),y,M(c')\backslash\{c\})$
- $M(v')\backslash\{v\}$ = {messages from all v'-nodes except node v}
- $r_{cv}(b) = \Pr(\text{check equation } c \text{ is satisfied }|\,v = b, M(v')\backslash\{v\})$
- L_{vc}: the extrinsic log-likelihood to be sent from v-node v to c-node c
- L_{cv}: the extrinsic log-likelihood to be sent from c-node c to v-node v

We are interested in computing the APP that a given bit in a transmitted codeword $v = [v_0\ v_1\ ...\ v_{n-1}]$ equals 1, given the received word $y = [y_0\ y_1\ ...\ y_{n-1}]$. Let us focus on decoding of the variable node v and we are concerned in computing LLR:

$$L(v) = \log\left[\frac{\Pr(v = 0\,|\,\boldsymbol{y})}{\Pr(v = 1\,|\,\boldsymbol{y})}\right], \ v \in \{v_0,...,v_{n-1}\} \tag{6.16}$$

The SPA, as indicated above, is an iterative decoding algorithm based on Tanner graph description of an LDPC code. We interpret the v-nodes as one type of processors and c-nodes as another type of processors, while the edges as the message paths for log-APPs.

The sub-graph illustrating the passing of messages (extrinsic information) from c-node to v-nodes in the check-nodes update half-iteration is shown in Fig. 6.5a. The information passed is the probability that parity-check equation c_0 is satisfied. The information passed form node c_0 to node v_1 represents the extrinsic information it had received form nodes v_0, v_4, and v_5 on the previous half-iteration. We are concerned in calculating the APP that a given bit v equals 1, given the received word y and the fact that all parity-check equations involving bit v are satisfied; with this APP denoted as $\Pr(v = 1|\boldsymbol{y}, E(v))$. Instead of APP, we can use log-APP or LLR, defined as $\log\left[\dfrac{\Pr(v = 0\,|\,y,E(v))}{\Pr(v = 1\,|\,y,E(v))}\right]$.

In the v-nodes update half iteration, the messages are passed in the opposite direction-from v-nodes to c-nodes, as depicted in the sub-graph of Fig. 6.5b. The information passed concerns $\log[\Pr(v_0 = 0|\boldsymbol{y})/\Pr(v_0 = 1|\boldsymbol{y})]$. The information being passed from node v_0 to node c_2 is information from the channel (via y_0) and extrinsic information node v_0 had received from nodes c_0 and c_1 on a previous half-iteration. This procedure is performed for all v-node/c-node pairs.

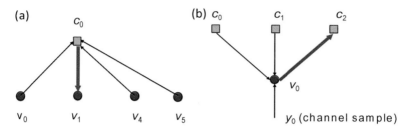

Figure 6.5 Illustration of half-iterations of the SPA: (a) sub-graph of bipartite graph corresponding to the H-matrix with the 0-th row [1 1 0 0 1 1 0...0], (b) sub-graph of bipartite graph corresponding to the H-matrix with the 0-th column [1 1 1 0...0].

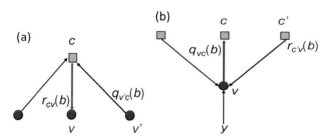

Figure 6.6 Illustration of calculation of: (a) extrinsic information to be passed from node c to variable node v, (b) extrinsic information (message) to be passed from variable node v to check node c regarding the probability that $v = b$, $b \in \{0, 1\}$.

The calculation of the probability that c-th parity-check equation is satisfied given $v = b$, $b \in \{0,1\}$, denoted as $r_{cv}(b)$, is illustrated in Fig. 6.6a. On the other hand, the calculation of the probability that $v = b$ given extrinsic information from all check nodes, except node c, and given channel sample y, denoted as $q_{vc}(b)$, is illustrated in Fig. 6.6b.

In the first half-iteration, we calculate the extrinsic LLR to be passed from node c to variable node v, denoted as $L_{vc} = L(r_{cv})$, as follows:

$$L_{cv} = \left(\prod_{v'} \alpha_{v'c}\right) \phi\left[\sum_{v' \in N(c)\backslash v} \phi(\beta_{v'c})\right]; \ \alpha_{vc} = \text{sign}[L_{vc}], \ \beta_{vc} = |L_{vc}|, \quad (6.17)$$

where with $\phi(x)$ we denoted the following function:

$$\phi(x) = -\log \tanh\left(\frac{x}{2}\right) = \log\left[\frac{e^x + 1}{e^x - 1}\right], \qquad (6.18)$$

which is plotted in Fig. 6.7.

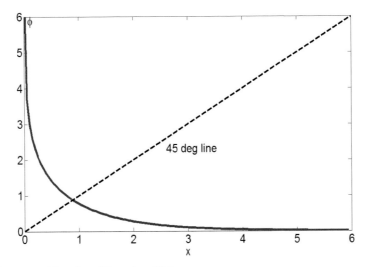

Figure 6.7 The plot of function $\phi(x)$.

In the second half-iteration, we calculate the extrinsic LLR to be passed from variable node v to function node c regarding the probability that $v = b$, denoted as $L_{vc} = L(q_{vc})$, as

$$L_{vc} = L(q_{vc}) = L_{ch}(v) + \sum_{c' \in N(v)\backslash c} L(r_{c'v}). \qquad (6.19)$$

Now we can summarize the *log-domain SPA* as follows.

(0) *Initialization*: For $v = 0, 1, ..., n-1$; initialize the messages L_{vc} to be sent from v-node v to c-node c to channel LLRs $L_{ch}(v)$, namely $L_{vc} = L_{ch}(v)$.

(1) *c-node update rule*: For $c = 0, 1, ..., n-k-1$; compute $L_{cv} = \boxed{+}_{N(c)\backslash\{v\}} L_{vc}$. The box-plus operator is defined by

$$L_1 \boxed{+} L_2 = \prod_{k=1}^{2} \text{sign}(L_k) \cdot \phi\left(\sum_{k=1}^{2} \phi(|L_k|)\right),$$

where $\phi(x) = -\log \tanh(x/2)$. The box operator for $|N(c)\backslash\{v\}|$ components is obtained by recursively applying 2-component version defined above.

(2) *v-node update rule*: For $v = 0, 1, ..., n-1$; set $L_{vc} = L_{ch}(v) + \sum_{N(v)\backslash\{c\}} L_{cv}$ for all c-nodes for which $h_{cv} = 1$.

(3) *Bit decisions*: Update $L(v)$ ($v = 0, ..., n-1$) by $L(v) = L_{ch}(v) + \sum_{N(v)} L_{cv}$ and set $\hat{v} = 1$ when $L(v) < 0$ (otherwise, $\hat{v} = 0$). If $\hat{v}\boldsymbol{H}^T = \boldsymbol{0}$ or pre-determined number of iterations has been reached then stop, otherwise go to step 1).

Because the *c*-node update rule involves log and tanh functions, it is computationally intensive, and there exist many approximations. The very popular is the *min-sum-plus-correction-term approximation* [42]. Namely, it can be shown that "box-plus" operator $\boxed{+}$ can also be calculated by

$$L_1 \boxed{+} L_2 = \prod_{k=1}^{2} \text{sign}(L_k) \cdot \min(|L_1|, |L_2|) + c(x, y), \qquad (6.20)$$

where $c(x,y)$ denotes the correction factor defined by

$$c(x, y) = \log(1 + e^{-|x+y|}) - \log(1 + e^{-|x-y|}), \qquad (6.21)$$

commonly implemented as a look-up table (LUT). Given the fact that $|c(x, y)| < 0.693$, very often this term can be ignored. Alternatively, the following approximation can be used:

$$c(x,y) \simeq \begin{cases} -d, |x-y|<2 \cap |x+y|>2|x-y| \\ d, |x+y|<2 \cap |x-y|>2|x+y| \\ 0, \text{otherwise} \end{cases} \tag{6.22}$$

with typical d being 0.5.

Another popular decoding algorithm is the *min-sum algorithm* in which we simply ignore the correction term in (6.20). Namely, the shape of $\phi(x)$, shown in Fig. 6.7, suggests that the smallest β_{vc} in the summation (6.17) dominates and we can write:

$$\phi\left[\sum_{v' \in N(c)\setminus v} \phi(\beta_{v'c})\right] \cong \phi(\phi(\min_{v'} \beta_{v'c})) = \min_{v'} \beta_{v'c} \tag{6.23}$$

Therefore, the min-sum algorithm is thus the log-domain algorithm with step (1) replaced by

$$L_{cv} = \left(\prod_{v' \in N(c)\setminus v} \alpha_{cv}\right) \cdot \min_{v'} \beta_{v'c} \tag{6.24}$$

6.3.3 Nonbinary LDPC Codes: Quasi-Cyclic Code Design and Decoding Algorithms

In this section, we first describe the nonbinary (NB) irregular QC-LDPC code design derived from pairwise balanced designs (PBDs), introduced in [12], defined over GF(q). Let a set V of size v represent a set of elements (points), with any subset being called a block. Then, a pairwise balanced design PBD(v, K, δ) is defined as a collection of blocks of different sizes taken from set K, such that every pair of points is contained in δ of the blocks. (For additional details on PBDs the interested reader is referred to [34]). The parity-check matrix of an irregular NB QC-LDPC codes based on PBDs is given by [12]:

$$H = \begin{bmatrix} \alpha^0 I\ (b_{00})I & \alpha^1\ (b_{10})I & \cdots & \alpha^{c-1} I\ (b_{c-1,0})I \\ \alpha^{c-1} I\ (b_{01})I & \alpha^0 I\ (b_{11})P^{s[1]} & \cdots & \alpha^{c-2} I\ (b_{c-1,1})P^{s[c-1]} \\ \alpha^{c-2} I\ (b_{02})I & \alpha^{c-1} I\ (b_{12})P^{2s[1]} & \cdots & \alpha^{c-3} I\ (b_{c-1,2})P^{2s[c-1]} \\ \vdots & \vdots & \ddots & \vdots \\ \alpha^{c-(r-1)} I\ (b_{0,r-1})I & \alpha^{c-r+2} I\ (b_{1,r-1})P^{(r-1)s[1]} & \cdots & \alpha^{c-r} I\ (b_{c-1,r-1})P^{(r-1)s[c-1]} \end{bmatrix}, \tag{6.25}$$

where I is $B \times B$ (B is a prime number) identity matrix, P is $B \times B$ permutation matrix given by $P = (p_{ij})_{B \times B}$, $p_{i,i+1} = p_{B,1} = 1$ (zero otherwise); and r and c represent the number of block-rows and block-columns in (6.25), respectively. In (6.25), $\{b_{ij}\}$ are points of the i-th block in PBD(r, K, δ), with the largest size of block k in set of sizes K satisfying the inequality $k \leq r$. The terms α^i are nonzero elements of Galois field of size q, denoted as GF(q). Finally, the term $\mathcal{I}(b_{ij})$ denotes the indicator function, which has the value 1 for the existing point within the i-th block, and 0 for the non-existing point. Therefore, only those submatrices for which the indicator function is 1 will be preserved from template, regular, QC-LDPC code design. Notice the repetition of PBD blocks in (6.25) is allowed. Given the fact that PBDs have regular mathematical structure that can be algebraically described, the irregular NB QC-LDPC codes derived from PBDs have the complexity comparable or lower to that of regular NB QC-LDPC code design. As an illustration, the irregular NB QC-LDPC code derived from PBD(5, {3,2,1}, 1) = {{0,1,3}, {1,2,4}, {1,2}, {0,4}, {3,4}, {0}, {1}} has the following form:

$$H = \begin{bmatrix} \alpha^0 I & 0 & 0 & \alpha^3 I & 0 & \alpha^5 I & 0 \\ \alpha^6 I & \alpha^0 P^{S[1]} & \alpha^1 P^{S[2]} & 0 & 0 & 0 & \alpha^5 P^{S[5]} \\ 0 & \alpha^6 P^{2S[1]} & \alpha^0 P^{2S[2]} & 0 & 0 & 0 & 0 \\ \alpha^4 I & 0 & 0 & 0 & \alpha^1 P^{3S[4]} & 0 & 0 \\ 0 & \alpha^4 P^{4S[1]} & 0 & \alpha^6 P^{4S[3]} & \alpha^0 P^{4S[4]} & 0 & 0 \end{bmatrix}. \quad (6.26)$$

Since both the identity matrix and the power of permutation matrix have a single 1 per row, the block size of the i-th block from PBD determines the i-th block-column weight. In the example above, the first two block-columns have column-weight 3; the third, fourth and fifth have the column-weight 2; and the last two block-columns have weight 1. Notice that for GF(4) = {0, 1, α, α^2}, we have that $\alpha^3 = 1$, $\alpha^4 = \alpha$, $\alpha^5 = \alpha^2$, and $\alpha^6 = \alpha^3$.

If all blocks in PBD are of the same size, the PBD becomes the balanced incomplete block design (BIBD) [34], and the corresponding nonbinary LDPC code given by Eq. (6.25) becomes regular.

The q-ary sum-product algorithm (qSPA), where q is the cardinality of the finite field, and a low-complexity version via fast Fourier transformation (FFT-qSPA) were proposed for decoding nonbinary LDPC codes [43, 44]. Log-domain qSPA (Log-qSPA) was presented in [45] where multiplication is replaced by addition. Later, a mixed-domain version of the FFT-qPSA (MD-FFT-qSPA) decoder was proposed [46], where large size look-up tables are required for the purpose of domain conversion. To avoid the complicated operations involved in the aforementioned decoding algorithms, extended min-sum algorithm [47] and min-max algorithm [48] were widely adopted.

Following the same notation in Section 6.3.2, let $L_{ch,v}(a)$, $L_{vc}(a)$, $L_{cv}(a)$, $L_{cv}^{k,l}$ and $L_v(a)$ represent the prior information of v-node v, the message from v-node v to c-node c, the message from c-node c to v-node v, and the APP value concerning symbol $a \in \mathrm{GF}(q)$, respectively. The *nonbinary min-max algorithm* can be summarized as follows.

(0) *Initialization*: For $v = 0, 1, ..., n - 1$; initialize the messages $L_{vc}(a)$ to be sent from v-node v to c-node c to channel LLRs $L_{ch,v}(a)$, namely $L_{vc}(a) = L_{ch,v}(a)$.

(1) *c-node update rule*: For $c = 0, 1, ..., n{-}k{-}1$; compute $L_{cv}(a) = \min_{a'_{v} \in I(c|a_v)}[\max_{v' \in N(c)\backslash v} L_{vc}(a'_{v'})]$, where $I(c/a_v)$ denotes the set of codewords such that $\sum_{v' \in N(c)\backslash v} h_{cv'}a'_{v'} = h_{cv}a_v$. Here h_{cv} denotes the c-th row, v-th column element of the parity-check matrix.

(2) *v-node update rule*: For $v = 0, 1, ..., n{-}1$; compute $L_{vc}(a) = L_{ch,v}(a) + \sum_{N(v)\backslash c} L_{cv}(a)$ for all c-nodes for which $h_{cv} \neq 0$.

(3) *Post processing*: $L_{vc}(a) = L_{vc}(a) - \min_{a' \in \mathrm{GF}(q)} L_{vc}(a')$, it is necessary for numerical reasons to ensure the nondivergence of the algorithm.

(4) *Symbol decisions*: Update $L_v(a)$ ($v = 0, ..., n - 1$) by $L_v(a) = L_{ch,v}(a) + \sum_{N(v)} L_{cv}(a)$ and set $\hat{v} = \arg\min_{a \in \mathrm{GF}(q)}(L_v(a))$. If $\hat{v}H^T = 0$ or pre-determined number of iterations has been reached then stop, otherwise go to step 1).

The bit-error rate (BER) vs. signal-to-noise ratio performances of the decoding algorithms are presented in Fig. 6.8. All the decoders have been simulated with floating point precision for GF(4) (8430, 6744) nonbinary LDPC code over binary input (BI) AWGN channel. The maximum number of iteration I_{max} is set to 20. The two decoding algorithms, Log-QSPA and MD-FFT-qSPA, achieve the best performance as they are based on theoretical derivations while the reduced-complexity decoding algorithms, min-max, max-log and min-sum algorithms, face the performance degradation at BER of 10^{-7} by 0.13 dB, 0.27 dB, and 0.31 dB, respectively.

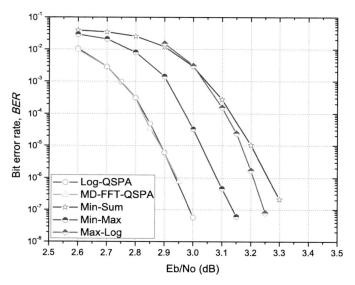

Figure 6.8 BER performance of NB-LDPC code over BI-AWGN.

6.3.4 Rate-Adaptive LDPC Coding Implementations in FPGA

The basic rate adaptation via either shortening or puncturing is widely used everywhere in communication systems, and can be introduced in both block and convolutional codes. In this subsection, we use shortening to achieve rate adaptive LDPC coding since it allows a wide range of rate adjustment with unified decoder architecture through a set of reconfigurable registers in an FPGA [49]. Because of the quasi-cyclic structure of our nonbinary LDPC codes, we shorten entire sub-block by adding the least number of logics' blocks. For example, we start from a (3,15)-regular nonbinary LDPC codes with rate of 0.8, and we can obtain a class of shortened regular nonbinary LDPC codes with column weight and row weight of {(3,14), (3,13), (3,12), (3,11), (3,10)}, which corresponding to code rates of {0.786, 0.77, 0.75, 0.727, 0.7}.

As shown in Fig. 6.9a, the binary LDPC decoder consists of two major memory blocks (one stores channel LLR and another stores APP messages), two processing blocks (variable node unit (VNU) and check node unit (CNU)), an early termination unit (ETU), and a number of mux blocks, wherein its selection of output signal can be software reconfigurable to adjust the shortening length. The memory consumption is dominated by the LLR message with size of $n \times W_L$ and APP messages of size $c \times n \times W_R$, where W_L and W_R denote the precisions used to represent LLR and APP messages. The logic consumption is dominated by CNU, as shown in Fig. 6.9b. The ABS-block first takes the absolute value of the inputs and the sign XOR array produces the output sign. In the two least minimums' finder block, we find the first minimum value via binary tree and trace back the survivors to find the second minimum value as well as the position of the first minimum value. This implies that we can write 3 values and r sign bits back to the APP memories instead of r values. However, we will not take advantage of memory reduction techniques for comparison in the following paragraphs.

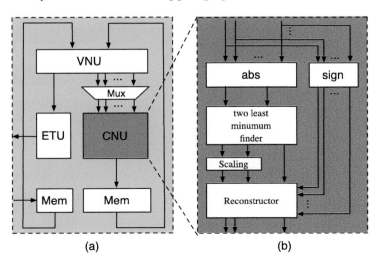

(a)   (b)

Figure 6.9 Architecture of rate adaptive binary LDPC decoder: (a) overall schematic diagram, (b) schematic diagram of CNU.

Similarly to the rate adaptive binary LDPC decoder architecture discussed above, the architecture of the *layered min-max algorithm* (LMMA)-based nonbinary LDPC decoder is presented in Fig. 6.10a. There are four types of memories used in implementation: memory $R_{c,v}$ with size of $c \times n \times q \times W_R$ stores the information from check nodes to variable nodes, memory for L_v with size of $n \times q \times W_L$ stores the initial log-likelihood ratios, memory for $\hat{c}$ with size $n \times \log_2 q$ stores the decoded bits, and memories inside each CNU store the intermediate values. The same notations are borrowed from previous subsection except that q denotes the size of Galois field. As shown in Fig. 6.10b, it is obvious that the CNU is the most complex part of the decoding algorithm, which consists of r inverse permutators, r BCJR-based min-max processors and r permutators, and two types of the first-in-first-out (FIFO) registers. The inverse permutator block shifts the incoming message vector cyclically. The first FIFO is used to perform the parallel-to-serial conversion. After the min-max processor, which is implemented by low-latency bidirectional BCJR algorithm, the processed data is fed into another FIFO block performing serial-to-parallel conversion, followed by the permutator block. Because of the high complexity of the CNU design and high memory requirements of the nonbinary decoder than that of binary decoder, reduced-complexity architectures and selective version of MMA can be further exploited.

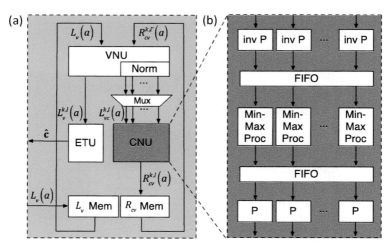

Figure 6.10 Architecture of rate adaptive nonbinary LDPC decoder: (a) overall schematic diagram, (b) schematic diagram of CNU.

We compare two rate-adaptive schemes based on binary LDPC codes and nonbinary LDPC codes. These architectures can be software-defined by initializing configurable registers in the FPGA. The resource utilization is summarized in Table 6.1. One can clearly notice that the LMMA-based nonbinary LDPC codes consumes 3.6 times larger memory size than the binary one because of large field size and high quantization precision, while the occupied number of slices is five times larger than that in binary case because of the higher complexity of the CNU.

Table 6.1 Logic resources utilization

Resources	Binary LDPC decoder	Nonbinary LDPC decoder
Occupied Slices	2,969 out of 74,400 (3%)	16,842 out of 74,400 (22%)
RAMB36E1	113 out of 1,064 (10%)	338 out of 1,064 (31%)
RAMB18E1	14 out of 2,128 (1%)	221 out of 2,128 (9%)

The BER vs. Q-factor performances of the rate adaptive binary and nonbinary LDPC code are presented in Figs. 6.11 and 6.12.

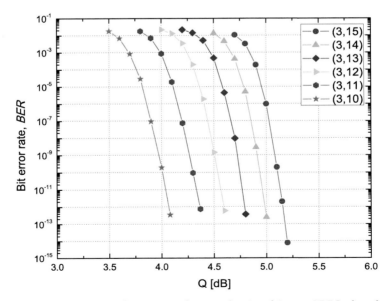

Figure 6.11 BER performance of rate-adaptive binary LDPC decoder implemented in an FPGA.

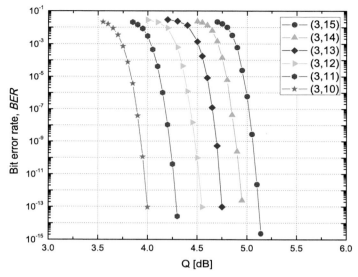

Figure 6.12 BER performance of rate-adaptive nonbinary LDPC decoder implemented in FPGA.

The FPGA-based emulation was conducted over binary (BI)-AWGN channel and 6 and 8 bits precision are used in binary and nonbinary LDPC decoder, respectively. A set of column weight and row weight configurations of {(3, 15), (3, 14), (3, 13), (3, 12), (3, 11), (3, 10)}, which corresponds to the code rates of {0.8, 0.786, 0.77, 0.75, 0.727, 0.7}, can be achieved by software-based reconfiguration of specific register in FPGA. The girth-10 regular (34635, 27710, 0.8) binary and nonbinary mother code can achieve a Q-limit of 5.2 dB and 5.14 dB at BER of 10^{-15}, which corresponds to NCG of 11.83 and 11.89 dB. The rate adaptive nonbinary LDPC codes outperform the binary LDPC codes by approximated 0.06 dB in all range of rate from 0.7~0.8. In addition, we believe this gap will be larger when combined with higher modulation schemes enabling 100 Gbits/s (with QPSK) 400 Gbits/s (with 16QAM) optical communication systems.

In the rest of this section, we discuss *other different strategies for rate adaptation*, in addition to shortening. The code rate adaptation is performed by *partial reconfiguration* of the decoder based on (6.25), for a fixed codeword length, by changing the size of the permutation matrix P while keeping the number of block rows constant, and/or by varying the number of employed block-rows while keeping P fixed. Another alternative to change the code rate is to use the same regular parity-check matrix as a template but employ the PBDs corresponding to different code rates. It is also possible to perform the puncturing of parity symbols in the original code (n, k, d) to obtain a linear block code $(n\text{-}p, k, d_p)$, $d_p \leq d$, where p is the number of removed parity symbols. Notice that this approach when applied to LDPC codes can introduce an early error floor phenomenon. This is the reason why the re-encoding approach has been introduced in [13]. To encode, the adaptive LDPC encoder encapsulates the last M symbols of the proceeding codeword and the incoming $K\text{-}M$ information symbols into a K-symbol vector. In other words, each codeword is generated continuously by re-encoding the last M symbols of its preceding codeword. (The first M symbols in the first codeword are set as known since no proceeding codeword exists and the first M re-encoded symbols of each codeword are not transmitted.) Therefore, the actual code rate is $R' = (K\text{-}M)/(N\text{-}M)$, where $0 < M < K$ and can be tuned easily in the range $(0, R]$ (R is the code rate of template code) by adjusting the re-encoded data size M. Notice that the template code should be systematic to leverage the advantage of re-encoding in decoding.

6.4 Coded Modulation for Optical Communications

The coded modulation (CM) has initially been proposed for wireless and wireline communications [50–54]. Different coded-modulation schemes recently proposed for use in optical communications include: (i) block-interleaved coded modulation [1–5], (ii) multilevel coding [1, 19], (iii) nonbinary LDPC-coded modulation [18], (iv) multidimensional coded modulation [12], (v) multilevel nonbinary LDPC-CM [55], and hybrid coded modulation [15]. For some other CM scheme, the interested reader is referred to [1, 19]. Using the coded modulation approach, modulation, coding and multiplexing are performed in a unified fashion so that, effectively, the transmission, signal processing, detection and decoding are done at much lower symbol rates, where dealing with fiber nonlinear effects is manageable.

6.4.1   Coded Modulation Fundamentals

Let d_{unc} denote the minimum Euclidean distance in the original, uncoded signal constellation diagram, and let d_{min} denote the minimum distance in sequence of symbols after coding. When the uncoded normalized average energy, denoted as E_{unc}, is different from the normalized average energy after coding, denoted as E_{coded}, the *asymptotic coding gain* can be defined as follows:

$$G_a = \frac{E_{unc}/d_{unc}^2}{E_{coded}/d_{min}^2} = \underbrace{\frac{E_{unc}}{E_{coded}}}_{G_C} \cdot \underbrace{\frac{d_{min}^2}{d_{unc}^2}}_{G_D} = G_C \cdot G_D, \tag{6.27}$$

where G_C is the *constellation expansion factor* and G_D is the *increased distance gain*. The generic coded modulation scheme, applicable to both block-coded modulation (BCM) and trellis coded modulation (TCM), is illustrated in Fig. 6.13, which is based on Forney's interpretation [56, 57] (see also [58]). Two-dimensional (2-D) modulator such as M-ary QAM (I/Q) modulator or M-ray PSK modulator, is used to impose the sequence of symbols to be transmitted. The N-dimensional lattice is used for code design. A *lattice* represents a discrete set of vectors in a real Euclidean N-dimensional space, which forms a group under ordinary vector addition, so the sum or difference of any two vectors in the lattice is also in the lattice. A sub-lattice is a subset of a lattice that is itself a lattice. The sequence space of uncoded signal is a sequence of points from an N-cube, obtained as a Cartesian product of 2-D rectangular lattices with points located at odd integers. Once the densest lattice is determined, we create its decomposition into partition subsets known as *cosets*. The k information bits are used is input to binary encoder, be block or convolutional. The binary encoder generates the codeword of length $k + r$, where r is the number of redundant bits. With $k + r$ bits we can select one out of 2^{k+r} possible cosets. Each coset contains 2^{n-k} points, and therefore, $n-k$ additional information bits are needed to select a point within the coset. Given that the lattice is defined in N-dimensional space while 2-D modulator imposes the data in 2-D space, $N/2$ consecutive transmissions are needed to transmit a point from the coset. Therefore, information bits are conveyed in two ways: (i) through the sequence of cosets from which constellation points are selected and (ii) through the points selected within each coset. Let d_{points} denote the minimum distance of points within the coset and d_{seq} denote the minimum distance among coset sequences, then the minimum distance of the code can be determined as d_{min} = min(d_{points}, d_{seq}). Since the fundamental volume per N dimensions is 2^r, where r is the number of redundant bits, the normalized volume per two dimensions would be $\sqrt[N/2]{2^r} = 2^{2r/N}$, the coding gain of this coded modulation scheme can be estimated as

$$G = \frac{d_{min}^2}{2^{2r/N}} = 2^{-2r/N} \, d_{min}^2. \tag{6.28}$$

Clearly, the normalized redundancy per two dimensions is equal to $r/(N/2) = 2r/N$. When the constellation is chosen to be N-sphere-like to reduce the average energy, we need to account for an additional gain, often called the *shaping gain*, denoted with G_s, which measures the reduction in energy. Calderbank and Sloane have shown that the shaping gain of an N-sphere over N-cube, when N is even, is given by [59]

$$G_s = \frac{\pi}{6} \frac{(N/2)+1}{\left[(N/2)!\right]^{1/(N/2)}}. \tag{6.29}$$

For instance for $N = 2$ and 4 the shaping gains are $\pi/3$ (0.2 dB) and $\pi/2^{3/2}$ (0.456 dB), respectively. The shaping gain in the limiting case, as N tends to infinity, is $\pi e/6$ (1.533 dB). This limiting shaping gain can also be achieved by nonuniform signaling [60, 61], also known as *probabilistic shaping* [62], in which different points from a signal constellation are transmitted with different probabilities. It has been shown in [60] that when the distribution of constellation points matches Maxwell-Boltzmann distribution, the ultimate shaping gain of 1.533 dB can be achieved. For additional details on probabilistic shaping, the interested reader is referred to Chapter 5.

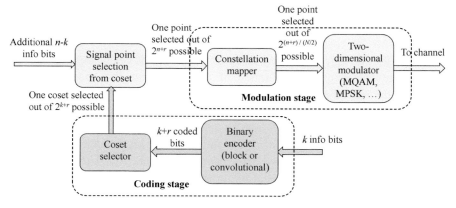

Figure 6.13 Generic coded modulation scheme.

Based on Forney's guidelines [56, 63], it is possible to generate maximum-density N-dimensional lattices by using a simple partition of a 2-D lattice with corresponding block/convolutional code, when $N = 4, 8, 16$, and 24; by applying the subset partition rules as shown in Fig. 6.14. The *4-D lattice* is obtained by taking all sequences of pair of points from the same subset, such as (A,A) or (B,B). The *8-D lattice* is obtained by taking all sequences of four points of either type-A or type-B. Each point in subset has a single subscript. Within each of four point subset, the point subscripts satisfy the parity check equation $i_1 + i_2 + i_3 + i_4 = 0$, so that the sequence subscripts must be codewords in the (4,3) parity-check code, which has a minimum Hamming distance of two. Therefore, three information bits and one parity bit are used to determine the lattice subset. The *16-D lattice* is obtained by taking all sequences of eight points of either type-A or type-B. Each point has two subscripts. The subscripts in the subset belong to a codeword from extended Hamming code (16,11) of min Hamming distance 4. Finally, the *24-D lattice* is obtained by taking all sequences of 12 points of either type-A or type-B. Each point has three subscripts (i, j, k). The subscripts (i,j) in 12-point subset form a codeword from the Golay (24,12) code of minimum Hamming distance 8. The third subscript k represents the overall parity-check; for B-points even parity-check is used and for A-points odd-parity-check is used (or vise versa). The corresponding coding gains are 1.5 dB, 3 dB, 4.5 dB, and 6 dB for lattice codes of dimensionalities 3, 8, 16, and 24, respectively [56].

· · ·	·	·	·	·	·	·	·	·	·	·	·	· · ·
· · ·	B_{000}	A_{000}	B_{110}	A_{110}	B_{000}	A_{000}	B_{110}	A_{110}	B_{000}	A_{000}	· · ·
· · ·	A_{101}	B_{010}	A_{010}	B_{101}	A_{101}	B_{010}	A_{010}	B_{101}	A_{101}	B_{010}	· · ·
· · ·	B_{111}	A_{111}	B_{001}	A_{001}	B_{111}	A_{111}	B_{001}	A_{001}	B_{111}	A_{111}	· · ·
· · ·	A_{011}	B_{100}	A_{100}	B_{011}	A_{011}	B_{100}	A_{100}	B_{011}	A_{011}	B_{100}	· · ·
· · ·	B_{000}	A_{000}	B_{110}	A_{110}	B_{000}	A_{000}	B_{110}	A_{110}	B_{000}	A_{000}	· · ·
· · ·	A_{101}	B_{010}	A_{010}	B_{101}	A_{101}	B_{010}	A_{010}	B_{101}	A_{101}	B_{010}	· · ·
· · ·	B_{111}	A_{111}	B_{001}	A_{001}	B_{111}	A_{111}	B_{001}	A_{001}	B_{111}	A_{111}	· · ·
· · ·	A_{011}	B_{100}	A_{100}	B_{011}	A_{011}	B_{100}	A_{100}	B_{011}	A_{011}	B_{100}	· · ·
· · ·	B_{000}	A_{000}	B_{110}	A_{110}	B_{000}	A_{000}	B_{110}	A_{110}	B_{000}	A_{000}	· · ·
· · ·	A_{101}	B_{010}	A_{010}	B_{101}	A_{101}	B_{010}	A_{010}	B_{101}	A_{101}	B_{010}	· · ·
· · ·	·	·	·	·	·	·	·	·	·	·	· · ·

Figure 6.14 Subset partition for up to 24-dimensional lattice.

6.4.2 Multilevel Coded Modulation and Unequal Error Protection

Multilevel coding (MLC) was initially proposed by Imai and Hirakawa in 1977 [50]. The key idea behind the MLC is to protect individual bits using different binary codes and use M-ary signal constellations, as illustrated in Fig. 6.15. The i-th component code, denoted as C_i, is $(n, k_i, d_i^{(H)})$ $(i = 1, 2, ..., m)$ linear block code of Hamming distance $d_i^{(H)}$. The length of the information word u_i for the i-th component code is k_i, while the corresponding codeword length n is the same for all codes. The bits-to-symbol mapper can be implemented with the help of block-interleaver, in which the codewords from different component codes are written in row-wise fashion into interleaver, and read out m bits at the time in column-wise fashion. With this interpretation, the symbol at j-th time instance ($j = 1, 2, ..., n$) is obtained after the mapping $\underline{s}_j = s(\mathbf{c}_j)$, where $\mathbf{c}_j = (c_{i,j} \ c_{i,2} \ ... \ c_{i,m})$ and $s(\cdot)$ is corresponding mapping rule. The 2^m-ary 2-D modulator, such as I/Q and PSK modulators, is used to impose the selected signal constellation point $\underline{s}_{i,} = (s_{i,I}, s_{i,Q})$, where $s_{i,j}$ and $s_{i,Q}$ are the corresponding in-phase and quadrature components, on the corresponding carrier signal. For instance, the signal index can be used as the address for a look-up-table (LUT) where the coordinates are stored. Once the coordinates are selected, the in-phase and quadrature components are used as inputs of corresponding pulse shapers, followed by driver amplifiers. For optical communications, the optical I/Q modulator is further used to perform electrical-to-optical conversion.

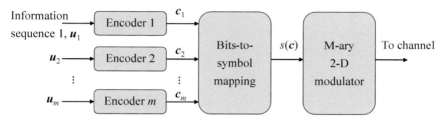

Figure 6.15 Multilevel encoder for MLC/MCM scheme.

The overall code rate of this MCM scheme, expressed in bits/symbol, is given by

$$R = \sum_{i=1}^{m} R_i = \sum_{i=1}^{m} k_i / n = \frac{\sum_{i=1}^{m} k_i}{n}. \tag{6.30}$$

The minimum squared Euclidian distance (MSED) of this scheme is lower-bounded by [50]

$$\text{MSED} \geq \min_{1 \leq i \leq m} \{d_i^{(H)} d_i\}, \tag{6.31}$$

where d_i is the Euclidean distance at the i-th level of set partitioning. As an illustration, the Euclidean distances at the i-th level of set partitioning (i = 1,2,3) have been provided in Fig. 6.16.

In set partition, a 2^m-ary signal constellation set, is partitioned into m-levels. At the i-th partition level ($0 \leq i \leq m - 1$), the signal S_i is split into two subsets $S_i(0)$ and $S_i(1)$ so that the intra-set squared Euclidean distance d_i^2 is maximized. With each subset choice $S_i(c_i)$ we associate a bit label c_i. The end result of this process is unique m-label $c_0 c_1 ... c_{m-1}$ associated with each constellation point $s(c_0 c_1 ... c_{m-1})$. By employing this Ungerboeck (standard) set partitioning of 2^m-ary signal constellation [51], the intra-set distances per levels are arranged in a nondecreasing order as the level index increases; in other words: $d_0^2 \le d_1^2 \le \cdots d_{m-1}^2$. For the set partitioning of 8-PSK with natural mapping rule $m = 3$, and in level three there are four subsets (cosets) with each subset containing two points. When the component codes are selected as follows: (8,1,8), (8,7,2), (8,8,1), the MSED for coded 8-PSK is lower-bounded by $\mathrm{MSED} \ge \min\{0.7654^2 \cdot 8, (\sqrt{2})^2 \cdot 2, 2^2 \cdot 1\} = 4$, indicating that the asymptotic coding gain with respect to uncoded QPSK is at least 3 dB. When properly designed LDPC codes are used as component codes, the near-capacity achieving performance can be obtained.

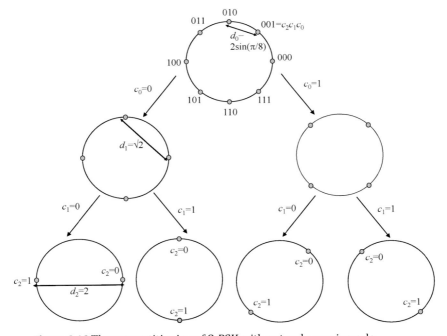

Figure 6.16 The set partitioning of 8-PSK with natural mapping rule.

Each component code can be represented using the trellis diagram description of the code. By taking the Cartesian product of individual trellises and then perform mapping based on set partitioning we obtain a trellis description of this MCM scheme. Now we can apply either Viterbi algorithm or BCJR algorithm to perform decoding on such a trellis. Unfortunately, the complexity of such decoding is too high to be of practical importance. Instead, the decoding is typically based on so-called *multistage decoding* (MSD) algorithm [64] in which the decisions from prior (lower) decoding stage are passed to next (higher) stages, which is illustrated in Fig. 6.17.

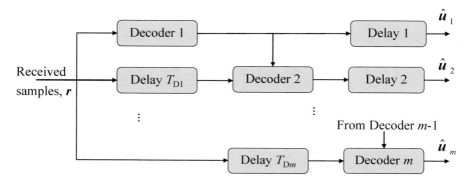

Figure 6.17 Multistage decoder for MLC/MCM scheme.

The i-th decoder operation (i = 2, ..., m) is delayed until the ($i–1$)-th decoder completes the decoding algorithm. Further, after the decoding, the i-th decoder (i = 1, ..., $m−1$) output is delayed until the m-th decoder decoding algorithm is completed. Clearly, even though the complexity of MSD algorithm is lower compared to Viterbi/BCJR algorithm operating on MLC trellis diagram, since the i-th component decoder operates on trellis of C_i code, the overall decoding latency is still high. Moreover, when the component codes are weak codes the error not corrected in prior decoding stage will affect the next stage and result in error multiplicity. On the other hand, when component codes are properly designed LDPC codes, this problem can be avoided [52, 65]. Moreover, it has been shown in [52], when the code rates and degree distribution are optimized for a given mapping rule, BER performance degradation when component LDPC decoders operate independently can be made arbitrary small compared to MSD algorithm. This decoding algorithm in which all components decoders operate independently of each other is commonly referred to as *parallel independent decoding* (PID) [52, 65]. The

PID decoding architecture is illustrated in Fig. 6.18, assuming that properly designed LPDC codes are used as component codes.

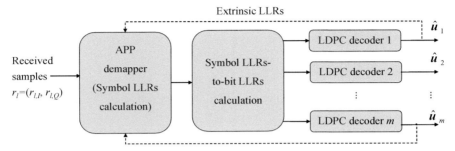

Figure 6.18 Parallel independent decoding (PID) for MLC/MCM scheme.

Clearly, in MLC different bits transmitted over the channel are differently protected, and as such the MLC is suitable for unequal error protection (UEP). The MLC channel encoder consists of m different binary error-correcting codes C_1, ..., C_m with decreasing codeword distances. The i-th priority bit stream is used as the input to the i-th encoder, which generates the coded bits c_i. The rest is the same as described above. As an illustration, in image compression, one type of channel code can be used for bits corresponding to the low-resolution reproduction of the image, whereas high resolution bits that simply refine the image can be protected by a different channel code. This is in particularly important in JPEG 2000 image compression standard, in which the image is decomposed into a multiple resolution representation. The JPEG 2000 provides efficient code-stream organizations that are progressive by pixel accuracy and by image resolution. Thanks to such organization, after a smaller portion of the whole file has been received, the viewer can see a lower quality version of the picture. The quality of the picture is then improved progressively through downloading additional data bits from the source.

For UEP, the *nested coding* can also be used. Nested coding can be interpreted as follows. Let us consider m different information vectors u_i ($i = 1$, ..., m) of length k_m. We would like to jointly encode these information vectors on such a way that each information vector is associated with a codeword from a different subcode. The i-th subcode C_i is represented by the generator matrix G_i of rate $R_i = k_i/n$. The overall generator matrix is given by

$$G = \begin{bmatrix} G_1 \\ G_2 \\ \vdots \\ G_m \end{bmatrix}. \tag{6.32}$$

The overall codeword c can be obtained as follows:

$$c^T = \begin{bmatrix} u_1^T & u_2^T & \cdots & u_m^T \end{bmatrix} G = \begin{bmatrix} u_1^T & u_2^T & \cdots & u_m^T \end{bmatrix} \begin{bmatrix} G_1 \\ G_2 \\ \vdots \\ G_m \end{bmatrix}$$

$$= u_1^T G_1 \oplus u_2^T G_2 \oplus \cdots \oplus u_m^T G_m, \tag{6.33}$$

where we use $\oplus$ to denote the bitwise XOR operation. If we are interested in unequal error correction, by setting $u_i = u$, by varying the number of generator matrices G_i we can achieve different levels of protection. The lowest level of protection would be to use only one generator matrix. The highest level of protection, corresponding to high priority bits, will be achieved by encoding the same information vector u m-times.

To improve the spectral efficiency of MLC scheme, the *time-multiplexed coded modulation* can be used. The key idea is to use different coded modulation schemes for different priority classes of information bits. Let T_i be the fraction of time in which the i-th priority code C_i is employed and let $R_{s,i}$ denote the corresponding symbol rate. The overall symbol rate R_s of this scheme would be

$$R_s = \frac{\sum_i R_{s,i} T_i}{\sum_i T_i}. \tag{6.34}$$

When all component codes in MCM scheme have the same rate, the corresponding scheme in [1–5], and related papers, is called *block-interleaved* coded modulation, as it contains the block-interleaver. This particular version, when combined with optimum signal constellation and optimum mapping rule designs [66–68], performs comparable to MLC, but has even lower complexity. At the same time it is suitable for implementation in hardware at high-speed, as m LDPC decoders operate in parallel at bit rate equal to symbol rate R_s. In conventional BICM scheme [54], described in the next section, a single LDPC decoder is required operating at rate m R_s, which can exceed the speed of existing electronics for fiber-optics communications, where the information symbol rate is typically 25 GS/s.

6.4.3   Bit-Interleaved Coded Modulation

In BICM [54], a single binary encoder is used followed by a pseudorandom bit interleaver, as shown in Fig. 6.19. The output of the bit interleaver is grouped in block of m bits, and during the signaling interval these m bits are used as input of corresponding mapper, such as Gray, anti-gray, natural mapping, and after the mapper, m bits from mapper are used to select a point from a 2^m-ary signal constellation. The rest of the transmitter is similar to that for MLC scheme. On the receiver side, the configurations of demodulator, APP demapper, and symbol LLRs-to-bit LLRs calculation block are similar to those for MLC scheme. After the deinterleaving of bit LLRs, corresponding bit reliabilities are forwarded to a single binary decoder. For this scheme, the choice of mapping rule is of high importance. For iterative decoding and

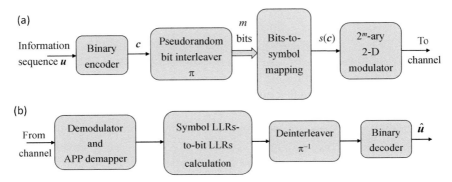

Figure 6.19 Bit-interleaved coded modulation (BICM) scheme: (a) BICM encoder and (b) BICM decoder configurations.

demapping, the Gray mapping rule is not necessarily the best mapping rule. The EXIT chart analysis should be used to match the APP demapper and LDPC decoder choice.

6.4.4 Hybrid Multidimensional Coded Modulation Scheme for High-Speed Optical Transport

It has been shown that multilevel coding with parallel independent decoding (MLC/PID) and iterative detection and decoding (IDD) perform very well under diverse optical fiber transmission scenarios [1–5]. Recently, nonbinary LDPC-coded modulation (NB-LDPC-CM) has been shown to outperform its corresponding MLC/PID+IDD scheme with performance gaps increasing as the underlying signal constellation size increases [18]. NB-LDPC-CM employs a single component code defined over an alphabet that matches in size to the underlying constellation size, which in return eliminates the need for iterative demapping-decoding but increases the decoding complexity. In this section, we described alternative, *hybrid multidimensional LDPC-coded modulation scheme*, which represents the generalization of hybrid coded modulation scheme introduced in [15] that lies essentially in between the two extremes described in [1–5]–employing only binary LDPC codes– and [18]–employing a single nonbinary LDPC code over a large finite field–in terms of decoding complexity and error correction performance. Having such alternatives at hand can provide the much-needed flexibility during link budget analysis of optical communication systems rather than forcing the system designer to opt into one of the two extremes. Additionally, various coded modulation schemes including BICM, MLC, nonbinary LDPC CM, multilevel nonbinary LDPC-coded modulation (NB-LDPC-CM) [55], to mention few, are just particular instances of hybrid coded modulation scheme.

A generic adaptive hybrid D-dimensional NB LDPC-coded modulation scheme to be used in combination with irregular/regular NB LDPC coding, is shown in Fig. 6.20. (To facilitate explanations, the details related to compensation of dispersion and nonlinear effects, carrier-frequency and carrier-phase estimations are omitted.) It is applicable to various spatial division multiplexing (SDM) fibers: few-mode fibers (FMFs), few-core fibers (FCFs), and few-mode-few-core fibers (FMFCFs). It is also applicable to single-mode fibers (SMFs), which can be considered as SDM fibers with only one fundamental mode and two polarization states. Both electrical and optical basis functions are applicable. Electrical basis functions include modified orthogonal polynomials and orthogonal prolate spheroidal wave functions (also known as Slepian sequences in discrete-time domain), while optical basis functions include spatial modes and polarization states. This scheme is very flexible and can be used in various configurations ranging from spatial multiplexing of various multidimensional signals to fully D-dimensional signaling.

Generic Tx configuration for the hybrid multidimensional LDPC-coded modulation scheme is presented in Fig. 6.20a. The MLC scheme, discussed earlier, is a degenerate case of the hybrid CM scheme and occurs when $b_m = 1$ for all m, while LDPC codes at different levels have different code rates. When all LDPC codes have the same code rate, we refer to this scheme as the *block-interleaved coded modulation*.

The conventional BICM is a serial concatenation of channel encoder, interleaver, and multilevel modulator [54]. This scheme is based on a single LDPC code, as discussed in previous section.

On the other hand, the NB-LDPC-CM of [18] is a particular instance of hybrid CM scheme employing a single level with an M-ary LDPC code and performing just a single pass through the detector and the decoder without any feedback. In hybrid multidimensional CM scheme, the number of levels is between 1 and $\log_2(M)$. The hybrid CM approach combines component codes defined over $GF(2^{b_i})$, $1 \le i \le m$ in a way that the condition $b_1 + \cdots + b_m = \log_2(M)$ is satisfied. Thus, the resulting NB-LDPC-coded modulation schemes are not homogenous in the field orders of their component codes, but are rather heterogeneous. They enable the interaction of component codes over various fields and hence the name *hybrid* LDPC-coded modulation scheme.

The configuration of D-dimensional modulator is shown in Fig. 6.20c. Clearly, conventional PDM scheme is just a special case corresponding to fundamental mode only by setting $N_1 = N_2 = 1$.

Transmitter side DSP provides coordinates for *D*-dimensional signaling. The 4-D modulator is composed of polarization beam splitter, two *I/Q* modulators corresponding to *x*- and *y*-polarizations, and polarization beam combiners. One spatial mode contains the data streams of N_1 power combined 4-D data streams. The N_2 data streams, corresponding to N_2 spatial modes are combined into a signal data stream by spatial division multiplexer and transmitted over SDM fiber. On the other hand, the *D*-dimensional demodulator, shown in Fig. 6.20d, provides the samples corresponding to signal constellation estimates. For additional details about this hybrid multidimensional coded modulation scheme, the interested reader is referred to [69].

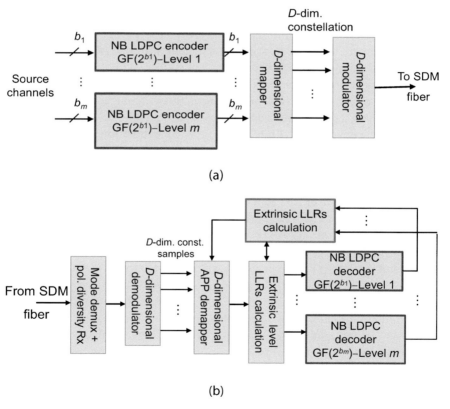

Figure 6.20 Hybrid multidimensional coded modulation: (a) Transmitter configuration. (b) Receiver configuration.

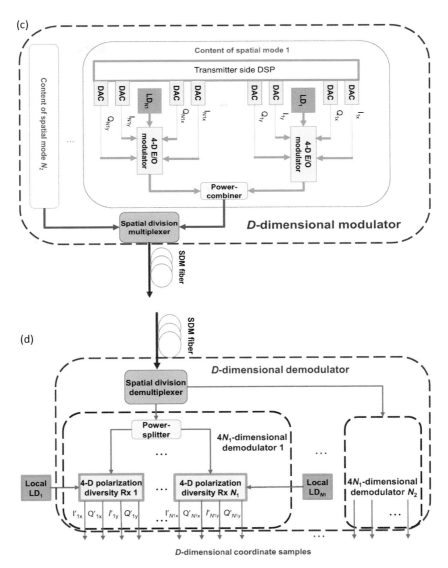

Figure 6.20 (continued) (c) Configuration of D-dimensional modulator. (d) D-dimensional demodulator configuration. SDM: spatial division multiplexing. SDM fibers could be FMF, FCF, FMFCF, or SMF. The corresponding signal space is $D = 4N_1N_2$-dimensional, where N_1 is number of sub-bands in a superchannel with orthogonal center frequencies and N_2 is number of spatial modes. The factor 4 originates from 2 polarization states and in-phase and quadrature channels.

6.5 LDPC Coded Modulation for Optical Communications Enabling Quasi-Single-Mode Transmission over Transoceanic Distances Using Few-Mode Fibers

High-speed optical systems rely solely on single-mode fibers for transmission beyond a few kilometers reach and, state-of-the art single-mode fibers are required to achieve spectral efficiencies beyond 6 bits/s/Hz over transoceanic distances. Recently, there has been a renewed interest in transmission over FMFs to increase the capacity over a single fiber strand by the multitude of the spatial modes. However, such an approach requires a radical transformation of optical communication infrastructure which has been built on the premise of transmission over a single spatial mode. It was proposed that FMFs can be integrated into the otherwise single-mode infrastructure by transmitting only in the fundamental mode, while taking advantage of their significantly larger effective area. The feasibility of using FMFs was demonstrated in a recent transmission experiment over 2600 km. In our recent paper [70], 101.6 km-long hybrid spans consisting of low-loss silica-core FMFs and single-mode fibers are used to achieve a new record spectral efficiency of 6.5 b/s/Hz in unidirectional transmission configuration over the transoceanic distance of 6600 km using EDFAs only. To the best of our knowledge, this is the first time FMFs are used to outperform state-of-the-art single-mode fibers in a transoceanic transmission experiment. To make the FMFs compatible with single-mode transmission, single-mode fiber jumpers are spliced to each end. This excites only the fundamental mode of the FMF at launch, and also strips any high-order modes excited during transmission at the exit splice. Such systems may suffer from multi-path interference (MPI) even with inherently low mode coupling FMFs. In effect, by using FMFs the nonlinear fiber impairments are traded for linear impairments from MPI which are easier to manage. While the improvements possible from mitigating nonlinear impairments by DSP is fundamentally limited in scope, and computationally costly, it was demonstrated recently that most of the penalty from the MPI can be mitigated with much less computational complexity by including an additional decision-directed least mean square (DD-LMS) equalizer to the standard digital coherent DSP. The differential modal group delay (DMGD) of a FMF can be of the order of 1 ns/km and the required number of taps in the DD-LMS equalizer may be several hundreds to thousands, which not only makes them less stable and computationally demanding, but also degrades its performance. In this work, instead of using standard single-carrier modulation (SCM), a digital multi-subcarrier modulation (MSCM) signal with 32 subcarriers at 1 GBaud is used to reduce the number of equalizer taps per subcarrier by more than an order of magnitude while increasing the temporal extent of MPI that can be mitigated. To further reduce MPI, a hybrid FMF-single-mode fiber span design is used for the first time with the FMF at the beginning of the span to reduce nonlinearity, and the single-mode fiber at the end to avoid excessive MPI. Hybrid spans also reduce the DSP complexity as they have smaller modal group delay per span compared to all FMF spans.

The experimental setup is shown in Fig. 6.21.

At the transmitter side, digital-to-analog converters (DACs) with a sampling rate of 64 GHz are used to generate WDM channels tuned to a 33 GHz grid. WDM channels are prepared in two groups. The first group consists of six neighboring channels generated with tunable ECLs. These six channels are tuned together across the C-band and only the center channel which uses a narrow linewidth (<1 kHz) laser is measured. Each of the even and odd subgroups of the six tunable channels are modulated separately by 4 independent streams of data for all in-phase and quadrature (I/Q) rails in both polarizations. The 111 loading channels are modulated with only independent I/Q rails followed by polarization multiplexing emulators. The tunable channels and the dummy channels are combined using a wavelength selective switch (WSS) with a 1 GHz grid resolution and a coupler. Using the DACs, either a 32 Gbaud, Nyquist-shaped SCM 16QAM, or a MSCM signal of 32×1 Gbaud 16QAM subcarriers with 10 MHz guard-band is generated per wavelength. The 2^{18}-1 PRBS binary data are encoded by 16-ary irregular quasi-cyclic nonbinary LDPC encoder of girth 10 to generate codewords at three overheads 14.3%, 20%, and 25%, to maximize the capacity across C-band. The codeword length is 13680 for 14.3% and 20% OHs, and 74985 for 25% OH. Encoder output is uniformly distributed into 32 sub-carriers of MSCM signal.

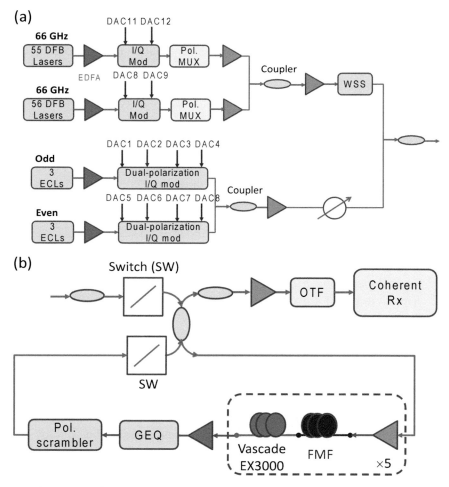

Figure 6.21 System setup: (a) Transmitter with six tunable channels, and 111 loading channels, on a 33 GHz grid. (b) Recirculating loop with 5 × 101.6 km-long hybrid spans of FMF and Vascade EX3000.

At the receiver side, the WDM channel under test is filtered, and captured with a standard offline coherent receiver, using a narrow-linewidth (<1 kHz) laser as the local oscillator and a real-time 80 GSa/s sampling scope. In the case of a SCM signal, after resampling and chromatic-dispersion compensation (CDC) the signal is fed into a multi-modulus algorithm for initial convergence of polarization de-multiplexing followed by carrier phase recovery. Subsequently a second stage of T-spaced equalizer using DD-LMS is applied to the 32 Gbaud SCM signal to mitigate MPI. In the case of MSCM, after frequency offset estimation and CDC each subcarrier is digitally filtered and processed individually in the same way as the single-carrier signal. The data of all subcarriers is combined and is further improved by the BCJR equalization and then sent to nonbinary LDPC decoder for detecting post-FEC bit errors.

The circulating loop test bed is composed of five hybrid spans obtained by splicing 51.3 km-long FMFs with 50.3 km-long Vascade® EX3000 fiber with average attenuations of 0.157 dB/km, and 0.153 dB/km, respectively. The FMFs support the LP11 mode albeit with a slightly higher attenuation and the differential modal delay is estimated to be 0.95 ns/km with the LP01 lagging. The average effective areas of the single-mode fiber and the FMF portions are 151 μm^2 and an estimated 200 μm^2, respectively. A short piece of Vascade® EX3000 is used as a bridge fiber between the input end of the FMF and the standard-single mode fiber pigtail of the EDFA output. The average span loss is 16.48 dB corresponding to an additional 0.7 dB incurred by the 4 splices and the two connectors. The mid-band chromatic dispersion is approximately 21 ps/nm/km. Span loss is compensated by C-band EDFAs and the accumulated amplifier gain tilt is compensated by a WSS each loop.

The Q-factor is measured across the C-band at 6600 km for a total of 111 WDM channels, from 1534.5 to 1563.6 nm. Q-factors averaged over 32 carriers, both polarizations, and over 10 separate measurements are plotted in Fig. 6.22 along with the received spectrum. The Q-factor varies from 4.5 to 5.7 dB from the shorter to the longer wavelengths. To maximize the capacity with such a variation of Q-factor three different FEC OHs at 14.3%, 20% and 25% are used for, 28, 64 and 19 WDM channels across the C-band, as shown in Fig. 6.22, resulting in an average spectral efficiency of 6.5 bits/s/Hz. For each channel, more than 25 million bits were processed for LDPC decoding and all the channels were decoded error-free in the LDPC decoder assisted with BCJR equalizer. The coding performance is further evaluated in the BTB configuration as a function of pre-FEC Q-factor derived from bit-error counting as shown in Fig. 6.23. To achieve sufficient confidence level, over 100 million bits are considered for each BER point, and exponential fitting is used for extrapolating the measured BER data and thus estimating the FEC limit as 5.5, 4.9, and 4.35 dB Q-factor for 14.3%, 20% and 25% OH, respectively.

To make direct performance comparisons, we next constructed 3 comparable loops from either all single-mode fiber, all FMF, or hybrid spans. Four of the hybrid spans are broken up and re-spliced to obtain either two spans of all-single-mode fiber spans or all-FMF spans with almost identical span lengths and losses. For the hybrid case, three spans are removed from the loop.

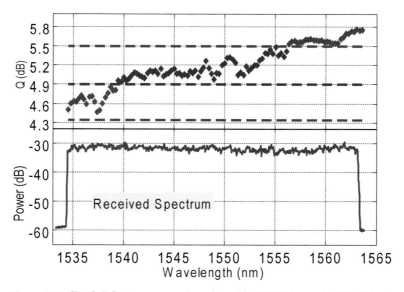

Figure 6.22 (Top) Q factor vs. wavelengths with 14.3% (green), 20% (blue), and 25% (red) FEC OH after 6600 km. (Bottom) Received spectrum. After ref. [70]; © IEEE 2015; reprinted with permission.

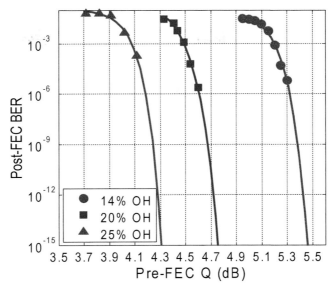

Figure 6.23 Back-to-back FEC coding performance. After ref. [70]; © IEEE 2015; reprinted with permission.

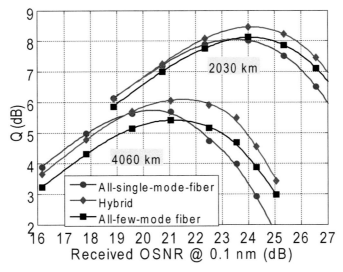

Figure 6.24 *Q*-factor versus OSNR at 2030 and 4060 km, all-single-mode fiber (blue), hybrid (red), and all-FMF (black). After ref. [70]; © IEEE 2015; reprinted with permission.

Figure 6.24 shows the comparison after 10 and 20 circulations using the same MSCM signals, and the same receiver DSP with 101-tap long CMA-based filer. Because the FMFs have larger effective area, the optimum OSNR is larger by about 1 dB compared to the all-SMF case. However, some of this improvement is lost because the impact of MPI cannot be removed altogether which is more evident in portions of the curves dominated by the linear noise, especially when all-FMF spans are used. This is most clear at 4060 km in the low OSNR regime. However, for the hybrid span case, the nonlinear improvement from the large effective area more than compensates for the penalty due to residual MPI affording 0.4 dB improvement over the all-single-mode fiber configuration. For additional details on this experiment, the interested reader is referred to [70].

6.6   Concluding Remarks

In this chapter, we have described different codes on graphs of interest for optical communications including turbo-product and LDPC codes. Considering the high potential of LDPC codes in enabling the next-generation high-speed optical transmission, we have focused our discussion on LDPC codes. Specifically, we described various binary and nonbinary LDPC decoding algorithms as well as the FPGA-based hardware implementation of rate-adaptive LDPC coding. In addition, we have presented a variety of advanced coded modulation schemes suitable for high-speed optical transmission. Finally, we have described an illustrative experiment in which the concepts described in this chapter have been used as key enabling technologies to improve the spectral efficiency and extend the transmission distance.

Acknowledgments

The author would like to thank his former students Yequn Zhang, Tao Liu, and Ding Zou as well as NEC Labs collaborators Ting Wang, Shaoliang Zhang, and Fatih Yaman for their involvement in joint research.

References

1. M. Cvijetic, I. B. Djordjevic, *Advanced Optical Communication Systems and Networks* (Artech House, USA), 2013.

2. I. Djordjevic, W. Ryan, B. Vasic, *Coding for Optical Channels* (Springer, USA), 2010.

3. I. B. Djordjevic, M. Arabaci, L. Minkov, Next generation FEC for high-capacity communication in optical transport networks, *J. Lightw. Technol.*, 27, pp. 3518–3530, 2009.

4. I. B. Djordjevic, *Optical Fiber Telecommunications VI* B, eds. I. Kaminow, T. Li, A. Willner, Chapter 6, Advanced coding for optical communication systems (Elsevier/Academic Press) pp. 221–296, 2013.

5. I. B. Djordjevic, On advanced FEC and coded modulation for ultra-high-speed optical transmission, *IEEE Commun. Surveys Tutorials*, PP (99), pp. 1–31, DOI: 10.1109/COMST.2016.2536726, 2016.

6. B. P. Smith, et al., Staircase codes: FEC for 100 Gb/s OTN, *J. Lightw. Technol.*, 30, pp. 110–117, 2012.

7. Y. Zhang, I. B. Djordjevic, Staircase rate-adaptive LDPC-coded modulation for high-speed intelligent optical transmission, *Proc. OFC/NFOEC 2014*, Paper M3A.6.

8. A. J. Felström, K. S. Zigangirov, Time-varying periodic convolutional codes with low-density parity-check matrix, *IEEE Trans. Inform. Theory*, 45(6), pp. 2181–2191, 1999.

9. S. Kudekar, T. J. Richardson, R. L. Urbanke, Threshold saturation via spatial coupling: Why convolutional LDPC ensembles perform so well over the BEC, *IEEE Trans. Inform. Theory*, 57(2), pp. 803–834, Feb. 2011.

10. K. Sugihara, et al., A spatially-coupled type LDPC code with an NCG of 12 dB for optical transmission beyond 100 Gb/s, *Proceedings OFC/NFOEC 2013*, Paper OM2B.4.

11. D. Chang, et al., LDPC convolutional codes using layered decoding algorithm for high speed coherent optical transmission, *Proceedings OFC/NFOEC 2012*, Paper OW1H.4.

12. I. B. Djordjevic, On the irregular nonbinary QC-LDPC-coded hybrid multidimensional OSCD-modulation enabling beyond 100 Tb/s optical transport, *J. Lightwave Technol.*, 31(16), pp. 2969–2975, Aug. 15, 2013.

13. M.-F. Huang, A. Tanaka, E. Ip, Y.-K. Huang, D. Qian, Y. Zhang, S. Zhang, P. N. Ji, I. B. Djordjevic, T. Wang, Y. Aono, S. Murakami, T. Tajima, T. J. Xia, G. A. Wellbrock, Terabit/s Nyquist superchannels in high capacity fiber field trials using DP-16QAM and DP-8QAM modulation formats, *J. Lightw. Technol.*, 32(4), pp. 776–782, Feb. 15, 2014.

14. M. Arabaci, I. B. Djordjevic, L. Xu, T. Wang, Nonbinary LDPC-coded modulation for rate-adaptive optical fiber communication without bandwidth expansion, *IEEE Photon. Technol. Lett.*, 23(18), pp. 1280–1282, Sept. 15, 2012.

15. M. Arabaci, I. B. Djordjevic, L. Xu, T. Wang, Hybrid LDPC-coded modulation schemes for optical communication systems, *Proceedings CLEO 2012*, Paper CTh3C.3.

16. M. Arabaci, I. B. Djordjevic, R. Saunders, R. M. Marcoccia, High-rate nonbinary regular quasi-cyclic LDPC codes for optical communications, *J. Lightwave Technol.*, 27(23), pp. 5261–5267, Dec. 1, 2009.

17. T. Liu, I. B. Djordjevic, On the optimum signal constellation design for high-speed optical transport networks, *Opt. Exp.*, 20(18), pp. 20396–20406, 27 August 2012.

18. M. Arabaci, I. B. Djordjevic, R. Saunders, R. M. Marcoccia, Polarization-multiplexed rate-adaptive non-binary-LDPC-coded multilevel modulation with coherent detection for optical transport networks, *Opt. Exp.*, 18(3), pp. 1820–1832, 2010.

19. L. Beygi, E. Agrell, P. Johannisson, M. Karlsson, A novel multilevel coded modulation scheme for fiber optical channel with nonlinear phase noise, *Proceedings IEEE GLOBECOM.*, 2010.

20. B. P. Smith, F. R. Kschischang, A pragmatic coded modulation scheme for high-spectral-efficiency fiber-optic communications, *J. Lightw. Technol.*, 30(13), pp. 2047–2053, July 2012.

21. J.-X. Cai, et al., Transmission over 9,100 km with a capacity of 49.3 Tb/s using variable spectral efficiency 16QAM based coded modulation, *Proceedings OFC/NFOEC 2014*, Paper Th5B.4, 2014.

22. C. Berrou, A. Glavieux, P. Thitimajshima, Near Shannon limit error-correcting coding and decoding: Turbo-codes, *Proceedings IEEE International Conference on Communications 1993 (ICC '93)*, 2, pp. 1064–1070, 23-26 May 1993.

23. B. P. Smith, F. R Kschischang, Future prospects for FEC in fiber-optic communications, *IEEE J. Sel. Top. Quantum Electron.*, 16(5), pp. 1245–1257, Sept.-Oct. 2010.

24. R. M. Pyndiah, Near optimum decoding of product codes, *IEEE Trans. Comm.*, 46, pp. 1003–1010, 1998.

25. O. A. Sab, V. Lemarie, Block turbo code performances for long-haul DWDM optical transmission systems, *Proceedings OFC 2001*, Paper ThS5.

26. T. Mizuochi et al., Forward error correction based on block turbo code with 3-bit soft decision for 10 Gb/s optical communication systems, *IEEE J. Selected Topics Quantum Electronics*, 10(2), pp. 376–386, Mar./Apr. 2004.

27. P. Elias, Error-free coding, *IRE Trans. Inform. Theory*, IT-4, pp. 29–37, Sep. 1954.

28. W. E. Ryan, Concatenated convolutional codes and iterative decoding, *Wiley Encyclopedia in Telecommunications,* J. G. Proakis, ed., John Wiley and Sons, 2003.

29. Y. Zhao, J. Qi, F. N. Hauske, C. Xie, D. Pflueger, G. Bauch, Beyond 100G optical channel noise modeling for optimized soft-decision FEC performance, *Proceedings OFC/NFOEC 2012* Paper OW1H.3.

30. F. Vacondio, et al., Experimental characterization of Gaussian-distributed nonlinear distortions, *Proceedings ECOC 2011*, Paper We.7.B.1.

31. R. G. Gallager, *Low Density Parity Check Codes.* Cambridge, MA: MIT Press, 1963.

32. D. J. C. MacKay, Good error correcting codes based on very sparse matrices, *IEEE Trans. Inform. Theory*, 45, pp. 399–431, 1999.

33. B. Vasic, I. B. Djordjevic, R. Kostuk, Low-density parity check codes and iterative decoding for long haul optical communication systems, *J. Lightwave Technol.*, 21(2), pp. 438–446, Feb. 2003.

34. I. Andersen, Combinatorial designs: Construction methods, *Mathematics and Its Applications.* Chichester, U.K.: Ellis Horwood, 1990.

35. L. Lan, L. Zeng, Y. Y. Tai, L. Chen, S. Lin, K. Abdel-Ghaffar, Construction of quasi-cyclic LDPC codes for AWGN and binary erasure channels: A finite field approach, *IEEE Trans. Inform. Theory*, 53, pp. 2429–2458, 2007.

36. I. B. Djordjevic, S. Sankaranarayanan, B. Vasic, Projective plane iteratively decodable block codes for WDM high-speed long-haul transmission systems, *IEEE/OSA J. Lightwave Technol.*, 22, pp. 695–702, March 2004.

37. J. L. Fan, Array-codes as low-density parity-check codes, *Second International Symposium On Turbo Codes*, Brest, France, pp. 543–546, Sept. 2000.

38. O. Milenkovic, I. B. Djordjevic, B. Vasic, Block-circulant low-density parity-check codes for optical communication systems, *IEEE/LEOS J. Selected Top. Quantum Electronics*, 10(2), pp. 294–299, March/April 2004.

39. Z.-W. Li, L. Chen, L.-Q. Zeng, S. Lin, W. Fong, Efficient encoding of quasi-cyclic low-density parity-check codes, *IEEE Trans. Commun.*, 54(1), pp. 71–81, Jan. 2006.

40. R. M. Tanner, A recursive approach to low complexity codes, *IEEE Tans. Information Theory*, IT-27, pp. 533–547, 1981.

41. W. E. Ryan, An introduction to LDPC codes, *CRC Handbook for Coding and Signal Processing for Recording Systems,* B. Vasic, ed., CRC Press, 2004.

42. J. Chen, A. Dholakia, E. Eleftheriou, M. Fossorier, X.-Y. Hu Reduced-complexity decoding of LDPC codes, *IEEE Trans. Comm.,* 53, pp. 1288–1299, 2005.

43. G. J. Byers, F. Takawira, Fourier transform decoding of non-binary LDPC codes, *Proceedings Southern African Telecommunication Networks and Applications Conference (SATNAC)*, Sept. 2004.

44. M. Davey, D. J. C. MacKay, Low-density parity check codes over GF(q), *IEEE Comm. Lett.*, 2(6), pp. 165–167, 1998.

45. L. Barnault, D. Declercq, Fast decoding algorithm for LDPC over GF($2q$), *Proceedings* ITW2003, pp. 70–73.

46. C. Spagnol, W. Marnane, E. Popovici, FPGA Implementations of LDPC over GF(2^m) Decoders, *IEEE Workshop on Signal Processing Systems* Shanghai, China, 2007, pp. 273–278.

47. D. Declercq, M. Fossorier, Decoding algorithms for nonbinary LDPC codes over GF(q), *IEEE Trans. Comm.*, 55(4), pp. 633–643, 2007.

48. V. Savin, Min-max decoding for non-binary LDPC codes, *Proceedings ISIT* 2008, pp. 960–964, Toronto, ON, Canada, 6–11 July 2008.

49. D. Zou, I. B. Djordjevic, FPGA implementation of advanced FEC schemes for intelligent aggregation networks, *Proceedings Photonics West 2016, OPTO, Optical Metro Networks and Short-Haul Systems VIII*, SPIE vol. 9773, pp. 977309-1– 977309-7, San Francisco, California United States (Invited Paper), 13–18 February 2016.

50. H. Imai, S. Hirakawa, A new multilevel coding method using error correcting codes, *IEEE Trans. Inf. Theory*, IT-23(3), pp. 371–377, May 1977.

51. G. Ungerboeck, Channel coding with multilevel/phase signals, *IEEE Trans. Inform. Theory*, IT-28, pp. 55–67, 1982.

52. J. Hou, P. H. Siegel, L. B. Milstein, H. D. Pfitser, Capacity-approaching bandwidth-efficient coded modulation schemes based on low-density parity-check codes, *IEEE Trans. Information Theory*, 49(9), pp. 2141–2155, 2003.

53. S. Benedetto, D. Divsalar, G. Montorsi, H. Pollara, Parallel concatenated trellis coded modulation, *Proceedings IEEE Int. Conf. Communications*, Dallas, TX, pp. 974–978, 1996.

54. G. Caire, G. Taricco, E. Biglieri, Bit-interleaved coded modulation, *IEEE Trans. Information Theory*, 44(3), pp. 927–946, May 1998.

55. Y. Zhang, I. B. Djordjevic, Multilevel nonbinary LDPC-coded modulation for high-speed optical transmissions, *Proceedings Asia Communications and Photonics Conference (ACP)*, Paper ATh1E.6, Shanghai, China, 11–14 November 2014.

56. G. D. Forney, R. G. Gallager, G. R. Lang, F. M. Longstaff, S. U. Qureshi, Efficient modulation for band-limited channels, *IEEE J. Sel. Areas Comm.*, SAC-2(5), pp. 632–647, 1984.

57. G. D. Forney, Coset codes-part I: Introduction and geometrical classification, *IEEE Trans. Infrom. Theory*, 34(5), pp. 1123–1151, Sept. 1988.

58. A. Goldsmith, *Wireless Communications*. Cambridge: Cambridge University Press, 2005.

59. A. R. Calderbank, N. J. A. Sloane, New trellis codes based on lattices and cosets, *IEEE Trans. Inform. Theory*, IT-33, pp. 177–195, 1987.

60. F. Kschischang, S. Pasupathy, Optimal nonuniform signaling for Gaussian channels, *IEEE Trans. Inform. Theory*, 39, 913–929, 1993.

61. T. Liu, Z. Qu, C. Lin, I. B. Djordjevic, Nonuniform signaling based LDPC-coded modulation for high-speed optical transport networks, *OSA Asia Communications and Photonics Conference (ACP) 2016*, Wuhan, China. (Invited Paper), Nov. 2–5 2016.

62. T. Fehenberger, D. Lavery, R. Maher, A. Alvarado, P. Bayvel, and N. Hanik, Sensitivity gains by mismatched probabilistic shaping for optical communication systems, *IEEE Photo. Technol. Lett.*, 28(7), 786–789, 2016.

63. G. D. Forney, Trellis shaping, *IEEE Trans. Inform. Theory*, 38(2), pp. 281–300, March 1992.

64. S. Lin, D. J. Costello, *Error Control Coding: Fundamentals and Applications*, 2nd ed., Pearson Prentice Hall, Upper Saddle River, NJ, 2004.

65. I. B. Djordjevic, B. Vasic, Multilevel coding in *M*-ary DPSK/differential QAM high-speed optical transmission with direct detection, *IEEE/OSA J. Lightw. Technol.*, 24(1), pp. 420–428, Jan. 2006.

66. T. Liu, I. B. Djordjevic, On the optimum signal constellation design for high-speed optical transport networks, *Opt. Exp.*, 20(18), pp. 20396–20406, 27 August 2012.

67. T. Liu, I. B. Djordjevic, Optimal signal constellation design for ultra-high-speed optical transport in the presence of nonlinear phase noise, *Opt. Express*, 22(26), pp. 32188–32198, 29 Dec 2014.

68. T. Liu, I. B. Djordjevic, Multi-dimensional optimal signal constellation sets and symbol mappings for block-interleaved coded-modulation enabling ultra-high-speed optical transport, *IEEE Photonics J.*, 6(4), Paper 5500714, Aug. 2014.

69. I. B. Djordjevic, M. Cvijetic, and C. Lin, Multidimensional signaling and coding enabling multi-Tb/s optical transport and networking, *IEEE Sig. Proc. Mag.*, 31(2), pp. 104–117, Mar. 2014.

70. F. Yaman, S. Zhang, Y.-K. Huang, E. Ip, J. D. Downie, W. A. Wood, A. Zakharian, S. Mishra, J. Hurley, Y. Zhang, I. B. Djordjevic, M.-F. Huang, E. Mateo, K. Nakamura, T. Inoue, Y. Inada, T. Ogata, First quasi-single-mode transmission over transoceanic distance using few-mode fibers, *Proceedings OFC Postdeadline Papers*, Paper Th5C.7, Los Angeles, CA, 22–26 March 2015.

Chapter 7

Nonlinear Fourier Transform-Based Optical Transmission: Methods for Capacity Estimation

Jaroslaw E. Prilepsky,[a] Stanislav A. Derevyanko,[b] and Sergei K. Turitsyn[a]

[a]*Aston Institute of Photonic Technologies, Aston University,*
B4 7ET Birmingham, UK
[b]*Department of Electrical and Computer Engineering,*
Ben-Gurion University of the Negev, Beer Sheva 84105, Israel

y.prylepskiy1@aston.ac.uk, stasd@bgu.ac.il, s.k.turitsyn@aston.ac.uk

7.1   Introduction

The ever-increasing deluge in the world's IP data traffic volume driven by the immense growth of versatile bandwidth-hungry on-line services, such as cloud computing, on-demand HD video streams, on-line business analytics and content sharing, the development of the Internet of Things, and many others, sets higher and higher requirements on the speed and quality

Optical Communication Systems: Limits and Possibilities
Edited by Andrew Ellis and Mariia Sorokina
Copyright © 2020 Jenny Stanford Publishing Pte. Ltd.
ISBN 978-981-4800-28-0 (Hardcover), 978-0-429-02780-2 (eBook)
www.jennystanford.com

characteristics of information flows interconnecting network constituents. It is currently well understood that increasing data rates in the core fibre communication systems are quickly approaching the limits of contemporary transmission technologies [1–3]. For *linear* channels the bounds on the transmission rate were established in works of Nyquist [4] and Shannon [5]. However, there is now a common belief that it is the inherent *nonlinear* transmission effects that serve as a major limiting factor in modern fibre-optic communication systems [3, 6–8]: These nonlinear properties make optical fibre channels very different from wireless and other linear ones [9, 10]. To reduce the nonlinearity impact on the transmission characteristics, there have been proposed a great number of nonlinearity compensation methods, including digital back-propagation (DBP) [11], digital [12] and optical [13, 14] phase conjugations (spectral inversion), and phase-conjugated twin waves [15], to mention a few important advances [16, 17]. Note that in the techniques itemized above, the fibre nonlinearity is treated as a "transmission degrading factor", and so the goal of each of the aforementioned methods is to suppress its negative effect at the expense of performing some additional processing operations. With regard to the latter aspect, the nonlinear Fourier transform (NFT) based transmission can be reckoned as a conceptually different approach: Here nonlinearity enters like a constructive part of processing and transmission, defining the features of the system architecture and its characteristics. Thereby the nonlinearity can play a rather positive role in defining the performance of the NFT-based system. At the same time, the major difficulty in NFT application lies in the fact that some quite common "linear" concepts may need to be reconsidered or appended with a new non-trivial meaning. For instance, in addition to the usual notions of frequency, spectrum, power, bandwidth, etc., one has to work with their nonlinear analogs that can be drastically different from the "ordinary" counterparts. Simultaneously, these new signal parameters can serve as well-defined and adjustable characteristics of our signal [18–20], which can be used for modulation and information transmission. The overall idea of using NFT for the communications is that inside the nonlinear Fourier (NF) domain the propagation of the so-called nonlinear spectral data (that play the role of Fourier spectrum for a special class of nonlinear problems) is uncoupled and linear. It is exactly this remarkable property of NF spectrum that makes the NFT-based transmission potentially free from the cross-talk and bandwidth related sources of the capacity degradation, that plague most of the conventional transmission systems [6–8], as the nonlinearity is effectively included into the NFT operations.

The original idea of using the NF spectrum for modulation and transmission was first proposed in the currently well-recognized work by Hasegawa and Nyu "Eigenvalue communication" [21]. After its having remained almost unattended for some 20 years, this direction experiences a boost of researchers' interest at present. The existing optical transmission methods based on NFT can be classified into three general groups. The first group [22, 23] employs NFT as a tool for solving NLSE backwards, very much alike the DBP approach [11]. The second group assumes using the parameters of nonlinear modes themselves (both discrete and continuous) for the modulation, data encoding, transmission and detection [18, 24–30]. Finally, the third direction attributes to the works of Bülow [31–33] and Osaka group and co-authors [34–36], demonstrating the possibility of the NFT-based optical transmission by a hybrid algorithm: while the data encoding takes place in the time domain, the detection is performed using the signal characteristics from the NF domain.

The purview of this chapter is to present the general framework of how to estimate the capacity for the channel defined inside the NF domain. First we derive the general asymptotically exact stochastic equations for the signal evolution for the NF spectrum affected by noise: This part is common and can be utilized in the study of each of three aforementioned NFT-based transmission methods. Then we address already specifically the transmission in the NF domain, and use these equations further for the definition of discrete input-output channel for the NF data and the forthcoming analysis of the resulting channel capacity, which generalizes the previous results [18, 37, 38]. However, since the treatment of the general problem with the modulation of all parts of NF spectrum is still under development, we provide some explicit example estimations for one specific method build on the NFT usage: for the so-called nonlinear inverse synthesis (NIS) [24–26], where only a part of the NF spectrum is used for the modulation [39]. This allows us to eventually obtain the analytical expression for the lower capacity per symbol bound in a closed form, which finalizes our study here.

7.2 Main Model and Basics of NFT

7.2.1 Nonlinear Schrödinger Equation (NLSE)

The principal master model for the electrical field $U(z, t)$ evolution inside a single-mode optical fibre with the account of amplification can be written as a noise-perturbed NLSE, where the noise inevitably arises due to the amplifier spontaneous emission (ASE) [3, 7–9]:

$$\frac{\partial U}{\partial z} + \frac{i}{2}\frac{\beta_2}{2}\frac{\partial^2 U}{\partial t^2} - i\gamma \, |U|^2 \, U = N(z,t), \qquad (7.1)$$

with z being a distance along the fibre, t is time in the frame co-moving with the velocity of the envelope. The parameter β_2 is the characteristic of chromatic dispersion, $\beta_2 < 0$ for the anomalous dispersion case considered here, with the typical value $|\beta_2| \approx 22$ ps^2/km; γ is the nonlinear Kerr coefficient, typically $\gamma = 1.27$ W^{-1} km^{-1}. For the quasi-lossless distributed Raman amplification (DRA) scheme [40, 41], we describe the signal evolution by the lossless NLSE (7.1) perturbed by the additive white Gaussian noise (AWGN) term $N(z, t)$ (having zero mean). The latter is fully characterized by the power spectral density (PSD) per unit of propagation distance, N_{ASE}, of its autocorrelation function [3, 8]:

$$E[N(z,t)\overline{N}(z',t')] = (N_{ASE}/L)\delta(t-t')\delta(z-z'), \qquad (7.2)$$

where here and in the chapter's sequel the overbar means the complex conjugate, $E[.\,.\,.]$ is the expectation value, $\delta(.\,.\,.)$ is the Dirac delta-function, and L is the propagation distance. In the case of ideal DRA we have: $N_{ASE}/L = h\nu_s K_T \alpha$, where $\alpha \approx 0.2$ dB/km at the is the fibre loss coefficient at the carrying wavelength λ_s = 1.55 μm, $K_T \approx 1$ is the temperature-dependent factor (related to the phonon-occupancy factor) that characterizes the Raman pump providing the distributed gain; v_s = 193.55 THz is the carrying frequency of the signal corresponding to λ_s. The lossless NLSE weakly perturbed by the AWGN term (7.1) suits best the NFT application, as it is quite close to the non-perturbed NLSE to which the NFT operations themselves are attributed. However, the NFT method can still be successfully applied to the erbium doped fibre amplifiers [26] or non-ideal DRA [42, 43] cases, where Eq. (7.1) with renormalized parameters can still be taken as a good leading approximation describing signal evolution.

Now we introduce the normalizations,

$$t/T_s \to t, \quad z/Z_s \to z, \quad U/\sqrt{P_0} \to q, \tag{7.3}$$

with $P_0 = (\gamma Z_s)^{-1}$. In Eq. (7.3) either of three parameters, T_s, Z_s, or P_0, can be taken for the normalization, but then the remaining two have to be properly adjusted. The noise PSD has to be normalized in accordance to (7.3): $N_{\text{ASE}} Z_s (P_0 T_s)^{-1}/L \to \varepsilon$ where L is in normalized units. The resulting NLSE form becomes

$$\frac{\partial q}{\partial z} - \frac{i}{2}\frac{\partial^2 q}{\partial t^2} - i\,|q|^2 q = N(z,t). \tag{7.4}$$

It was first shown in the seminal work by Zakharov and Shabat [44] that Eq. (7.4) without noise belongs to the class of the so-called *integrable nonlinear systems*. The advanced mathematical method, widely known as the inverse scattering transform (IST), can be applied to find the solution of integrable nonlinear equations, in particularly, to the noiseless NLSE. Mathematically, this can be treated as an effective linearization of the nonlinear evolution. In other terms, the integrability here means that any solution of the NLSE can be presented in a basis of nonlinear modes that propagate without interacting with each other, and that is how the name NFT emerged, underlining the similarity with the application of Fourier transform for the linear problems. There exists a vast amount of literature where the integrability notion is elucidated in very details [18, 44–47]. In the subsection below we provide the explicit description of NFT.

7.2.2 NFT Operations

The NFT decomposes the signal $q(z, t)$ at a fixed location $z = z_0$ and returns the corresponding NF spectrum. The INFT reverses this process, i.e., given a NF spectrum it returns the signal $q(z_0, t)$.

Direct NFT operation. The direct NFT is computed from specific (auxiliary) solutions $v_{1,2}(t, \zeta) = v_{1,2}(t, \zeta; z_0)$ to the Zakharov–Shabat problem (ZSP)

$$\frac{dv_1}{dt} = q(z_0, t)v_2 - i\zeta v_1, \quad \frac{dv_2}{dt} = -\bar{q}(z_0, t)v_1 + i\zeta v_2, \tag{7.5}$$

for different values of the complex spectral parameter $\zeta = \xi + i\eta$ which plays the role of the nonlinear analog of frequency (we address specifically the anomalous dispersion case). The signal $q(z_0, t)$ acts as an effective potential. Under the assumption that $q(z_0, t)$ decays at least exponentially for $t \to \pm\infty$, specific solutions (the so-called Jost functions) $\phi_{1,2}(t, \xi)$ and $\psi_{1,2}(t, \xi)$ to the ZSP can be obtained from the "boundary conditions"

$$\phi_1(t,\xi) \to e^{-i\xi t}, \quad \phi_2(t,\xi) \to 0 \text{ for } t \to -\infty,$$
$$\psi_1(t,\xi) \to 0, \quad \psi_2(t,\xi) \to e^{i\xi t} \text{ for } t \to +\infty. \tag{7.6}$$

In practical realization of the transmission schemes, the pulse $q(t)$ is truncated to have a finite duration (a symbol duration), i.e. we operate in the so-called burst mode [25], and thus one sets the initial conditions at the trailing and leading ends of the finite-extent pulse. The pairs $\tilde{\phi}_1 = -\bar{\phi}_2$ and $\tilde{\phi}_2 = \bar{\phi}_1$ as well $\tilde{\psi}_1 = -\bar{\psi}_2$ and $\tilde{\psi}_2 = \bar{\psi}_1$ also solve the ZSP (7.5). These different ZSP solutions are linearly dependent as follows:

$$[\phi_1 \ \phi_2] = a(\xi)[\tilde{\psi}_1 \ \tilde{\psi}_2] + b(\xi)[\psi_1 \ \psi_2], \tag{7.7}$$

$$[\tilde{\phi}_1 \ \tilde{\phi}_2] = -\tilde{a}(\xi)[\psi_1 \ \psi_2] + \tilde{b}(\xi)[\tilde{\psi}_1 \ \tilde{\psi}_2]. \tag{7.8}$$

The functions $a(\xi)$ and $b(\xi)$ forming the core of the NFT signal decomposition, are known as the *Jost scattering coefficients*. Due to the boundary conditions (7.6), we have

$$a(\xi) = \lim_{t \to \infty} \phi_1(t,\xi) e^{i\xi t}, \qquad b(\xi) = \lim_{t \to \infty} \phi_2(t,\xi) e^{-i\xi t}. \tag{7.9}$$

The scattering coefficients satisfy: $|a(\xi)|^2 + |b(\xi)|^2 = 1$. The NF spectrum of the signal $q(z_0, t)$ consists of two parts. The first (continuous) part is given either by the left and or right reflection coefficient (RC), respectively:

$$l(\xi) = \overline{b}(\xi)/a(\xi), \qquad r(\xi) = b(\xi)/a(\xi), \qquad \xi \in R \tag{7.10}$$

The second part of the NF spectrum consists of the discrete complex numbers $\zeta_n = \zeta_n + i\eta_n$, which are the eigenvalues of ZSP having a positive imaginary part $\eta > 0$, and the associated left/right norming constants, which are defined by the residue of $l(\zeta)$ (or $r(\zeta)$) at the point ζ_n:

$$l_n = [b(\zeta_n)a'(\zeta_n)]^{-1}, \qquad r_n = b(\zeta_n)/a'(\zeta_n), \tag{7.11}$$

where the prime designates the derivative with respect to ζ. The complete (left or right) NF spectrum of the decaying (truncated) signal $q(z_0, t)$ is given by

$$\sum\nolimits_l = \{l(\xi), [\zeta_n, l_n]_{n=1}^{N}\}, \qquad \sum\nolimits_r = \{r(\xi), [\zeta_n, r_n]_{n=1}^{N}\}, \tag{7.12}$$

where N is the total number of solitons in the signal, which can be equal to zero. The NF spectrum characterizes the signal $q(z_0, t)$ completely and can be used to recover it.

The z-dependence of the NFT data, $\sum_{l,r}(z)$, is given by the following expressions. First, the eigenvalues ζ_n are independent on z, and for the rest of the quantities we have:

$$l(\xi, z) = l(\xi, z_0)\, e^{-2i\xi(z-z_0)},\; l_n(z) = l_n(z_0)\, e^{-2i\zeta_n(z-z_0)},$$

$$r(\xi, z) = r(\xi, z_0)e^{2i\xi(z-z_0)},\; r_n(z) = r_n(z_0)e^{2i\zeta_n(z-z_0)}. \tag{7.13}$$

INFT operation. The inverse NFT (INFT) maps the scattering data $\sum_{l,r}$ (7.12) onto the field $q(t)$: This is achieved via the solution of the so-called Gelfand–Levitan–Marchanko equations (GLME) [44–47]. However, we do not present here the explicit expressions and refer a reader to the aforementioned references for more details.

7.3 General Expressions for Noise Autocorrelation Functions inside NF Domain

7.3.1 Perturbed Evolution of NF Spectrum

According to the IST theory [48, 49], the perturbed z-evolution of the RC and $2N$ complex soliton parameters (eigenvalues and norming constants) is given by the following equations:

$$\frac{\partial a}{\partial z} = I[\phi, \psi], \quad \frac{\partial b}{\partial z} - 2i\xi^2 b = -I[\phi, \tilde{\psi}], \tag{7.14a}$$

$$\frac{\partial l}{\partial z} + 2i\xi^2 l = -\frac{1}{a^2}I[\psi, \psi], \quad \frac{\partial r}{\partial z} - 2i\xi^2 r = -\frac{1}{a^2}I[\phi, \phi], \tag{7.14b}$$

$$\frac{\partial \zeta_k}{\partial z} = -r_k\, I[\psi, \psi]\big|_{\zeta_k}, \tag{7.14c}$$

$$\frac{\partial r_k}{\partial z} = 2i\zeta_k^2 r_k + \frac{a_k''}{(a_k')^3}I[\phi, \phi]\bigg|_{\zeta_k}$$

$$-\frac{1}{(a_k')^2}\int\limits_{-\infty}^{\infty} dt\{N(t,z)\varphi_2^k + \bar{N}(t,z)\varphi_1^k\}, \tag{7.14d}$$

$$\frac{\partial l_k}{\partial z} = -2i\zeta_k^2 l_k + \frac{a_k''}{(a_k')^2 a_k} I[\psi,\psi]\Big|_{\zeta_k}$$

$$-\frac{1}{(a_k')^2}\int_{-\infty}^{\infty} dt\{N(t,z)\gamma_2^k + \overline{N}(t,z)\gamma_1^k\}, \tag{7.14e}$$

where $I[u, v]$ is the projection of the noisy perturbation $N(t, z)$ on the corresponding unperturbed squared Jost functions:

$$I[u,v,\xi] = \int_{-\infty}^{\infty} dt[N(t,z)u_2(\xi,t,z)v_2(\xi,t,z)$$

$$+ \overline{N}(t,z)u_1(\xi,t,z)v_1(\xi,t,z)], \tag{7.15}$$

and we have introduced the simplified denotations $\varphi_i^k \equiv \varphi_i^k(t,z) = \partial_\zeta \varphi_i^2(\zeta,t,z)$ and $\gamma_i^k \equiv \gamma_i^k(t,z) = \partial_\zeta \psi_i^2(\zeta,t,z)$, both evaluated at $\zeta = \zeta_k$. The system (7.14) is incomplete unless one specifies the z-dependence of Jost functions. Since our treatment is a perturbative one, it is sufficient to evaluate $\psi(\zeta, t; z)$ and $\phi(\zeta, t; z)$ in the zeroth order (i.e. in the absence of perturbation) taking into account that Eq. (7.15) already contains a small parameter via $N(t, z)$. Note that since the perturbation terms in the R.H.S. of Eq. (7.14) are *linear* with respect to noise term, the effective noise in NF domain is Gaussian albeit its correlation properties depend on the input data and their unperturbed evolution.

7.3.2 Noise Autocorrelation Functions for the Continuous Part of NF Spectrum

In this subsection we present the evolution of the NF spectrum continuous part given by either of Eq. (7.14b), assuming that *no solitons* are embedded into our time-domain waveform. These expressions will be utilized in our example capacity estimates in Section 7.4. First we address the evolution of the right RC $r(\xi, z)$ and rewrite the corresponding equation from (7.14b) introducing a new effective Gaussian noise term $\Gamma(z, \xi)$:

$$\frac{\partial r}{\partial z} - 2i\xi^2 r = \Gamma(z,\xi). \tag{7.16}$$

One finds that the average of the new noise is zero, as same as it is for the progenitor noise in the space-time domain $N(z, t)$, and the autocorrelation functions are given through the unperturbed Jost functions and scattering amplitudes as follows:

$$E[\Gamma(z,\xi)\overline{\Gamma}(z',\xi')] = \frac{\varepsilon\delta(z-z')}{a^2(\xi)\overline{a}^2(\xi')}$$

$$\times \int_{-\infty}^{\infty} dt[\phi_2^2(\xi,t,z)\overline{\phi}_2^2(\xi',t,z)$$

$$+ \phi_1^2(\xi,t,z)\overline{\phi}_1^2(\xi',t,z)]$$

$$\equiv \varepsilon A(z;\xi,\xi')\delta(z-z'), \tag{7.17a}$$

$$E[\Gamma(z,\xi)\Gamma(z',\xi')] = \frac{\varepsilon\delta(z-z')}{a^2(\xi)a^2(\xi')}$$

$$\times \int_{-\infty}^{\infty} dt[\phi_2^2(\xi,t,z)\phi_1^2(\xi',t,z)$$

$$+ \phi_2^2(\xi',t,z)\phi_1^2(\xi,t,z)]$$

$$\equiv \varepsilon B(z;\xi,\xi')\delta(z-z'), \tag{7.17b}$$

where, for the convenience of forthcoming analysis we have introduced the functions $A(\xi, \xi'; z)$ and $B(\xi, \xi'; z)$. Already from these expressions, one notices the following distinctions from the progenitor AWGN term in the space-time domain: (i) The noise inside the NF domain loses its circular symmetry, and (ii) this new noise is neither homogeneous nor uncorrelated. The most important distinctive property of Γ is, however, that (iii) it depends on the initial spectrum, $r(\xi)$. From the information theory perspective, the latter means that Eqs. (7.16), (7.17) define an *input-dependent Gaussian channel with memory* [50].

For the completeness of this section we also write down the expressions associated with the dynamics of left RC $\ell(\xi; z)$ from Eq. (7.14b). We rewrite the evolution equation introducing the effective new noise term $\Delta(z, \xi)$, as

$$\frac{\partial l}{\partial z} + 2i\xi^2 l = \Delta(z,\xi), \tag{7.18}$$

where the new Gaussian noise $\Delta(z, \xi)$, as before, has zero average.

The autocorrelator functions of Δ are very similar to Eq. (7.17) where the squares of Jost functions ϕ_i^2 are to be swapped for ψ_i^2.

For our example below, we will use just the expression presented in this subsection. However, the theory can be generalised to the case when both the continuous and discrete NF spectrum parts coexist. The corresponding cumbersome expressions for the autocorrelation functions between the NF spectrum components are summarized in the Appendix.

7.4 Capacity Estimates for the Nonlinear Inverse Synthesis NFT-Based Method

7.4.1 NIS Basics and Continuous Input-Output Channel Model

The NIS scheme based on the modulation of the continuous part of the NF signal spectrum, was proposed recently in Refs. [24–26]: The data within the NIS approach are encoded on and retrieved from the RC, say $r(\xi)$. The evolution of $r(\xi)$ in the ideal (noise-free) NLSE channel amounts to the dispersive phase rotation, so that the orthogonality of nonlinear normal modes is preserved during the signal propagation. The INFT maps the encoded scattering data Σ at the transmitter onto the field $q(t)$. Then, after the fibre propagation, at the receiver one reads the waveform $q(t, L)$ and retrieves the NF spectrum $r(\xi; L)$ by using the NFT (7.5). Unrolling the accumulated dispersion in the NF domain we recover the initial data, and this completes the NIS scheme (Fig. 7.1). The thorough assessment of the NIS performance in terms of the Q-factor behaviour in dependence on the input power for the distributed ideal DRA case is given in [25] for the orthogonal frequency division multiplexing (OFDM) [52, 53] and Nyquist [8, 53] modulations of NF spectrum. The first experimental realisation and real conditions assessment of the NIS-based transmission system was presented in [30].

We start from a general encoded wavelength-division-multiplexing (WDM) sequence in effective "nonlinear time" domain:

$$q_{in}(\tau) = \sum_{\sigma=0}^{N_b-1} \sum_{k=0}^{N_{ch}-1} c_{\sigma k} s(\tau - \sigma T_s) e^{i\Omega_k \tau}, \tag{7.19}$$

Figure 7.1 Simplified flow-chart of the NIS transceiver scheme, involving the NFT-based synthesis operation at the transmitter and NFT-based information recovery at the receiver.

where N_b is the length of the symbol sequence (i.e. burst), N_{ch} is the number of WDM channels or, alternatively, OFDM subcarriers, $s(t)$ is the base wave-shape defining the modulation format, T_s is the symbol width, Ω_k is an individual channel frequency or an OFDM subcarrier frequency. The parameter τ in (7.19) is not the true time but the variable Fourier-conjugated to the nonlinear "frequency" ξ from (7.5). It is the discrete set of M coefficients $c_{\sigma k}$ that now bears our informational content, with $M = N_b \times N_{ch}$. Within the NIS scheme we use the linear spectrum of (7.19) and take it as the nonlinear spectrum of a new optical signal $q(z = 0, t)$ that, in turn, is to be launched into the fibre. To define a particular mapping between the linear spectrum $q_{in}(\omega)$ and the quantity $r(\xi)$ from the NF domain, the convenient choice (proposed in Refs. [24, 25]) is to keep the correspondence between the initial waveform and the generated NIS waveform in the linear limit, i.e. when $|q(t)| \sim |q(\omega)| \to 0$. We have the following relation between the nonlinear and linear spectra in the limit of low powers [24, 45, 47]: $r(\xi) \to -\bar{q}|_{\omega=-2\xi}$ when $|q| \to 0$. Thus it is convenient to map the linear spectrum to the continuous NF spectrum (the continuous NF channel input) $X_\xi \equiv r(\xi, 0)$ via the following rule [24]: $r_{in}(\xi) = X_\xi = \bar{q}_{in}(\omega)|_{\omega=-2\xi}$, and then for X_ξ we have

$$X_\xi = r(\xi,0) = -\sum_{\sigma=0}^{N_b-1} \sum_{k=0}^{N_{ch}-1} \bar{c}_{\sigma k} \bar{s}(-2\xi - \Omega_k) e^{2i\xi\sigma T_s}. \tag{7.20}$$

After applying the INFT to the RC (7.20), the generated pulse is launched into the fibre and propagates towards the receiver.

For the information-theoretic analysis, the continuous NF channel model given by Eqs. (7.16), (7.17) should be formulated as an input-output probabilistic model, i.e., the conditional probability density function (PDF) of the channel output given the channel input X_ξ. We define the continuous *channel output* Y_ξ as the solution of (7.16) at the receiver located at distance $z = L$ with the compensated phase rotation and filtering:

$$Y_\xi \equiv \mathcal{H}(\xi)e^{-2i\xi^2 L}r(\xi,L) = X_\xi + N(\xi,X_\xi), \qquad (7.21)$$

where $\mathcal{H}(\xi)$ is the rectangular bandpass filtering function in the NF domain, that selects a given channel of interest (COI). The filtered noise $N(\xi, X_\xi)$ (with zero mean) has the following correlation properties:

$$E[N(\xi,X_\xi)\overline{N}(\xi',X_{\xi'})] = \varepsilon \int_0^L e^{-2i(\xi^2 - \xi'^2)z} \mathcal{A}(z;\xi,\xi')dz,$$

$$E[N(\xi,X_\xi)N(\xi',X_{\xi'})] = \varepsilon \int_0^L e^{-2i(\xi^2 + \xi'^2)z} \mathcal{B}(z;\xi,\xi')dz. \qquad (7.22)$$

Due to filtering these relations hold within the channel of interest (COI) only. To evaluate these correlation functions, one needs to know the full z-evolution of the unperturbed Jost functions $\varphi_i(z; t, \xi)$, and this problem does not have a closed form solution in the general case. However, in the regime of a long fibre system one can use the asymptotic solutions of ZS problem (7.5) [54], assuming large L, and then can obtain a simpler expression that explicitly depends only on the initial spectral data (RC):

$$E[N(\xi,X_\xi)\overline{N}(\xi',X_{\xi'})] \approx N_{\mathrm{ASE}}\pi\delta(\xi - \xi')E_1(\xi),$$

$$E[N(\xi,X_\xi)N(\xi',X_{\xi'})] \approx N_{\mathrm{ASE}}\pi\delta(\xi - \xi')E_2(\xi), \qquad (7.23)$$

where $E_1(\xi) = 1 + |X_\xi|^2 + |X_\xi|^4$ is an effective PSD, $E_2(\xi) = X_\xi^2$ is the effective pseudo-PSD, $N_{\mathrm{ASE}} = \varepsilon L$ is the normalized accumulated noise variance per sample, and we have omitted the non-diagonal terms of order unity as small compared to those $\sim L$.

Two important observations regarding the properties of noise $N(\xi, X_\xi)$ should be made. First, within the nonlinear bandwidth of COI the noise PSD in the NFT domain, $E_1(\xi)$, grows nonlinearly with the spectral power of the input. Second, the channel model (7.21), (7.23) is *local* in nonlinear "frequency" ξ.

So, for example, in the case of dense WDM one can simply match the nonlinear bandwidth of the filter with that of the COI and prevent both direct and noise-induced channel cross-talks without losing any of the informational content of the message, since the signal dependent nonlinear spectral broadening is virtually absent. It is this remarkable property of the nonlinear spectrum that makes the NIS-based transmission being potentially free from the cross-talk and bandwidth related sources of capacity degradation.

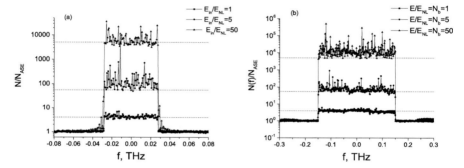

Figure 7.2 The simulated PSD of the nonlinear noise (normalized to its linear counterpart N_{ASE}) for different input burst energies for OFDM (a) and Nyquist (b) sequences. Theoretical predictions of Eq. (7.23) are shown as dotted horizontal lines. For the explanations of the energy scales meaning, E_{in} and E_{NL}, see Subsection 7.4.3. From [39].

Let us compare the theoretical predictions with the results of numerical simulations for OFDM and Nyquist modulations used in (7.19) for the NIS scheme. We plot the numerically extracted ratio of the NF domain noise PSD to its linear value N_{ASE} [i.e. the function E_1 from Eq. (7.23)] as a function of linear frequency for different values of the energy ratio $E_{in}/E_{NL} = (S/P_s)N_b$ (for the explanations of the energy scales meaning, E_{in} and E_{NL}, see Subsection 7.4.3 below) where $P_s = |\beta_2|/(\gamma T_s^2)$ is a typical nonlinear power in the r.w.u. In all the simulations the propagation distance L was fixed at 500 km. We have used $N_{ch} = 3$ channels for Nyquist and $N_{ch} = 112$ subcarriers for OFDM modulation. For the OFDM transmission the length of the burst was set constant, $N_b = 1$, and we changed the energy ratio by increasing the signal power inside the NF domain, S. For the Nyquist case we fixed $S = P_s$ and changed the number of symbols in the burst to achieve the desired energy ratio. The averaging was performed both over noise and over Gaussian-sampled set of the input symbols c_{ok}. From Fig. 7.2 one can see that the spectrum-averaged PSD confirms reasonably well with theoretical predictions, Eq. (7.23).

7.4.2 Discrete Input-Output Model

In digital communications the signal is modulated and sampled in the time or frequency domain, and for each time sample the information is encoded via discrete complex amplitude levels sets [8], i.e. via the coefficients c_{ok} from Eqs. (7.19), (7.20). As before, we consider two closely related standard frequency multiplexed schemes, namely dense Nyquist WDM [8] and OFDM [52, 53] both adapted to the NIS scheme. Since the channel in the NF domain is characterized by additive Gaussian input-dependent noise, the discrete NF channel has the same property:

$$c_{ok}(L) = c_{ok}(0) + N_{ok}, \qquad (7.24)$$

where N_{ok} is the projection of the NF spectral noise $N(\xi, X_\xi)$ onto the corresponding subcarrier in the OFDM case and the Nyquist-sampled noise vectors for the COI in the WDM case. Introducing $2M$ real and imaginary parts of c_{ok} as discrete real-valued input and output, **X** and **Y** correspondingly, one gets for the input-output conditional PDF the multivariate Gaussian distribution with the $2M \times 2M$ quadrature correlation matrix $\hat{\Sigma}$ whose elements are obtained from the correlation functions (7.22). Since the noise autocorrelation functions (7.23) depend on **X**, so does the correlation matrix: $\hat{\Sigma} = \hat{\Sigma}(\mathbf{X})$. Using input (7.19) and asymptotic expressions (7.23), one obtains for the components of

$$\hat{\Sigma}(N_{ok} = N_{ok}^R + iN_{ok}^I):$$

$$E[N_{ok}^{R/I} N_{o'k'}^{R/I}] = \frac{\mathcal{N}_{ASE}}{4\pi} \mathrm{Re}\left[\int d\omega \left\{ e^{i\omega(\sigma - \sigma')T_\sigma} f_k(\omega)\overline{f}_{k'}(\omega) \right.\right.$$

$$\left.\left. \times E_1\left(-\frac{\omega}{2}\right) \pm e^{i\omega(\sigma + \sigma')T_s} f_k(\omega)f_{k'}(\omega)\overline{E}_2\left(-\frac{\omega}{2}\right) \right\}\right], \qquad (7.25a)$$

$$E[N_{\sigma k}^R N_{\sigma' k'}^I] = \frac{N_{ASE}}{4p} \mathrm{Im} \left[\int d\omega \left\{ -e^{i\omega(\sigma-\sigma')T_\sigma} f_k(\omega)\overline{f}_{k'}(\omega) \right. \right.$$

$$\left. \left. \times E_1\left(-\frac{\omega}{2}\right) + e^{i\omega(\sigma+\sigma')T_s} f_k(\omega)f_{k'}(\omega)\overline{E}_2\left(-\frac{\omega}{2}\right) \right\} \right], \qquad (7.25b)$$

where $E_{1,2}(\xi)$ were defined below Eq. (7.23). Coefficients $f_k(\omega)$ are the format dependent form-factors: $f_k(\omega) = P_k(\omega)$ for the Nyquist modulation, where $P_k(\omega) = 1$ when the frequency is inside the k-th band and zero otherwise; for the OFDM modulation $f_k(\omega) = e^{i(\omega-\Omega_k)T_s/2} \mathrm{sinc}[(\omega - \Omega_k)T_s/2]$. In as much as the channels do not overlap, the noise components from different channels are *uncorrelated*. This is the consequence of the asymptotic absence of channel cross-talk in the continuous model (7.21), (7.23).

7.4.3 Capacity Estimates for WDM/OFDM NIS Transmission

For an arbitrary vector information channel the input-output *mutual information* $I(\mathbf{X}, \mathbf{Y})$ is defined as [5, 8, 50]:

$$I(\mathbf{X}; \mathbf{Y}) = H(\mathbf{Y}) - H(\mathbf{Y}|\mathbf{X}),$$

$$H(\mathbf{Y}) = -\int d^{2M}\mathbf{Y} P_Y(\mathbf{Y})\log_2[P_Y(\mathbf{Y})], \qquad (7.26)$$

where H designates the entropy. The *Shannon capacity per symbol*, C, is the maximum of $I(\mathbf{X}; \mathbf{Y})/M$ over the input distribution $P_X(\mathbf{X})$ subject to the average power per sample constraint $E[|\mathbf{X}|^2]/M \le S$. For any additive Gaussian channel the expression for the channel entropy $H(\mathbf{Y}|\mathbf{X})$ is obtained by averaging the determinant of the conditional correlation matrix $\Sigma(\mathbf{X})$ over the input distribution. Our channel (7.24) with the non-diagonal input-dependent correlation matrix (7.25) obviously has memory which makes the direct optimization of the mutual information functional extremely difficult. Therefore we make do only with a *lower bound* for the capacity by making a plausible guess of the optimal statistics of input distribution. A standard approach is to use Gaussian input

$\mathbf{X}_G$ with independently distributed real quadrature samples each having the variance $S/2$, which in the continuous limit corresponds to a Gaussian process with constant spectral density S [8]. However, the analytical expressions for the mutual information for the channel given by Eqs. (7.24)–(7.25) are generally intractable even with the Gaussian i.i.d. input. Thus even this lower bound for capacity has to be further relaxed to achieve a tractable analytical result. Here we use the effective Gaussian input-output model and Pinsker's formula [6, 55] for the capacity to lower bound it further. It can be shown [39] that in the limit of large effective SNR defined by the equation below, the Pinsker lower bound for the capacity, C_G, in bits per symbol in the r.w.u. is

$$C_G = \log_2\left[\frac{ST_s/N_{\mathrm{ASE}}}{2(E_{\mathrm{in}}(S)/E_{\mathrm{NL}})^2 + (E_{\mathrm{in}}(S)/E_{\mathrm{NL}})^2 + 1}\right]$$
$$\equiv \log_2(SNR_{\mathrm{eff}}), \tag{7.27}$$

where the second line is the definition of the effective SNR, $E_{\mathrm{in}}(S) = ST_s N_b N_{\mathrm{ch}}$ is the average energy of the initial optical burst (before the NIS module) and $E_{\mathrm{NL}} = |\beta_2|N_{\mathrm{ch}}/(\gamma T_s)$ is a typical nonlinear energy. This formula holds for both OFDM and Nyquist-based NIS transmission. It is accurate up to the terms of order $O[1/SNR_{\mathrm{eff}}]$; the general applicability criteria are discussed in the next subsection. Note that for a fixed propagation distance L and symbol rate, it displays a characteristic *peaky behaviour* in both average input power, S, and burst energy, E_{in}, which is a common feature to many Gaussian-based lower-bound estimates for conventional transmission formats [6–8].

We can now put the obtained results into perspective by considering a fixed distance, the number of channels/subcarriers, burst duration and symbol rate. The argument of log in (7.27) is a monotonically growing function of E_{in} (or S) up to $E_{\mathrm{in}}^* = E_{\mathrm{NL}}/\sqrt{2}$ which corresponds to the maximum (Pinsker) capacity:

$$C_G^{\mathrm{max}} = \log_2\left(\frac{2\sqrt{2}-1}{7}\frac{E_{\mathrm{NL}}}{N_{\mathrm{ASE}}N_b N_{\mathrm{ch}}}\right)$$
$$= \log_2\left(\frac{|\beta_2|}{K_T h\nu_s \alpha\gamma T_b L}\right) - 1.94 \text{ bits/symbol}, \tag{7.28}$$

where $T_b = T_s N_b$ is the duration of the burst. From Eq. (7.28) it is seen that the estimate deteriorates slowly (logarithmically) with the product $L \times T_b$. On the other hand it does not depend on the channel number N_{ch}, which is a direct consequence of the absence of the channel cross-talk.

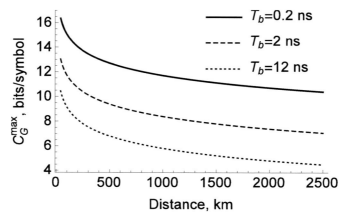

Figure 7.3 Lower bound for the maximum achievable capacity (7.28) vs. transmission distance at 500 Gbaud at different burst sizes. From [39].

Now we deal with two typical examples of Nyquist and OFDM systems with the parameters that are close to those reported both in conventional [7, 8] and NIS based [25, 26] systems. For the Nyquist transmission we pick 5 channels with individual band width $B_0 = T_s^{-1} = 100$ GHz, corresponding to the overall symbol rate of 500 Gbaud (in all channels) and for the OFDM we pick 100 subcarriers with the spacing of $B_0 = 5$ GHz and the same effective baud rate. The optimal initial energy is then $E_{in}^* = 6$ pJ for both Nyquist and OFDM cases. The required power levels in optical domain can be estimated if one fixes the burst size, N_b. For example if we use one OFDM supersymbol of length $T_s = 0.2$ ns as a burst, i.e. $N_b = 1$, $T_b = T_s$, the corresponding Nyquist burst of the same length and total baud rate would contain $N_b = 20$ symbols. The optimal average power in the optical domain Sopt can be estimated by the averaging of NIS-generated patterns. One obtains: $S_{opt} = 13.32$ dBm in the OFDM, and $S_{opt} = 12$ dBm in the Nyquist case. For such parameters it is possible to achieve the capacity of at least ~10.7 bits/symbol at 500 Gbaud at 2000 km. The result deteriorates with the burst size for both Nyquist and OFDM transmission as predicted by (7.28). Note, however, that for a fixed symbol rate the penalty of going from, say, 1000 to 2000 km is only 1 bit/symbol. All this is illustrated in Fig. 7.3, where the estimate (7.28) is plotted as a function of distance for different burst sizes. For the Nyquist case, varying the burst size is equivalent to changing the number of symbols in the burst while keeping the bandwidth (symbol rate) fixed, whereas for the OFDM one needs to change the number of subcarriers to keep the baud rate fixed.

7.4.4   Applicability of Results

In the limit of small power and short burst, when $E_{in} \ll E_{NL}$, the definition of the SNR_{eff} coincides with the linear one. However, in the nonlinear regime the effective SNR deteriorates if one considers either a high power regime or a very long burst. The overall consistency criteria for combining perturbation theory and asymptotic analysis can be written as $SNR_{eff} \gg 1$, so that the validity condition for Eq. (7.16) is met, and the propagation distance L must be much greater than the dispersion length (defined via the total transmission bandwidth $B = N_{ch}B_0$), in order to assume the diagonal form of the correlation functions (7.23) that was further used to get Eq. (7.25). For the fixed fibre and burst parameters considered in Section 7.4, and the given symbol length T_s, assuming the optimal energy $E^*_{in} = E_{NL}/\sqrt{2}$ corresponding to the optimal power launch level $S_* = |\beta_2|/\sqrt{2}\gamma T^2_s N_b$), all the above conditions turn into a restricted window of distances in the r.w.u.:

$$\frac{1}{B^2|\beta_2|} \ll L \ll \frac{|\beta_2|}{K_T h v_s \chi \gamma T_b}. \tag{7.29}$$

For B_0 = 100 GHz and 5 WDM channels with the total bandwidth $B = 5 B_0 = 0.5$ THz, and the burst size $T_b = 12$ ns in a standard telecom fibre, the above reads as $0.2 \ll L$ [km] $\ll 2.1 \times 10^5$, and this condition is easily met in all realistic implementations.

Now let us study what are the theoretical restrictions on the input power inside the NF domain, S. In the nonlinear regime, when $E_{in} \gtrsim E_{NL}$, the quadratic term in the denominator of SNR_{eff} in Eq. (7.27) dominates and the condition $SNR_{eff} \gg 1$ is equivalent to the following restriction:

$$S \ll \frac{|\beta_2|^2}{\gamma^2 T_s T_b^2 N_{ASE}}.$$

Thus, for the Nyquist modulation with five channels each having the bandwidth $B_0 = T^{-1} = 100$ GHz, $T_b = 12$ ns and $L = 2000$ km considered above, the perturbative approach is valid up to the optical power levels $S_{opt} \sim 11$ dBm.

7.5 Conclusion

In this chapter we described a theoretical approach based on the perturbation theory for the NF data for the asymptotical estimate of capacity limits of the NFT-based optical transmission. We underline that the intermediate results of our analysis, i.e. the expressions for the autocorrelator functions for the signal in the NF domain, are novel and quite important for further NFT-transmission SE and capacity studies. For the NIS transmission, considered as an example, Eq. (7.16) describes a continuous channel model for a generic NFT-based system in the presence of inline noise, while Eqs. (7.21), (7.22), (7.25) develop it further, introducing a discrete time channel model for the NFT-based transmission. The model predicts the absence of the nonlinear spectral broadening and channel cross-talk, which makes it applicable to multi-user routed optical transmission systems. Using very conservative lower bound estimate we showed that the channel capacity per symbol at 500 km exceeds the values of ~12 bits/symbol (corresponding to the SE of ~4.5 bits/symbol/Hz) in a multi-user environment and the result does not depend on the number of subcarriers/channels used. Moreover we believe that both the peaky estimates for the capacity and the relatively modest values of the spectral efficiency obtained in the current paper can be improved. Of course our original channel model (7.24) is only valid in the limit of large effective SNR. As shown in subsection 7.4.4, for a fixed burst size this puts the upper bound on the possible input power which is inversely proportional to the square of the burst length, N_b. For the spectral parameters of our system at 2000 km for bursts with $N_b \sim 10^3$, the model is valid up to the power levels of $S_{max} \approx 12 \div 13$ dBm, which is still way above the values considered in our example study. This means that even if the Pinsker estimate (7.27) fails at large power levels and long taps, the genuine SE is expected to grow all the way up to S_{max} where the integrability is already lost. Moreover, if one is able to obtain the capacity estimates that do not deteriorate with the burst size T_b, the problem of low spectral efficiency can be effectively circumvented by considering large bursts $T_b \gg T_G$.

To sum up, the burst mode of the NFT-based methods brings about the noticeable degradation of the effective SE of a system measured in the optical domain, which means that the NFT-based systems require further development and optimization. The straightforward way is to incorporate the remaining nonlinear degrees of freedom (solitons) into the NIS ideology for the sake of increasing the amount of information embedded into the same time and spectral volume. The first example of such a technique was recently proposed in [29]. Yet another way is to optimize further the time-domain occupation of the NIS-generated pulses using non-traditional formats tailored specially for the NFT-based transmission purposes. One more, even less trivial solution could be the use of different variants of inverse scattering method itself [56]. However, we note that the effective absence of the inter-mode cross-talk inside the nonlinear domain and other results of our study demonstrate the great potential of the NFT application, where the main problem now lies in the optimization and further elaboration of the approach rather than in some inevitable performance degradation sources, like the aforementioned nonlinear cross-talk.

Appendix: Random Walk of Discrete and Continuous NF Spectrum Parameters

First we note that in the context of conventional soliton-based transmission [46], when only a single eigenvalue is present and the continuous NF spectrum is absent, our perturbation theory reduces to the same level of approximation that was used for the derivation of the celebrated Gordon–Haus effect [51]. Here we take the analysis some steps further and consider multiple eigenvalue case in the presence of continuous NF spectrum.

The perturbation terms in the R.H.S. of Eqs. (7.14c)–(7.14e) form the components of $3N$ dimensional complex noise vector Γ_{kj} ($k = 1, \ldots, N, j = \zeta, r_k, l_k$). Following the same procedure as in Subsection 7.3.2, one can arrive at the autocorrelation functions for this vector. Let us start from the "diagonal" terms. For convince we introduce some more simplified denotations: $\psi_{i,k}^2 \equiv \psi_{i,k}^2(\xi, t, z) = \psi_i^2(\zeta_k, \xi|t, z)$, $\phi_{i,k}^2 \equiv \phi_{i,k}^2(\xi, t, z) = \phi_i^2(\zeta_k, \xi|t, z)$. For the autocorrelator functions of the continuous NF part we have:

$$E[\Gamma_{k\zeta}(z)\overline{\Gamma}_{k'\zeta}(z')] = \varepsilon\delta(z - z')r_k\overline{r}_{k'}\int_{-\infty}^{\infty} dt[\psi_{2,k}^2\overline{\psi}_{2,k'}^2 + \psi_{1,k}^2\overline{\psi}_{1,k'}^2], \qquad (7.30a)$$

$$E[\Gamma_{k\zeta}(z)\Gamma_{k'\zeta}(z')] = \varepsilon\delta(z - z')r_k r_{k'}\int_{-\infty}^{\infty} dt[\psi_{2,k}^2\psi_{1,k'}^2 + \psi_{1,k}^2\psi_{2,k'}^2]; \qquad (7.30b)$$

For the autocorellator functions of the discrete NF part of the spectrum:

$$E[\Gamma_{kr}(z)\overline{\Gamma}_{k'r}(z')] = \varepsilon\delta(z - z')$$
$$\times \left\{ \frac{a_k''}{(a_k')^3}\frac{\overline{a}_{k'}''}{(\overline{a}_{k'}')^3}\int_{-\infty}^{\infty} dt[\phi_{2,k}^2\overline{\phi}_{2,k'}^2 + \phi_{1,k}^2\overline{\phi}_{1,k'}^2] \right.$$
$$+ \frac{1}{(a_k')^2}\frac{1}{(\overline{a}_{k'}')^2}\int_{-\infty}^{\infty} dt[\varphi_2^k\overline{\varphi}_2^{k'} + \varphi_1^k\overline{\varphi}_1^{k'}]$$
$$\left. - \left(\frac{a_k''}{(a_k')^3}\frac{1}{(\overline{a}_{k'}')^2}\int_{-\infty}^{\infty} dt[\phi_{2,k}^2\overline{\varphi}_2^{k'} + \phi_{1,k}^2\overline{\varphi}_1^{k'}] + \text{h.c.} \right) \right\}, \qquad (7.30c)$$

$$E[\Gamma_{kr}(z)\Gamma_{k'r}(z')] = \varepsilon\delta(z - z')$$
$$\times \left\{ \frac{a_k''}{(a_k')^3}\frac{a_{k'}''}{(a_{k'}')^3}\int_{-\infty}^{\infty} dt[\phi_{2,k}^2\phi_{1,k'}^2 + \phi_{1,k}^2\phi_{2,k'}^2] \right.$$
$$+ \frac{1}{(a_k')^2}\frac{1}{(a_{k'}')^2}\int_{-\infty}^{\infty} dt[\varphi_2^k\varphi_1^{k'} + \varphi_1^k\varphi_2^{k'}]$$
$$\left. - \left(\frac{a_k''}{(a_k')^3}\frac{1}{(a_{k'}')^2}\int_{-\infty}^{\infty} dt[\phi_{2,k}^2\varphi_1^{k'} + \phi_{1,k}^2\varphi_1^{k'}] + \text{t.c.} \right) \right\}, \qquad (7.30d)$$

$$E[\Gamma_{kl}(z)\overline{\Gamma}_{k'l}(z')] = \varepsilon\delta(z - z')$$

$$\times \left\{ \frac{a_k''}{(a_k')^2 a_k} \frac{\overline{a}_{k'}''}{(\overline{a}_{k'}')^2 \overline{a}_{k'}} \int_{-\infty}^{\infty} dt[\psi_{2,k}^2\overline{\psi}_{2,k'}^2 + \psi_{1,k}^2\overline{\psi}_{1,k'}^2] \right.$$

$$+ \frac{1}{(a_k')^2} \frac{1}{(\overline{a}_{k'}')^2} \int_{-\infty}^{\infty} dt[\gamma_2^k\overline{\gamma}_1^{k'} + \gamma_1^k\overline{\gamma}_1^{k'}]$$

$$\left. - \left(\frac{a_k''}{(a_k')^2 a_k} \frac{1}{(\overline{a}_{k'}')^2} \int_{-\infty}^{\infty} dt[\psi_{2,k}^2\overline{\gamma}_1^{k'} + \psi_{1,k}^2\overline{\gamma}_1^{k'}] + \text{h.c.} \right) \right\}, \tag{7.30e}$$

$$E[\Gamma_{kl}(z)\Gamma_{k'l}(z')] = \varepsilon\delta(z - z')$$

$$\times \left\{ \frac{a_k''}{(a_k')^2 a_k} \frac{a_{k'}''}{(a_{k'}')^2 a_{k'}} \int_{-\infty}^{\infty} dt[\psi_{2,k}^2\psi_{1,k'}^2 + \psi_{1,k}^2\psi_{2,k'}^2] \right.$$

$$+ \frac{1}{(a_k')^2} \frac{1}{(a_{k'}')^2} \int_{-\infty}^{\infty} dt[\gamma_2^k\gamma_1^{k'} + \gamma_1^k\gamma_1^{k'}]$$

$$\left. - \left(\frac{a_k''}{(a_k')^2 a_k} \frac{1}{(a_{k'}')^2} \int_{-\infty}^{\infty} dt[\psi_{2,k}^2\gamma_1^{k'} + \psi_{1,k}^2\gamma_1^{k'}] + \text{t.c.} \right) \right\}, \tag{7.30f}$$

where "h.c." means the Hermitian conjugation (index swap $k \leftrightarrow k'$ followed by the complex conjugation) and "t.c." stands for conjugate transpose (i.e. just the index swap). These expressions should be supplemented with the "cross-correlation" functions between discrete and continuous NF spectrum components:

$$E[\Gamma_{k\zeta}(z)\overline{\Gamma}_{k'r}(z')] = \varepsilon\delta(z - z')$$

$$\times \left\{ -r_k \frac{\overline{a}_{k'}''}{(\overline{a}_k')^3} \int_{-\infty}^{\infty} dt[\psi_{2,k}^2\overline{\phi}_{2,k'}^2 + \psi_{1,k}^2\overline{\phi}_{1,k'}^2] \right.$$

$$\left. + \frac{r_k}{(\overline{a}_{k'}')^2} \int_{-\infty}^{\infty} dt[\psi_{2,k}^2\overline{\varphi}_2^{k'} + \psi_{1,k}^2\overline{\varphi}_1^{k'}] \right\}, \tag{7.31a}$$

$$E[\Gamma_{k\zeta}(z)\Gamma_{k'r}(z')] = \varepsilon\delta(z - z')$$

$$\times \left\{ -r_k \frac{a_{k'}''}{(a_k')^3} \int_{-\infty}^{\infty} dt[\psi_{2,k}^2\phi_{1,k'}^2 + \psi_{1,k}^2\phi_{2,k'}^2] \right.$$

$$\left. + \frac{r_k}{(a_{k'}')^2} \int_{-\infty}^{\infty} dt[\psi_{2,k'}^2\varphi_1^{k'} + \psi_{1,k}^2\varphi_2^{k'}] \right\}, \tag{7.31b}$$

$$E[\Gamma_{k\zeta}(z)\overline{\Gamma}_{k'r}(z')] = \varepsilon\delta(z-z')$$

$$\times\left\{-r_k\frac{\overline{a}_{k'}''}{(\overline{a}_{k'}')^2\overline{a}_{k'}}\int_{-\infty}^{\infty}dt[\psi_{2,k}^2\overline{\psi}_{2,k'}^2 + \psi_{1,k}^2\overline{\psi}_{1,k'}^2]\right.$$

$$\left.+\frac{r_k}{(\overline{a}_{k'}')^2}\int_{-\infty}^{\infty}dt[\psi_{2,k}^2\overline{\gamma}_2^{k'} + \psi_{1,k}^2\overline{\gamma}_1^{k'}]\right\}, \tag{7.31c}$$

$$E[\Gamma_{k\zeta}(z)\Gamma_{k'l}(z')] = \varepsilon\delta(z-z')$$

$$\times\left\{-r_k\frac{a_{k'}''}{(a_{k'}')^2 a_{k'}}\int_{-\infty}^{\infty}dt[\psi_{2,k}^2\psi_{1,k'}^2 + \psi_{1,k}^2\psi_{2,k'}^2]\right.$$

$$\left.+\frac{r_k}{(a_{k'}')^2}\int_{-\infty}^{\infty}dt[\psi_{2,k}^2\gamma_1^{k'} + \psi_{1,k}^2\gamma_2^{k'}].\right\} \tag{7.31d}$$

Since we are using only one of the two sets of coefficients, either r_k or l_k, their cross-correlation functions are not needed.

Acknowledgements

This work was supported by the UK EPSRC Programme Grant UNLOC EP/J017582/1, ERC project ULTRALASER and Grant of the Ministry of Education and Science of the Russian Federation (agreement No. 14.B25.31.0003). SD was supported by the Israeli Science Foudation (grant No 466/18).

References

1. D. J. Richardson, Filling the light pipe, *Science*, **330**, 327–328 (2010).

2. P. J. Winzer, Scaling Optical Fiber Networks: Challenges and Solutions, *Optics & Photonics News*, **26**, 28–35 (2015).

3. R. J. Essiambre, R. W. Tkach, and R. Ryf, Fiber nonlinearity and capacity: Single mode and multimode fibers, in *Optical Fiber Telecommunications, Vol. VIB: Systems and Networks*, 6th ed., Chapter 1, eds. I. Kaminow, T. Li, and A. E. Willner, pp. 1–43 (Academic Press, 2013).

4. H. Nyquist, Certain factors affecting telegraph speed, *Bell Syst. Tech. J.*, **3**, 324–346 (1924).

5. C. E. Shannon, A mathematical theory of communication, *Bell Syst. Tech. J.*, **27**, 379–423 (1948).

6. P. P. Mitra and J. B. Stark, Nonlinear limits to the information capacity of optical fiber communications. *Nature*, **411**, 1027–1030 (2001).

7. R.-J. Essiambre, G. J. Foschini, G. Kramer, and P. J. Winzer, Capacity limits of information transport in fiber-optic networks, *Phys. Rev. Lett.*, **101**, 163901 (2008).

8. R.-J. Essiambre, G. Kramer, P. J. Winzer, G. J. Foschini, and B. Goebel, Capacity limits of optical fiber networks, *J. Lightwave Technol.*, **28**, 662–701 (2010).

9. G. P. Agrawal, *Fiber-Optic Communication Systems*, 4th ed. (Wiley-Blackwell, 2010).

10. E. Agrell, A. Alvarado, and F. R. Kschischang, Implications of information theory in optical fibre communications, *Phil. Trans. R. Soc. A*, **374**, 20140438 (2016).

11. E. Ip and J. Kahn. Compensation of dispersion and nonlinear impairments using digital backpropagation, *J. Lightwave Tech.*, **26**, 3416–3425 (2008).

12. C. Xi, L. Xiang, S. Chandrasekhar, B. Zhu, and R. W. Tkach, Experimental demonstration of fiber nonlinearity mitigation using digital phase conjugation, in *Technical Digest of Optical Fiber Communication Conference and Exposition (OFC/NFOEC) and the National Fiber Optic Engineers Conference*, paper OTh3C.1 (2012).

13. S. L. Jansen, et al. Optical phase conjugation for ultra long-haul phase-shift-keyed transmission, *J. Lightwave Technol.*, **24**, 54–64 (2006).

14. I. Phillips, et al. Exceeding the nonlinear-Shannon limit using Raman laser based amplification and optical phase conjugation, in *Optical Fiber Communication Conference* (OFC 2014), Los Angeles, USA, paper M3C.1.

15. A. R. C. X. Liu, P. J. Winzer, R. W. Tkach, and S. Chandrasekhar, Phase-conjugated twin waves for communication beyond the Kerr nonlinearity limit, *Nat. Photonics*, **7**, 560–568 (2013).

16. L. B. Du, et al. Digital Fiber Nonlinearity Compensation: Toward 1-Tb/s transport, *IEEE Signal Proc. Mag.*, **31**, 46–56 (2014).

17. D. Rafique, Fiber nonlinearity compensation: Commercial applications and complexity analysis, *J. Lightwave Technol.*, **34**, 544–553 (2016).

18. M. I. Yousefi and F. R. Kschischang, Information transmission using the nonlinear Fourier transform, Parts I–III, *IEEE Trans. Inform. Theory*, **60**, 4312–4369 (2014).

19. J. E. Prilepsky, S. A. Derevyanko, and S. K. Turitsyn, Nonlinear spectral management: Linearization of the lossless fiber channel, *Opt. Express*, **21**, 24344–24367 (2013).

20. J. E. Prilepsky and S. K. Turitsyn, Eigenvalue communications in nonlinear fiber channels, in *Odyssey of Light in Nonlinear Optical Fibers: Theory and Applications*, Chapter 18, eds. K. Porsezian and R. Ganapathy, pp. 459–490 (CRC Press, 2015).

21. A. Hasegawa and T. Nyu, Eigenvalue communication, *J. Lightwave Technol.*, **11**, 395–399 (1993).

22. E. G. Turitsyna and S. K. Turitsyn, Digital signal processing based on inverse scattering transform, *Opt. Lett.*, **38**, 4186–4188 (2013).

23. S. Wahls, S. T. Le, J. E. Prilepsky, H. V. Poor, and S. K. Turitsyn, Digital backpropagation in the nonlinear Fourier domain, in *Proceedings of IEEE 16th International Workshop in Signal Processing Advances in Wireless Communications* (SPAWC 2015), Stockholm, Sweden, pp. 445–449.

24. J. E. Prilepsky, S. A. Derevyanko, K. J. Blow, I. Gabitov, and S. K. Turitsyn, Nonlinear inverse synthesis and eigenvalue division multiplexing in optical fiber channels, *Phys. Rev. Lett.*, **113**, 013901 (2014).

25. S. T. Le, J. E. Prilepsky, and S. K. Turitsyn, Nonlinear inverse synthesis for high spectral efficiency transmission in optical fibers, *Opt. Express*, **22**, 26720–26741 (2014).

26. S. T. Le, J. E. Prilepsky, and S. K. Turitsyn, Nonlinear inverse synthesis technique for optical links with lumped amplification, *Opt. Express*, **23**, 8317–8328 (2015).

27. S. Hari, F. Kschischang, and M. Yousefi, Multi-eigenvalue communication via the nonlinear Fourier transform, in *27th Biennial Symposium on Communications (QBSC)*, Kingston, ON, Canada, pp. 92–95, (2014).

28. Z. Dong, et al., Nonlinear frequency division multiplexed transmissions based on NFT, *IEEE Photon. Tech. Lett.*, **27**, 1621–1623 (2015).

29. I. Tavakkolnia and M. Safari, Signalling over nonlinear fibre-optic channels by utilizing both solitonic and radiative spectra, in *IEEE European Conference on Networks and Communications (EuCNC)*, Paris, France, pp. 103–107, (2015).

30. S. T. Le, et al., Demonstration of nonlinear inverse synthesis transmission over transoceanic distances, *J. Lightwave Tech.*, **34**, 2459–2466 (2016).

31. H. Bülow, Experimental assessment of nonlinear Fourier transformation based detection under fiber nonlinearity, in *European Conference on Optical Communications (ECOC)*, Cannes, France, paper We.2.3.2, (2014).

32. H. Bülow, Experimental demonstration of optical signal detection using nonlinear Fourier transform, *J. Lightwave Technol.*, **33**, 1433–1439 (2015).

33. H. Bülow, Nonlinear Fourier transformation based coherent detection scheme for discrete spectrum, in *Optical Fiber Communication Conference (OFC)*, Los Angeles, USA, paper W3K.2, (2015).

34. H. Terauchi and A. Maruta, Eigenvalue modulated optical transmission system based on digital coherent technology, *The 10th Conference on Lasers and Electro-Optics Pacific Rim, and the 18th OptoElectronics and Communications Conference/Photonics in Switching (CLEO-PR & OECC/PS)*, Kyoto, Japan, Paper WR2-5, (2013).

35. A. Maruta, Eigenvalue modulated optical transmission system (invited), *The 20th OptoElectronics and Communications Conference (OECC)*, Shanghai, China, Paper JThA.21, (2015).

36. A. Maruta, Y. Matsuda, H. Terauchi, and A. Toyota, Digital coherent technology-based eigenvalue modulated optical fiber transmission system, in *Odyssey of Light in Nonlinear Optical Fibers: Theory and Applications*, Chapter 19, eds. K. Porsezian and R. Ganapathy, pp. 491–505 (CRC Press, 2015).

37. E. Meron, M. Feder, and M. Shtaif, On the achievable communication rates of generalized soliton transmission systems, arXiv:1207.0297 (2012).

38. N. A. Shevchenko, J. E. Prilepsky, S. A. Derevyanko, A. Alvarado, P. Bavel, and S. K. Turitsyn, A lower bound on the per soliton capacity of the nonlinear optical fibre channel, in *IEEE Information Theory Workshop (ITW)*, Jeju Island, Korea, pp. 104–108 (2015).

39. S. A. Derevyanko, J. E. Prilepsky, and S. K. Turitsyn, Capacity estimates for optical transmission based on the nonlinear Fourier transform, *Nat. Commun.*, **7**, 12710 (2016).

40. J. D. Ania-Castañón, Quasi-lossless transmission using second-order Raman amplification and fibre Bragg gratings, *Opt. Express*, **12**, 4372–4377 (2004).

41. J. D. Ania-Castañón, T. J. Ellingham, R. Ibbotson, X. Chen, L. Zhang, and S. K. Turitsyn, Ultralong Raman fibre lasers as virtually lossless optical media, *Phys. Rev. Lett.*, **96**, 023902 (2006).

42. S. T. Le, et al. Modified nonlinear inverse synthesis for optical links with distributed Raman amplification, in *41st European Conference on Optical Communications (ECOC)*, Valencia, Spain, paper Tu 1.1.3 (2015).

43. S. T. Le, J. E. Prilepsky, P. Rosa, J. D. Ania-Castañón, and S. K. Turitsyn, Nonlinear inverse synthesis for optical links with distributed Raman amplification, *J. Lightwave Technol.*, **34**, 1778–1785 (2016).

44. V. E. Zakharov and A. B. Shabat, Exact theory of 2-dimensional self-focusing and one-dimensional self-modulation of waves in nonlinear media, *Sov. Phys. JETP*, **34**, 62–69 (1972).

45. N. J. Ablowitz, D. J. Kaup, A. C. Newell, and H. Segur, H. The inverse scattering transform-Fourier analysis for nonlinear problems, *Stud. Appl. Math.*, **53**, 249–315 (1974).

46. A. Hasegawa and Y. Kodama, *Solitons in Optical Communications* (Oxford University Press, 1995).

47. M. J. Ablowitz and H. Segur, *Solitons and the Inverse Scattering Transform* (SIAM, 1981).

48. D. J. Kaup, Perturbation expansion for Zakharov–Shabat inverse scattering transform, *SIAM J. Appl. Math.*, **31**, 121–133 (1976).

49. D. J. Kaup and A. C. Newell, Solitons as particles, oscillators, and in slowly changing media—singular perturbation-theory, *Proc. Roy. Soc. Lond. A Mat.*, **361**, 413–446 (1978).

50. T. M. Cover and J. A. Thomas, *Elements of Information Theory*, 2nd ed. (Wiley, 2006).

51. J. P. Gordon and H. A. Haus, Random walk of coherently amplified solitons in optical fiber transmission, *Opt. Lett.*, **11**, 665–667 (1986).

52. W. Shieh, H. Bao, and Y. Tang, Coherent optical OFDM: Theory and design, *Opt. Express*, **16**, 841–859 (2008).

53. R. Schmogrow, et al, Real-time Nyquist pulse generation beyond 100 Gbit/s and its relation to OFDM, *Opt. Express*, **20**, 317–337 (2012).

54. V. E. Zakharov and S. V. Manakov, Asymptotic behavior of non-linear wave systems integrated by the inverse scattering method, *Sov. Phys. JETP*, **44**, 106 (1976).

55. M. S. Pinsker, *Information and Informational Stability of Random Variables and Processes*, (Holden Day, 1964).

56. M. Kamalian, J. E. Prilepsky, S. T. Le, and S. K. Turitsyn, Periodic nonlinear Fourier transform for fiber-optic communications, Part I: theory and numerical methods, *Opt. Express*, **24**, 18353–18369 (2016).

Chapter 8

Spatial Multiplexing: Technology

Yongmin Jung, Qiongyue Kang, Shaif-ul Alam, and David J. Richardson

Optoelectronics Research Centre, University of Southampton, Southampton, SO17 1BK, UK

ymj@orc.soton.ac.uk

Space division multiplexing (SDM), utilizing few-mode fibres or multi-core fibres supporting multiple spatial channels, is currently under intense investigation as an efficient approach to overcome the current capacity limitations of high-speed long-haul transmission systems based on single mode optical fibres. In order to realize the potential energy and cost savings offered by SDM systems, the individual spatial channels should be simultaneously multiplexed, transmitted, amplified and switched with associated SDM components and subsystems. In this chapter, a review of recent progress on the implementation of various SDM technologies is provided and the latest strategies and trends are presented. In particular, integrated SDM amplifier technologies will be discussed in detail and the prospect of further scaling with respect to

Optical Communication Systems: Limits and Possibilities
Edited by Andrew Ellis and Mariia Sorokina
Copyright © 2020 Jenny Stanford Publishing Pte. Ltd.
ISBN 978-981-4800-28-0 (Hardcover), 978-0-429-02780-2 (eBook)
www.jennystanford.com

the number of spatial channels that can be amplified will be described.

8.1 Introduction

Over the past few decades, the demand for internet capacity has grown exponentially at an average rate of about 40–60% year-on-year and there are fears that future growth of the internet will be constrained due to fundamental capacity limits in optical fibres due to their intrinsic nonlinearity. To date operators have kept up with the increasing data traffic through a sequence of technical innovations (as labelled in Fig. 8.1) associated with ever better exploitation of the transmission capacity of single mode fibre (SMF) which has formed the essential physical fabric of our global communication systems for the past ~40 years. However, this situation is set to change: in the laboratory the fundamental limit to transmission capacity imposed by optical nonlinearity in SMF is rapidly being approached (the so-called "Nonlinear Shannon Limit" [1, 2] which constrains transmission capacity over 1000 km length scale to ~100 Tbit/s). Given the relentless growth in data traffic and the fact that 10 Tbit/s systems are already being installed, there are concerns of a future "Capacity Crunch" in the coming decade, where the need to install new fibres to take up additional traffic growth (at an effectively constant cost-per-bit as further capacity is added) threatens to constrain future internet growth. As a consequence, radical innovation is now required to overcome this threat and this will almost certainly dictate the need to revisit our choice of transmission fibres and corresponding optical amplifiers. The only physical signalling dimension not fully exploited in current system is "space" and consequently Space Division Multiplexing (SDM) [3–5] has emerged as the most promising solution to the "capacity crunch" problem. In SDM transmission system, independent data streams can be transmitted in parallel spatial channels and various SDM fibre strategies have been proposed. In the following section, we first summarize the state-of-the-art for various SDM approaches under investigation, tabulating progress in the development of SDM transmission fibres, SDM multiplexers/demultiplexers and the associated partner SDM amplifiers.

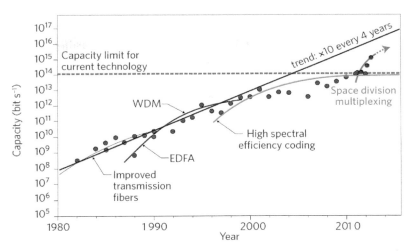

Figure 8.1 The historical evolution of transmission capacity in optical fibres [3, 5].

8.2   SDM Transmission Fibres

Several different SDM transmission fibres have been proposed and investigated to overcome the capacity limit of SMFs and Fig. 8.2 shows the classification of the primary SDM fibre types reported to date. There are two main SDM approaches for achieving multiple spatial channels within a single fibre: (1) multi-core fibre (MCF) whereby multiple cores are incorporated in a single fibre cladding, and (2) few-mode fibre (FMF) that utilizes multiple spatial modes in a large core fibre to define the distinct spatial information channels. Both of these approaches are being investigated intensively around the globe and an approximate 10-fold increase in overall fibre capacity (using a 12-core fibre) was achieved [6] in little more than 5 years. Moreover, at the most recent OFC'16 conference, MCFs with up to 32-cores were demonstrated providing a high spatial core density and low crosstalk supporting transmission over distance in excess of 1000 km [7], while a 15 spatial mode FMF has also been demonstrated [8], supporting 15 channel transmission over a distance of 22.8 km. In order to further increase the fibre capacity for future SDM networks, the number of spatial channels needs to be further scaled and the introduction of a hybrid SDM fibre structure (e.g. few-mode multi-core fibre, FM-MCF) can help further increase the capacity by combining the two basic spatial multiplexing schemes (FMF and MCF). At the OFC'15 conference, two state-of-the-art hybrid SDM fibres were presented one with 36 cores × 3 modes [9] and the other with 19 cores × 6 modes [10]. Initial hybrid SDM fibre transmission experiments in both fibres were presented highlighting the potential for 100-fold capacity increases with respect to standard single mode optical fibre. In addition, several other types of SDM transmission fibre have also been proposed offering spatial channels with interesting features. For example, single radial mode but azimuthally multi-mode transmission in ring-core fibres (RCFs) [11], orbital angular momentum (OAM) mode transmission in OAM fibres [12], low latency air-core guidance in hollow-core photonic bandgap fibres (HC-PBGFs) [13] and fibre bundle like multi-element fibres (MEFs) [14] that provide for easy access to the cores within the individual fibre elements. In the following sections, the detailed fibre design concepts of the two mainstream SDM fibres (i.e. FMFs and MCFs) will be provided.

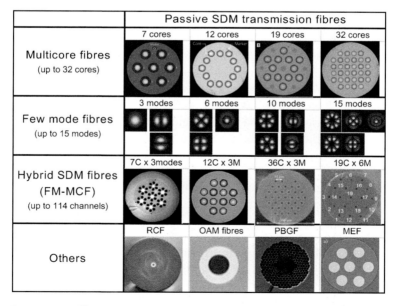

Figure 8.2 Different types of SDM transmission fibre reported so far; MCF [6, 7], FMF [8, 19, 36], FM-MCF [9, 10, 23] and others [11–14].

8.2.1 Few-Mode Fibres

A few-mode fibre (FMF) is a multi-mode fibre that has been designed to support a limited but carefully controlled number of spatial modes. The normalized frequency parameter of FMFs (the so-called the V-number) needs to be greater than 2.405 (the cut-off value for SM fibre operation), which means that the fibre has either to have a larger core diameter (>10 µm) and/or a larger refractive index differences (>0.005) between the core and cladding than in conventional telecom SMFs. Since the spatial modes are orthogonal, it is in principle possible to increase the transmission capacity in accordance with the number of modes supported by the FMFs. The distinguishable spatial data pathways required for SDM are defined either by the particular modes supported by the fibre, or alternatively by orthogonal combinations of these fibre modes. Since all spatial modes in FMFs have significant spatial overlap, the data signals are prone to couple randomly between spatial channels during propagation. Thus it is generally necessary to employ multiple input multiple output (MIMO) digital signal processing (DSP) at the receiver end in order to mitigate the transmission impairments from mode coupling. To fully compensate for the effects of mode coupling using MIMO, the equalization filter length needs to be longer than the impulse response spread of an individual symbol. Therefore, the modal dispersion (i.e. differential group delay (DGD)) is one of the most important factors affecting the transmission performance in FMFs.

Currently two different primary types of refractive index profile have been considered for the design of FMFs as shown in Table 8.1: high-DGD step-index FMFs and low-DGD graded-index FMFs [15, 16]. In the ray-optic analogy, light rays travelling down fibres with a step-index profile follow different optical paths along the axis of the fibre. Generally higher-order modes (HOMs) have larger propagation angles and therefore travel greater distances (or at slower group velocity) than the lower-order modes (LOMs), resulting in a significant amount of intermodal dispersion. The typical DGD value for a step-index FMF is a few ns/km and the number of taps required for MIMO processing (which is proportional to the DGD induced spread of a single data bit)

Table 8.1 Two different types of index profiles used in FMF design

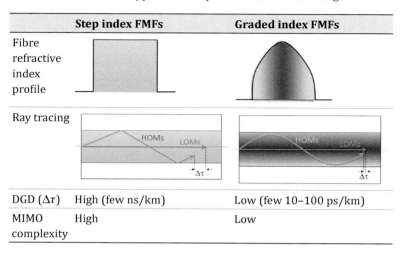

	Step index FMFs	Graded index FMFs
Fibre refractive index profile		
Ray tracing		
DGD ($\Delta\tau$)	High (few ns/km)	Low (few 10–100 ps/km)
MIMO complexity	High	Low

becomes increasingly challenging for transmission distances much above 10 km. Controlling the overall DGD in FMFs is therefore a primary design consideration and a large body of research work has been devoted to develop FMF designs providing substantially reduced DGD. Using a graded-index FMF design featuring an accurately parabolic refractive index profile, the DGD value can be efficiently reduced. As shown in Table 8.1 (right), the HOMs travelling in outer regions of the core (i.e. relatively low refractive index regions) will travel faster than the LOMs travelling in the central region of the core (i.e. high refractive index region). Therefore, the transit time for both HOMs and LOMs can be similar and the DGD value (or modal dispersion) can be significantly reduced. The index profile of the graded-index FMF is commonly characterized by a power-law index profile (i.e. α-profile governing the shape of the graded-index core) and the optimum α-value (i.e. nearly parabolic profile, $\alpha \sim 2$) can be selected to get the lowest possible DGD between the mode groups. Typical DGD values in graded-index FMFs are around a few 10–100 ps/km, which is one or two orders of magnitude smaller than those of step-index FMFs. However, it is very challenging to fabricate low DGD graded-index FMFs because the fibre DGD value is known to be very sensitive to small index profile variations during the fabrication process.

Therefore, an alternative (or complementary) approach of DGD compensation has recently been demonstrated. DGD compensation is implemented by constructing transmission lines comprising lengths of FMFs with opposite signs (positive and negative) of DGD. Link averaged net values of DGD as low as approximately 5 ps/km have been achieved in this way. In addition, trench assisted index profiles (i.e. incorporation of a low index trench surrounding the core) can be added to FMF designs, effectively improving light confinement within the core and thereby improving the macro-bend performance and optical loss of the HOMs. With the trench assisted FMF design, the mode dependent loss (MDL) is almost negligible and it opens up the possibility of transmission over more than 1000 km length scales when combined with the use of a suitable SDM optical amplifier.

8.2.2 Multi-Core Fibres

A multi-core fibre incorporates multiple cores in a single fibre cladding to provide multiple spatial channels within a single fibre. In a MCF-based SDM transmission, the inter-core crosstalk (XT) is one of the most important design considerations and sufficiently low XT MCF is crucial so that the individual cores can be considered as effectively independent information channels over the desired propagation distances and for the particular modulation formats used. Considerable research effort has been devoted to realize such fibres and different techniques have been used to minimize the XT for different MCF structures [17, 18], as summarized in Fig. 8.3. The simplest MCF approach (referred to as homogeneous MCF) uses an identical design for each core (Fig. 8.3a and the core density is dominated by the core-to-core distance (i.e. core pitch, Λ)). Since the phase velocity of each core is identical, coupling between adjacent cores is very strong and this can only be suppressed by keeping the cores relatively well separated (to minimize field overlap between neighbouring cores). Typical core spacing in homogenous MCFs with standard SMF core designs are more than 45 μm and it is difficult to substantially increase the number of independent cores within practical fibres of this form beyond 7. However, using trench assisted index profiles (i.e. similar to those discussed above in relation to FMFs) can be used within homogeneous MCFs to enhance mode confinement and to help suppress inter-core XT. More than 20–30 dB XT reduction can be readily achieved by engineering the trench volume. Due to the reduced XT, the core spacing can be reduced to 30–40 µm in trench-assisted homogeneous MCFs (Fig. 8.3b) and impressively low XT levels (e.g. less than –90 dB/km) have been demonstrated, thereby enabling independent parallel transmission over multi-1000 km length scales. Another common approach for reducing the XT of MCF is to employ a heterogeneous core arrangement with different types of cores, each having slightly different effective indices but similar effective mode areas. Due to the dissimilar phase matching conditions between neighbouring cores, the XT can be significantly reduced using heterogeneous core arrangements (Fig. 8.3c); however, it comes at the expense of increased fibre fabrication complexity. Another MCF fibre design consideration is the cladding thickness and the outer diameter of the fibre. Generally, the outermost cores in MCFs experience additional attenuation caused by leakage of the light field into the polymer coating as the fibre cladding thickness is reduced. Typically the cladding thickness should be greater than 30 µm due to micro-bend loss considerations. The choice of maximum fibre outer diameter is related to fibre failure probability (i.e. mechanical fibre reliability) and cladding diameters beyond approximately 250 µm are not considered practical for MCF-based SDM transmission. The research to date seems to indicate that the maximum number of independent cores that one can practically envisage using for long-haul transmission lies somewhere in the range 12–19, although fibres with as many as 32 cores and impressively low levels of XT of less than –40 dB over 100 km have just recently been demonstrated exploiting heterogeneous core arrangements [7]. Note that the current frontier in SDM research is to combine multiple SDM approaches to achieve much higher levels of spatial channel count (referred to as channel multiplicity). For example, both MCF and FMF technologies can be combined together (e.g. few-mode multi-core fibres) to support a total of $M \times N$ spatial channels. Excellent progress has been made and just recently data transmission at a record spectral efficiency of 345 bit/s/Hz was reported through a 9.8 km few-mode MCF containing 19 cores with each core supporting six spatial modes, providing a total of 114 distinguishable spatial channels [10].

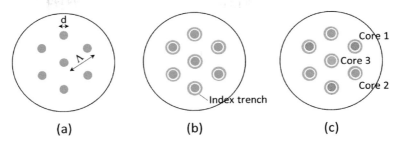

Figure 8.3 Different types of multi-core fibres to reduce the inter-core crosstalk: (a) Homogeneous MCFs, (b) Trench-assisted homogeneous MCFs. (c) Trench-assisted heterogeneous MCFs.

8.3 SDM Multiplexers and Demultiplexers

Other key optical components for SDM system are efficient and stable SDM multiplexers (MUXs) and demultiplexers (DEMUXs), which allow for individual spatial modes (or cores) to be accessed with minimal loss and crosstalk. A variety of technologies and approaches are being explored to combine and separate all spatial channels in SDM fibres. The following two subsections provide a brief summary of the most common techniques used for the two major classes of SDM fibre: (1) mode MUXs/DEMUXs for few-mode fibres and (2) fan-in/fan-out devices for multi-core fibres.

8.3.1 Mode MUXs/DEMUXs for Few-Mode Fibres

Mode MUXs can either couple light from multiple single mode fibres directly to specific guided modes of a FMF, or couple light to a linear combination of modes. The options for mode MUXs depend on the coupling amongst the guided modes. For transmission systems relying on weak mode coupling, selective mode excitation methods such as phase plates, spatial light modulators or mode selective fibre couplers are commonly used. For transmission systems where mode mixing can be tolerated, photonic lanterns, 3D waveguides and multi-spot options are suitable. The most commonly used mode MUX options are summarized in Table 8.2.

Table 8.2 The most commonly used mode multiplexers for FMFs

	Phase plates	3D waveguides	Photonic lanterns
Schematics			
Fabrication complexity	Simple	Moderate	Simple
Insertion loss	High	Moderate	Low
MDL	High	Moderate	Low
Scalability	Difficult	Good	Good
Footprint size	Large	Compact	Compact
Cost	Expensive	Moderate	Good

The first example illustrates a phase plate based mode multiplexer. Binary bulk phase plates are placed between two free space collimators to modify the spatial phase distribution of the input Gaussian beams originating from SMFs in order to efficiently excite higher-order modes in FMF. It is one of the easiest ways to enable the desired higher-order mode excitation and the modal purity is typically greater than 25 dB [19]. However, the MUX includes a number of 50/50 beam splitters to combine beams and the overall device is thus inherently bulk with a high insertion loss that scales badly with increased mode count. The second approach is a three-dimensional (3D) waveguide based mode MUX. A tightly focused femtosecond laser beam is used to modify the refractive index of a transparent bulky glass sample to enable 3D waveguide geometries with subwavelength precision by direct laser writing [20]. In this type of mode MUX, the mode transformation is accomplished by transitioning from a 1D array of single mode waveguides into a 2D coupled-core arrangement of waveguides supporting super-modes that can then be coupled to the FMF. In principle, the approach relies upon the same concept as the multi-spot launcher [21] where each beam spot excites a linear combination of modes but a special geometrical spot arrangement is required to excite a near-orthogonal combination of modes. These spot launchers can scale to a large number of modes without a significant increase in insertion loss. The third example is a photonic lantern [22]. Here, a bundle of N separate single mode fibres is inserted into a low index glass capillary tube and tapered together to obtain a multi-mode fibre structure supporting N spatial modes. Due to the adiabatic nature of the taper transition, this device is theoretically lossless and can scale to a large number of modes with negligible MDL.

8.3.2 Fan-In/Fan-Out Devices for Multi-Core Fibres

In the case of MCF, spatial multiplexers/demultiplexers are referred to as fan-in/fan-out (FIFO) devices and are used to efficiently couple light from individual SMFs to each core of the MCF and vice versa. Various configurations have been reported so far but three primary techniques are widely used as listed in Table 8.3. Note that FIFO device fabrication processes are very similar to those of the mode MUX/DEMUX. A critical feature here is that

Table 8.3 The most commonly used fan-in/fan-out devices for MCFs

	3D waveguides	Fused taper	Free space optics
Schematics			
Fabrication complexity	Moderate	Simple	Complex
Insertion loss	Moderate	Low	Low
Crosstalk	Moderate	Moderate	Low
Scalability	Good	Good	Moderate
Footprint size	Compact	Compact	Moderate
Cost	Moderate	Good	Expensive

at high core counts some level of cross-talk becomes inevitable and an effective means to achieve cross-talk reduction in FIFO devices is one of the most important factors limiting MCF transmission performance. Moreover, hybrid SDM multiplexers (i.e. for few-mode multi-core fibre) have also recently been reported to further increase spatial multiplicity by integrating both technologies [23].

8.4   SDM Optical Amplifiers

While various SDM transmission fibres have been demonstrated so far, the question naturally arises, what progress has been made in the suitable partner amplifiers required to enable long haul transmission? It is perhaps not surprising but fair to say that SDM amplifier research has lagged one step behind the advances in passive SDM transmission fibre technology as shown in Fig. 8.4—as soon as SDM amplifier research starts to address the particular optical amplification needs of one form of SDM fibre, a new generation of high-density SDM transmission fibres appears on the horizon. In the case of multi-core EDFAs (MC-EDFAs), both core-pumped and cladding-pumped 7-core EDFAs [24] have been demonstrated in 2011 and 2012, respectively. The core pumped MC-EDFA was then scaled to 19-cores in 2103 and the cladding pumped MC-EDFA to 32-cores in 2017 [25], which represents the highest core-density MC-EDFA to date. Improved sharing of the optical components and significant device integration was also demonstrated as required to obtain the anticipated cost reduction benefits of SDM. For example, a single free-space isolator was used to simultaneously prevent unwanted reflections from all 32 cores and two side pump couplers were used to couple the pump light into each individual active MCF core. In the case of few-mode erbium doped fibre amplifiers (FM-EDFAs), both core-pumped [26] and cladding-pumped 6-mode EDFAs [27] were demonstrated in 2014. More recently a cladding pumped 10-mode EDFA [28] was reported representing the highest mode count FM-EDFA reported so far. The cladding-pumped FM-EDFA, which was originally end-pumped using free space optics, has recently been upgraded to incorporate a fully fiberized side-pumping configuration and this functionality makes it possible to construct a fully integrated SDM amplifier capable of

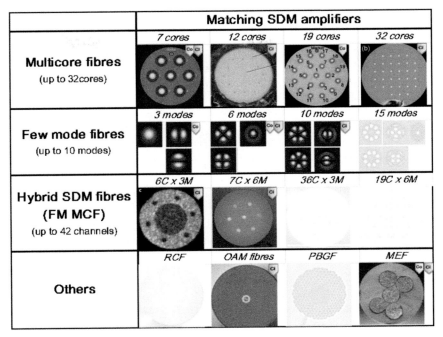

	Matching SDM amplifiers			
Multicore fibres (up to 32cores)	*7 cores*	*12 cores*	*19 cores*	*32 cores*
Few mode fibres (up to 10 modes)	*3 modes*	*6 modes*	*10 modes*	*15 modes*
Hybrid SDM fibres **(FM MCF)** (up to 42 channels)	*6C x 3M*	*7C x 6M*	*36C x 3M*	*19C x 6M*
Others	*RCF*	*OAM fibres*	*PBGF*	*MEF*

Figure 8.4 The corresponding SDM optical amplifiers. Faded images are used to indicate instances where suitable partner amplifier solutions have yet to be demonstrated.

providing stable modal amplification without the need for free-space optics. In order to realize a high-capacity hybrid SDM system a hybrid SDM amplifier supporting few-mode multi-core fibres represents an essential subsystem and a six-mode seven-core SDM amplifier [29] was recently demonstrated supporting 42 spatial channels. Other novel forms of SDM amplifier, e.g. ring-core fibre amplifiers [30], OAM fibre amplifiers [31] and MEF amplifiers [14], have been introduced. Most interestingly, ring-core fibre amplifiers can be designed to support single radial modes (i.e. LP_{m1} modes where m is an integer) and can provide exceptionally well-equalized modal gain (nearly identical gain) for all guided signal modes. These amplifiers are very attractive as few-mode EDFAs in terms of differential modal gain (DMG) mitigation. A fibre optic OAM amplifier capable of robustly amplifying modes with a helical phase front has also recently been demonstrated [31] and has attracted significant scientific interest not only in optical communications but also for optical trapping, laser material processing and quantum information processing. There is no report yet of matching optical amplifiers for HC-PBGFs due to the air-hole structure. Both core-pumped and cladding-pumped MEF amplifiers [14] have also been demonstrated with up to 7 fibre elements in a single assembly. In the following section, we will focus on the development of few-mode EDFAs offering low DMG between all supported modes. In FM-EDFAs, DMG directly affects the system outage probability and can limit system reach and we will discuss in detail methods to minimize it. We will concentrate in particular on the case of the 6-mode EDFA and will discuss DMG control for both core and cladding-pumped configurations. Finally, we review the results of a comprehensive theoretical study to show that it may be possible to achieve lower DMG over the full C-band of EDFAs supporting 10 spatial modes by optimizing the erbium dopant distribution of the active fibre in a cladding pumped configuration.

8.4.1 Strategies to Minimize Differential Modal Gain in Few-Mode EDFA

In FM-EDFAs, the most important factor is the DMG control required to ensure that all spatial channels are equally amplified. As shown in Fig. 8.5, the DMG is a function of the overlap integrals between the doping distribution, the pump mode profile and the signal mode profile. If we assume that the input signal mode profiles are fixed

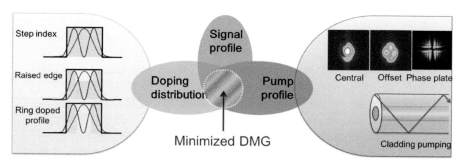

Figure 8.5 Strategies to minimize DMG in few-mode EDFAs. The DMG can be engineered by (i) tailoring the radial erbium-doping concentration profile of the erbium-doped fibre, (ii) controlling the pump field intensity distribution, and (iii) controlling the signal mode.

(i.e. the refractive index profile of the fibre remains constant) then there are two main strategies to minimize the DMG: (i) tailoring the radial erbium-doping concentration profile of the erbium doped fibre and (ii) controlling the pump field intensity distribution [32, 33]. With respect to the approach of controlling the doping distribution within the active fibre, we begin by considering a simple uniformly doped step-index EDF. While such a fibre can amplify multiple modes, it preferentially amplifies the LP_{01} mode compared to the HOMs. Indeed >10 dB MDG has been reported for uniformly doped step-index 3M-EDFAs [32]. To address this both raised-edge and ring-doped erbium dopant profiles have been introduced as a means to reduce the overlap of the dopant with the field of the fundamental mode in the central core region [26, 34]. Confining the erbium ions to lie within a ring around the fibre axis can be particularly effective and can drastically enhance the HOM gain. Pump mode control can also be used to control the DMG. On axis, launching of pump light into the active fibre generally creates an LP_{01}-like pump intensity profile which also tends to result in preferential amplification of the LP_{01} signal mode. In contrast launching the pump offset from the fibre axis or launching the pump light into well-defined HOMs tends to result in preferential gain for HOMs. More recently, the cladding pumped configuration has been employed to minimize the impact of the pump mode profile and to offer a cost-effective alternative to core-pumped variants. Cladding pumping allows low cost, high power multi-mode pumps to be used, and offers performance improvements, scalability and simplicity to FM-EDFA design. In practice, a combination of both strategies (i.e. dopant distribution control and pump mode profile engineering) is generally required in the case of core-pumped FM-EDFAs. Dopant distribution control is the only possible option for controlling DMG in cladding-pumped FM-EDFAs.

8.4.2 Core Pumped 6-Mode EDFA

Figure 8.6a shows a schematic of a core-pumped 6-mode EDFA (6M-EDFA) [26] for simultaneous amplification of the six lowest order spatial modes (i.e. LP_{01}, $LP_{11a,b}$, $LP_{21a,b}$ and LP_{02}).

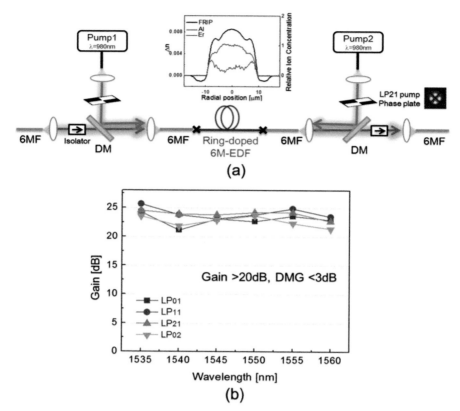

Figure 8.6 (a) The schematic of a core-pumped 6M-EDFA and (b) its gain performance.

Two dichroic mirrors are used for combining 980 nm pump light with the 1550 nm input signals and two polarization-insensitive free-space isolators are included to prevent unwanted parasitic lasing within the 6M-EDFA. In order to mitigate the DMG, a tailored ring-doped EDF was fabricated to provide dopant distribution control and bi-directional HOM pumping (e.g. a LP_{21} pump mode in this particular experiment) was used to provide pump mode control. As shown in Fig. 8.6b, the amplifier provides more than 20 dB signal gain for all six spatial modes with less than 3 dB DMG across the C-band. Note that the use of HOM pump profiles enables us to mitigate the DMG of the FM-EDFA. The pump coupling losses to the 6-mode fibre (6MF) depend on the particular pump mode under consideration with HOMs experiencing higher coupling loss. In our experiment an LP_{21} pump mode profile was used and a 3.5 dB coupling loss was observed. Therefore a single pump LD with a maximum output power of 750 mW cannot provide enough gain for all the spatial modes meaning we need to use two pump LDs in a bidirectional pump configuration to get acceptable signal gain. As the number of spatial modes is further scaled up to 10 and beyond, it will become increasingly challenging to meet the associated pump power requirements with single-mode pump diodes. For this reason, a cladding-pumped scheme is attractive to address such issues.

8.4.3   Cladding Pumped 6-Mode EDFA

Figure 8.7a shows a schematic of our cladding-pumped 6M-EDFA [27]. A D-shaped double-clad 6M-EDF was fabricated and coated with a low index polymer such that it guides pump light in the glass cladding. A side-pumping scheme was adopted to realize a fully fiberized integrated SDM amplifier. A multi-mode pump delivery fibre was tapered down to a central uniform waist of 10–20 µm and this taper was then wound around a short stripped section of the active fibre with a slight tension to couple pump light efficiently into the glass cladding. A pump coupling efficiency of more than 70% can easily be achieved. A UV-curable low-index acrylate polymer was then applied to the tapered section to ensure robust stable optical contact within the pump combiner. Residual pump power at the output of the amplifier was removed by applying high-index polymer to a further section of stripped fibre and removing the associated heat generated. As shown in Fig. 8.7b, more than 20 dB of average modal gain was successfully achieved across the C-band with a DMG of ~3 dB amongst the mode groups and a corresponding noise figure of 6–7 dB. The amplifier performance could be further improved by optimizing the core dopant distribution and by reducing the core-to-clad area ratio. We consider this to be an important step in increasing the mode scalability of the few-mode EDFA, which offers cost effective and efficient amplification of a large number of spatial data channels in a single device.

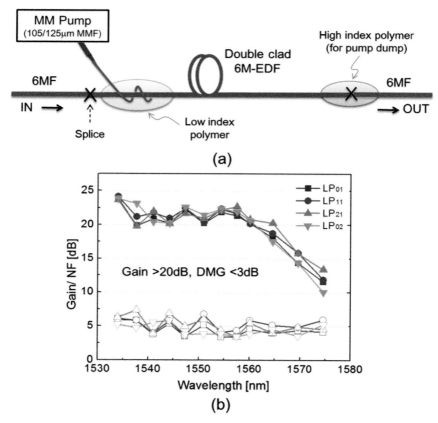

Figure 8.7 (a) The schematic of a cladding-pumped 6M-EDFA and (b) the associated gain performance.

8.4.4 Future Prospects to Increase the Number of Spatial Modes in FM-EDFAs

Scalability to a larger number of modes is an essential feature for the penetration of SDM in next generation optical transmission systems. In this section, we present a cladding pumped FM-EDFA design supporting 10 spatial modes with a multi-layer ring-doped erbium profile [35]. A Genetic Algorithm was employed to minimize the DMG over all supported signal modes (at the same wavelength). The optimal EDFA designs found through the algorithm provided less than 1 dB DMG across the C-band (1530–1565 nm) while achieving more than 20 dB gain per mode. As shown in Fig. 8.8a, the optimum 10M-EDF design exhibits a step-index fibre refractive index profile and a complex four-layer erbium dopant profile. For the purpose of reducing fabrication complexity, the number of erbium-doped ring layers should be minimized but we found that at least four-layer multiple ring structures are required in order to achieve a DMG of less than 1 dB across the C-band.

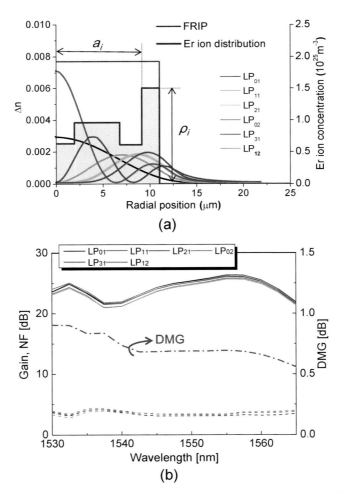

Figure 8.8 (a) Erbium doping distribution and refractive index profile of the cladding-pumped 10M-EDF and (b) the calculated gain and noise figure spectra.

The diameter of the inner-cladding was chosen to be 70 µm, compatible with our preferred choice of pigtailed pump diode. As shown in Fig. 8.8b, the averaged modal gain is well above 20 dB and NFs are found to be less than 5 dB. Note that FM-EDFA designs supporting more than 10 spatial modes will certainly become more complex and it will be very challenging to realize multiple ring-doped profiles with accurate radial control of the dopant distribution using conventional modified chemical vapour deposition and solution doping techniques. Also, dopant diffusion is another limiting factor in the fabrication of such a complex multi-layer structure because some degree of dopant diffusion is inevitable during the fibre fabrication process due to the heat treatment during tube collapsing and subsequent fibre drawing. In this respect, rather than targeting complex fibre doping profiles with low DMG (<1 dB), adopting a less complicated erbium doping profile offering moderate DMG (2–5 dB) and using this in conjunction with a spatial mode filtering device offering controllable modal loss is perhaps a more realistic approach. The filter (fixed or preferably dynamic) could be inserted in the middle of a dual-stage FM-EDFA and the total DMG can effectively be improved without noticeable NF degradation in an analogous manner to the gain-flattening filters used in many current single mode amplifiers.

8.5 Conclusion

Over the past few years, major advances in SDM technology have been made in both few-mode and multi-core fibres. Both of these approaches are being investigated around the globe and a 10-fold increase in overall fibre capacity has already been achieved. Furthermore, a heterogeneous SDM fibre structure (e.g. few-mode multi-core fibre) has been introduced to further increase the spectral efficiency in a single fibre by combining both technologies and has demonstrated up to 100-fold capacity increase. However, although these point-to-point transmission results are highly encouraging, in order for SDM to be exploited commercially, several key subsystems and optical components are still required to provide network functionality, while component integration will be essential to provide practical, field-deployable devices.

Acknowledgements

The authors acknowledge many helpful discussions with their collaborators on the European Union Framework 7 (FP7) funded project MODEGAP (258033) and H2020 program SAFARI (642928), and the UK Engineering and Physical Science Research Council (EPSRC) funded projects HYPERHIGHWAY (EP/I01196X/1) and COMIMO (EP/EP/J008591/1), which have helped them develop their understanding of the various aspects of SDM technology.

References

1. Ellis, A. D., Jian Zhao, and Cotter, D. (2010). Approaching the non-linear Shannon limit. *Journal of Lightwave Technology*, 28(4), pp. 423–433.

2. Essiambre, R.-J., Kramer, G., Winzer, P. J., Foschini, G. J., and Goebel, B. (2010). Capacity limits of optical fiber networks. *Journal of Lightwave Technology*, 28(4), pp. 662–701.

3. Richardson, D. J., Fini, J. M., and Nelson, L. E. (2013). Space-division multiplexing in optical fibres. *Nature Photonics*, 7, pp. 354–362.

4. Winzer, P. J. (2014). Making spatial multiplexing a reality. *Nature Photonics*, 8(5), pp. 345–348.

5. Richardson, D. J. (2016). New optical fibres for high-capacity optical communications Subject Areas: *Philosophical Transactions of the Royal Society A*, 374, pp. 20140441.

6. Takara, H., Sano, A., Kobayashi, T., Kubota, H., Kawakami, H., Matsuura, A., ... Morioka, T. (2012). 1.01-Pb/s (12 SDM/222 WDM/ 456 Gb/s) crosstalk-managed transmission with 91.4-b/s/Hz aggregate spectral efficiency. In *European Conference and Exhibition on Optical Communication* (p. Th.3.C.1). Washington, D.C.: OSA.

7. Mizuno, T., Shibahara, K., Ono, H., Abe, Y., Miyamoto, Y., Ye, F., ... Yamada, M. (2016). 32-core dense SDM unidirectional transmission of PDM-16QAM signals over 1600 km using crosstalk-managed single-mode heterogeneous multicore transmission line. In *Optical Fiber Communication Conference Postdeadline Papers* (p. Th5C.3). Washington, D.C.: OSA.

8. Fontaine, N. K., Ryf, R., Chen, H., Benitez, A. V., Guan, B., Scott, R., ... Amezcua-Correa, R. (2015). 30×30 MIMO transmission over 15 spatial

modes. In *Optical Fiber Communication Conference Post Deadline Papers* (p. Th5C.1). Washington, D.C.: OSA.

9. Sakaguchi, J., Klaus, W., Delgado Mendinueta, J. M., Puttnam, B. J., Luis, R. S., Awaji, Y., ... Kobayashi, T. (2015). Realizing a 36-core, 3-mode fiber with 108 spatial channels. In *Optical Fiber Communication Conference Post Deadline Papers* (p. Th5C.2). Washington, D.C.: OSA.

10. Igarashi, K., Souma, D., Wakayama, Y., Takeshima, K., Kawaguchi, Y., Tsuritani, T., ... Suzuki, M. (2015). 114 space-division-multiplexed transmission over 9.8-km weakly-coupled-6-mode uncoupled-19-core fibers. In *Optical Fiber Communication Conference Post Deadline Papers* (p. Th5C.4). Washington, D.C.: OSA.

11. Feng, F., Guo, X., Gordon, G. S., Jin, X., Payne, F., Jung, Y., ... Wilkinson, T. D. (2016). All-optical mode-group division multiplexing over a graded-index ring-core fiber with single radial mode. In *Optical Fiber Communication Conference* (p. W3D.5). Washington, D.C.: OSA.

12. Bozinovic, N., Yue, Y., Ren, Y., Tur, M., Kristensen, P., Huang, H., ... Padgett, M. J. (2013). Terabit-scale orbital angular momentum mode division multiplexing in fibers. *Science (New York, N.Y.)*, 340(6140), pp. 1545–1548.

13. Sleiffer, V. A. J. M., Jung, Y., Baddela, N. K., Surof, J., Kuschnerov, M., Veljanovski, V., ... de Waardt, H. (2014). High capacity mode-division multiplexed optical transmission in a novel 37-cell hollow-core photonic bandgap fiber. *Journal of Lightwave Technology*, 32(4), pp. 854–863.

14. Jain, S., Rancaño, V. J. F., May-Smith, T. C., Petropoulos, P., Sahu, J. K., and Richardson, D. J. (2014). Multi-element fiber technology for space-division multiplexing applications. *Optics Express*, 22(4), pp. 3787–3796.

15. Grüner-Nielsen, L., Sun, Y., Nicholson, J. W., Jakobsen, D., Jespersen, K. G., Lingle, R., and Pálsdóttir, B. (2012). Few mode transmission fiber with low DGD, low mode coupling, and low loss. *Journal of Lightwave Technology*, 30(23), pp. 3693–3698.

16. Sillard, P. (2015). Next-generation fibers for space-division-multiplexed transmissions. *Journal of Lightwave Technology*, 33(5), pp. 1092–1099.

17. Saitoh, K., and Matsuo, S. (2016). Multicore fiber technology. *Journal of Lightwave Technology*, 34(1), pp. 55–66.

18. Mizuno, T., Takara, H., Sano, A., and Miyamoto, Y. (2016). Dense space-division multiplexed transmission. *Journal of Lightwave Technology*, 34(2), pp. 582–592.

19. Ryf, R., Randel, S., Gnauck, A. H., Bolle, C., Sierra, A., Mumtaz, S., ... Lingle, R. (2012). Mode-division multiplexing over 96 km of few-mode fiber using coherent 6×6 MIMO processing. *Journal of Lightwave Technology*, 30(4), pp. 521–531.

20. Thomson, R. R., Bookey, H. T., Psaila, N. D., Fender, A., Campbell, S., MacPherson, W. N., ... Kar, A. K. (2007). Ultrafast-laser inscription of a three dimensional fan-out device for multicore fiber coupling applications. *Optics Express*, 15(18), pp. 11691–11697.

21. Ryf, R., Mestre, M. A., Gnauck, A., Randel, S., Schmidt, C., Essiambre, R., ... Lingle, R. (2012). Low-loss mode coupler for mode-multiplexed transmission in few-mode fiber. In *Optical Fiber Communication Conference* (p. PDP5B.5). Washington, D.C.: OSA.

22. Leon-Saval, S. G., Argyros, A., and Bland-Hawthorn, J. (2013). Photonic lanterns. *Nanophotonics*, 2(5-6), pp. 429–440.

23. van Uden, R. G. H., Correa, R. A., Lopez, E. A., Huijskens, F. M., Xia, C., Li, G., ... Okonkwo, C. M. (2014). Ultra-high-density spatial division multiplexing with a few-mode multicore fibre. *Nature Photonics*, 8(11), pp. 865–870.

24. Abedin, K. S., Taunay, T. F., Fishteyn, M., DiGiovanni, D. J., Supradeepa, V. R., Fini, J. M., ... Dimarcello, F. V. (2012). Cladding-pumped erbium-doped multicore fiber amplifier. *Optics Express*, 20(18), pp. 20191–20200.

25. Jain, S., Castro, C., Jung, Y., Hayes, J., Sandoghchi, R., Mizuno, T., ... Richardson, D. J. (2017). 32-core erbium/ytterbium-doped multicore fiber amplifier for next generation space-division multiplexed transmission system. *Optics Express*, 25(26), 32887–32896.

26. Jung, Y., Kang, Q., Sahu, J. K., Corbett, B., O'Callagham, J., Poletti, F., ... Richardson, D. J. (2014). Reconfigurable modal gain control of a few-mode EDFA supporting six spatial modes. *IEEE Photonics Technology Letters*, 26(11), pp. 1100–1103.

27. Jung, Y., Lim, E. L., Kang, Q., May-Smith, T. C., Wong, N. H. L., Standish, R., ... Richardson, D. J. (2014). Cladding pumped few-mode EDFA for mode division multiplexed transmission. *Optics Express*, 22(23), pp. 29008–29013.

28. Fontaine, N. K., Huang, B., Sanjabieznaveh, Z., Chen, H., Jin, C., Ercan, B., ... Amezcua Correa, R. (2016). Multi-mode optical fiber amplifier supporting over 10 spatial modes. In *Optical Fiber Communication Conference* (p. Th5A.4). Washington, D.C.: OSA.

29. Jung, Y., Wada, M., Sakamoto, T., Jain, S., Davidson, I. A., Barua, P., ... Richardson, D. J. (2019) High Spatial Density 6-Mode 7-Core Multicore

L-Band Fiber Amplifier. *In Optical Fiber Communication Conference* (p. Th1B.7). Washington, D.C.: OSA.

30. Kang, Q., Lim, E., Jun, Y., Jin, X., Payne, F. P., Alam, S., and Richardson, D. J. (2014). Gain equalization of a six-mode-group ring core multimode EDFA. In *2014 The European Conference on Optical Communication (ECOC)* (p. P.1.14). IEEE.

31. Jung, Y., Kang, Q., Yoo, S., Raghuraman, S., Ho, D., Gregg, P., ... Richardson, D. J. (2016). Optical orbital angular momentum amplifier based on an air-core erbium doped fiber. In *Optical Fiber Communication Conference Postdeadline Papers* (p. Th5A.5). Washington, D.C.: OSA.

32. Jung, Y., Alam, S., Li, Z., Dhar, A., Giles, D., Giles, I. P., ... Richardson, D. J. (2011). First demonstration and detailed characterization of a multimode amplifier for space division multiplexed transmission systems. *Optics Express*, 19(26), pp. B952–B957.

33. Bai, N., Ip, E., Wang, T., and Li, G. (2011). Multimode fiber amplifier with tunable modal gain using a reconfigurable multimode pump. *Optics Express*, 19(17), pp. 16601–16611.

34. Jung, Y., Kang, Q., Sleiffer, V. A. J. M., Inan, B., Kuschnerov, M., Veljanovski, V., ... Richardson, D. J. (2013). Three mode Er^{3+} ring-doped fiber amplifier for mode-division multiplexed transmission. *Optics Express*, 21(8), pp. 10383–10392.

35. Kang, Q., Lim, E.-L., Jung, Y., Poletti, F., Baskiotis, C., Alam, S., and Richardson, D. J. (2014). Minimizing differential modal gain in cladding-pumped EDFAs supporting four and six mode groups. *Optics Express*, 22(18), pp. 21499–21507.

36. Sleiffer, V. A. J. M., Jung, Y., Veljanovski, V., van Uden, R. G. H., Kuschnerov, M., Chen, H., ... de Waardt, H. (2012). 73.7 Tb/s (96 × 3 × 256-Gb/s) mode division multiplexed DP-16QAM transmission with inline MM-EDFA. *Optics Express*, 20(26), pp. B428–B438.

Chapter 9

Spatial Multiplexing: Modelling

Filipe Ferreira, Christian Costa, Sygletos Stylianos, and Andrew Ellis

Aston Institute of Photonic Technologies University,
Aston University, Birmingham, B47ET, UK

f.ferreira@aston.ac.uk

Spatial-division multiplexing has been proposed as a next-generation solution to overcome the imminent exhaustion of the capacity of current single-mode fiber based systems. However, these systems presented additional challenges such as the overall group-delay spread due to differential spatial-path delay and linear spatial-path coupling, and inter-mode nonlinear effects. The accurate modelling of these effects is preponderant on the performance optimization of mode-division multiplexing systems. This chapter reviews a method for the semi-analytical modelling of linear mode coupling. Simulations using this model matched the analytical predictions for the statistics of group-delays in few-mode fiber links, considering different coupling regimes with and without mode delay management. Furthermore, this chapter reviews the study of nonlinear performance of few-mode fiber

Optical Communication Systems: Limits and Possibilities
Edited by Andrew Ellis and Mariia Sorokina
Copyright © 2020 Jenny Stanford Publishing Pte. Ltd.
ISBN 978-981-4800-28-0 (Hardcover), 978-981-4800-28-0 (eBook)
www.jennystanford.com

links operating in all different linear coupling regimes and mode delay maps. The optimum link configurations minimizing the nonlinear penalty at practical levels of equalization complexity are presented. Finally, the limits of the extension of the Manakov approximation to the multi-mode case are accurately validated against a fully stochastic model developed considering distribution linear mode coupling.

9.1 Introduction

Mode-division multiplexing (MDM) over few-mode fibers (FMFs) is emerging as an attractive solution to overcome the capacity limit of single-mode fibers (SMFs) [1, 2]. However, the multitude of guided modes introduces new impairments that have to be addressed in order to reach FMFs' full capacity, namely: group delay (GD) spread [3–6] given the interplay between differential mode delay (DMD) and linear mode coupling (XT), and inter-mode nonlinear effects [7–10].

To correctly estimate the GD spread and the performance of a MDM equalizer, the mode coupling arising from the waveguide imperfections [3], need to be correctly modelled. Thereby, intense research has been accomplished to study the statistics of GDs analytically [4, 11–14] and numerically [15–19]. A significant number of works assume systems operating in the strong mode coupling regime, e.g. [11, 14], and consider a multi-section model where mode coupling is introduced through random unitary matrices each section, where the length of each section must be longer than the correlation length. However, few-mode fibers ([20–23]) usually operate in the weak or intermediate coupling regime for transmission distances 100–1000 km. Even though strong mode coupling can be assumed within groups of degenerate modes [24], the mode coupling between groups of nondegenerate modes cannot either be considered negligible or strong. Note that, nonlinear simulation requires a step-size much smaller than nonlinear effective length (~20 km) [25]; thus, the generation of coupling matrices with the appropriate level for 10–100 m is required. Therefore, models considering random unitary matrices do not cover many of the cases of interest.

To model systems operating in the weak and intermediate coupling regime, the introduction of coupling in the form of misaligned fiber splices in each section of a multi-section model was proposed [16]. In this case, the mode coupling matrices are obtained using an overlap integral approach. However, the matrix elements obtained this way present two limitations. First, even though the coefficients are effective in describing the mode power distribution, they fail to consider phase effects thereby appropriate only for incoherent sources 0. Second, the coupling elements inevitably include mode dependent loss given the nature of the overlap integral. Even though splices do introduce mode dependent loss, splices are here being used as a discrete representation of continuous imperfections which may introduce or not introduce mode dependent loss. Therefore, a model able to separate mode coupling from mode dependent loss is preferable. Recently, the authors have presented a semi-analytical model capable of describing the linear mode coupling for fibers operating in the intermediate coupling regime [17, 18]. Such method was demonstrated to match the analytical predictions for group-delay in few-mode fiber links [26, 27].

In the nonlinear regime, it has been shown that MDM systems performance can be dominated by inter-mode interactions for low DMD and low XT [28], and that high XT significantly reduces intermodal nonlinear [29]. However, in the intermediate coupling regime, the nature of the dependence of the nonlinear distortion on the (distributed) XT strength and DMD has only recently started being studied [10, 30] for mode delay uncompensated and compensated links. Recently, the authors have modified the single-mode split-step Fourier method to include semi-analytical solutions for linear mode coupling of arbitrary strength. Using such model, the authors were able to accurately study for the first time the nonlinear distortion in FMFs operating in the intermediate coupling regime [10, 30].

This chapter presents the derivation of a semi-analytical solution method for the linear mode coupling equations, and validate the group-delay spreading predictions for different coupling regimes and different link configurations. A validation of a multi-section model using semi-analytical solutions is presented for non-GD-managed links by matching different analytical predictions for the statistics of the GDs, namely: standard deviation, probability density function, and cumulative distribution function. Furthermore, a validation of the multi-section model for GD managed links is presented by matching the analytical predictions for standard deviation of the GDs. Finally, this chapter reviews the proposed few-mode split-step Fourier method. The nonlinear performance of FMFs is studied for all different linear coupling regimes and mode delay maps, in order to find the optimum link configuration minimizing the nonlinear penalty for practical equalization complexities. Moreover, the regimes under which the extension of the single-mode Manakov approximation to the multi-mode case [8, 29] is valid are assessed, considering the transmission of wavelength multiplexed channels in each of the polarization modes over the fully stochastic model presented including distributed XT.

9.2   Coupled-Mode Theory for Few-Mode Fibers

The linear mode coupling in few-mode fibers is due to refractive-index inhomogeneities or small deviations of the core-cladding boundary caused by perturbations introduced during the fabrication process or by mechanical stresses imposed on the fiber in the field. Figure 9.1a shows a fiber dielectric waveguide with distorted core-cladding boundary. These imperfections cause the modes of the fiber to couple among each other. When exciting a pure mode at the fiber beginning, some of its power is transferred to other guided modes. This power transfer results in signal distortion because each guided mode travels at its own characteristic group velocity. Therefore, the equalization of the received signal must span over a time window that covers all the significant distortions undergone by a given information symbol.

Mode coupling may even be a desirable effect. The mode delay spread can be reduced by introducing a significant amount of distributed coupling among all guided modes which introduces a sufficiently strong averaging effect of the different mode group velocities, see Section 9.5. However, in mode delay compensated fiber links, mode coupling may or may not be desirable, as discussed in Section 9.4.

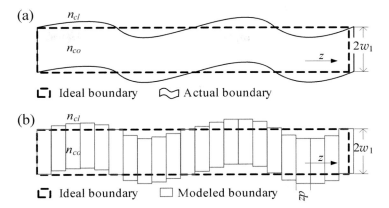

Figure 9.1 Fiber dielectric waveguide with distorted core-cladding boundary.

9.2.1 Coupled-Mode Equations

The perturbations that couple the ideal modes of the ideal waveguide can be described by variations of the dielectric tensor. That is, the perturbed dielectric tensor (ε_p) as a function of the space coordinates is written as

$$\varepsilon_p(x,y,z) = \varepsilon_u(x,y) + \Delta\varepsilon(x,y,z), \tag{9.1}$$

where $\varepsilon_u(x,y)$ is the unperturbed part of the dielectric tensor, thereby invariant with the fiber longitudinal coordinate z, and $\Delta\varepsilon(x,y,z)$ represents the dielectric perturbation, which in the general case varies with all space coordinates. Equation (9.1) can adequately describe the core-cladding perturbations in Fig. 9.1a.

If an arbitrary field of frequency ω is excited at $z = 0$, the propagation of this field in the unperturbed dielectric waveguide can be expressed as a linear combination of the ideal modes:

$$\mathbf{E}(x,y,z,t) = \Sigma A_m(z,t)\,\mathbf{E}_m(x,y)e^{j(wt-\beta_m z)}, \tag{9.2}$$

where m is the mode index, $A_m(z,t)$ is the slowly varying mode field envelope, β_{0m} is the mode propagation constant at ω, and $\mathbf{E}_m(x,y)$ is the electric field distribution.

In the presence of a dielectric perturbation $\Delta\varepsilon(x,y,z)$, the coupling between the ideal modes are described by the following coupled-mode equations [3, 4, 31]:

$$\left[\partial_z + \beta_{1m}\partial_t - j\frac{\beta_{2m}}{2}\partial_t^2 + j\frac{\beta_{3m}}{6}\partial_t^3 \cdots\right]A_m(z,t) = -j\sum_n C_{m,n}(z)A_n(z,t)e^{j(\beta_{0m}-\beta_{0n})z} \quad (9.3)$$

$$C_{m,n}(z) = \frac{\omega\varepsilon_0}{4}\int\int_{-\infty}^{+\infty}[\Delta\varepsilon(x,y,z)]\mathbf{E}_m^* \cdot \mathbf{E}_n\,dxdy, \quad (9.4)$$

where β_{lm} is the l-th order coefficient of a Taylor series expansion of $\beta_m(\omega)$ centered at the carrier frequency ω. $C_{m,n}$ are the coupling coefficients given by the area integral of the dot product of the electrical fields of mode m and mode n, over the area where the permittivity perturbation $\Delta\varepsilon(x,y,z) \neq 0$.

For the general case, where $\Delta\varepsilon$ is varying continuously with z, so is $C_{m,n}$, the solution of the coupling operator in (9.1) can only be achieved using numerical methods, e.g. Runge–Kutta method. However, the usage of these methods is computationally inviable for simulation of long-haul transmission links. To overcome such limitation, we propose a model that discretizes the core-cladding fluctuations by dividing the fiber in multiple sections, each with a random displacement of the core center position constant along the section. In this case, the dielectric tensor is given by

$$\varepsilon_p(x,y,z) = \varepsilon_{r0}(x+\delta x(z), y + \delta y(z), z), \quad (9.5)$$

where δx and δy are the random displacement of the abscissa and ordinate coordinates, respectively. Figure 9.1b shows a diagram of the discretization of the core-cladding fluctuations given the proposed method. In this case, each section has constant coupling coefficients. Therefore, in theory it should be possible to find (semi-) analytical solutions for the coupling operator present in (9.3).

Assuming the fiber section length is much shorter than both the dispersion length $L_D = T_0^2/|\beta_{2m}|$ and the walk-off length $L_W = T_0/|\beta_{1m} - \beta_{1n}|$, where T_0 is a measure of the pulse width, an approximate solution of (9.3) can be obtained by assuming the dispersive effects and linear coupling effects act independently. In the following, we will focus on finding a (semi-)analytical solution for the coupling operator, that is, we will be trying to solve

$$\partial_z A_m(z,t) = -j \sum_n C_{m,n}(z) A_n(z,t) e^{j(\beta_{0m} - \beta_{0n})z}.$$

(9.6)

9.2.2 Coupled-Mode Equations Solution for Two-Mode Fibers

The simple case of a two-mode fiber, where only the coupling between the LP_{01} mode ($m = 1$) and the LP_{11} mode ($n = 2$) is present, (9.6) can be solved analytically in each section [3]:

$$A_1(z) = e^{j(\Delta\beta/2)z} \left[\cos sz - j\frac{\Delta\beta}{2}\frac{\sin sz}{s} \quad -j\kappa\frac{\sin sz}{s} \right] \begin{bmatrix} A_1(0) \\ A_2(0) \end{bmatrix}$$

(9.7)

$$A_2(z) = e^{-j(\Delta\beta/2)z} \left[-j\kappa^*\frac{\sin sz}{s}\cos sz + j\frac{\Delta\beta}{2}\frac{\sin sz}{s} \right],$$

(9.8)

where $\Delta\beta = \beta_{01} - \beta_{02}$, $s^2 = \kappa\kappa^* + (\Delta\beta/2)^2$, and $\kappa = C_{12} = C_{21}^*$. From (9.7) and (9.8), it can be concluded that the coupling strength depends on the relation between $|\kappa|^2$ and $\Delta\beta^2$.

Figures 9.2a,b show the mode powers $|A_1|^2$ and $|A_2|^2$ as functions of the interaction distance z, for $\Delta\beta = 0$ and for $\Delta\beta = 4|\kappa|$, respectively, with $\kappa = \pi/2$. Figure 9.2 shows that the coupling efficiency is 100% when the phase mismatch is zero, a full power swap happens for every $|\kappa|z$ odd multiple of $\pi/2$. However, if the phase mismatch is different from zero the coupling is no longer the power coupling is incomplete, for $\Delta\beta = 2|\kappa|$ the maximum coupling is ½.

For higher number of modes, the dependence of the coupling strength on the phase mismatch and on the coupling coefficient should follow similar dependencies. Next section presents a solution method for higher number of modes.

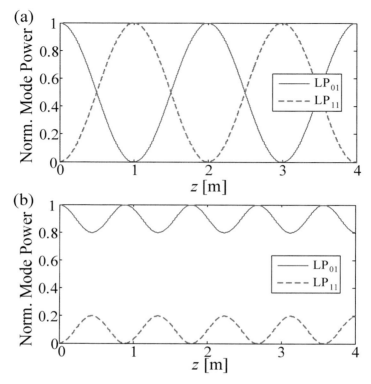

Figure 9.2 Fiber dielectric waveguide with distorted core-cladding boundary.

9.3 Semi-Analytical Solutions for Higher-Order Modes

For the simple case of a fiber with two modes, (9.6) can be easily solved by hand; however, this method becomes endless for higher number of modes. The use of a numerical method for the solution of (9.6), such as the Runge–Kutta–Fehlberg (RK45) method, is also not an option since it would be necessary to solve these equations for each fiber section with different fiber displacements, thus leading to computation times that are unaffordable in most applications. Therefore, an analytical approach is desirable.

Our approach starts by taking the Fourier Transform of (9.6) on z to avoid the complex exponentials, obtaining

$$A_m(w_z) = -\frac{j}{w_z} \sum_n C_{m,n} A_n(w_z - \Delta\beta_{m,n}), \tag{9.9}$$

where w_z is the spatial frequency and $\Delta\beta_{m,n} = \beta_{0m} - \beta_{0n}$. The system of Eq. (9.9) can be solved by substitution, thereby eliminating each A_p from all the equations for all $p \neq m$, obtaining an equation for A_m which can be written as

$$A_m(w_z) \cdot \left[a_{m,R}(jw_z)^R + \cdots + a_{m,0}(jw_z)^0 \right] = 0, \tag{9.10}$$

where R is equal to $2(M - 1)$, with M equal to the number of modes, and $a_{m,r}$ are functions of $\Delta\beta_{mn}$ and C_{mn}. The process described must be performed for $m = 1, \ldots, M$ to obtain equations like (9.10) for each mode. Now, by applying the inverse Fourier transform to (9.10), a differential equation with constant coefficients is obtained which can be solved using the method of the characteristic polynomial, obtaining

$$A_m(z) = b_{m,1} e^{s_{m,1} z} + \cdots + b_{m,R} e^{s_{m,R} z}, \tag{9.11}$$

where $s_{m,r}$ are the characteristic polynomial roots and $b_{m,r}$ are constants that can be determined from the initial conditions at $z = 0$ obtained by differentiating (9.6), $(d_z^i A_m)_{z=0}$, and equating the results. Finally, the coefficients $b_{m,r}$ are the solutions of the linear equations system:

$$\begin{bmatrix} s_{m,1} & \cdots & s_{m,N} \\ \vdots & \ddots & \vdots \\ s_{m,1}^{2(M-1)} & \cdots & s_{m,N}^{2(M-1)} \end{bmatrix} \begin{bmatrix} b_{m,1} \\ \vdots \\ b_{m,R} \end{bmatrix} = \begin{bmatrix} \left(d_z^0 A_m \right)_{z=0} \\ \vdots \\ \left(d_m^{2(M-1)} A_m \right)_{z=0} \end{bmatrix} \tag{9.12}$$

The solution method described is easily applied using a software tool with symbolic computation capability. We have used the Symbolic Math Toolbox from Matlab® to generate equations for $a_{m,r}$ and $(d_z^i A_m)_{z=0}$ as function of $\Delta\beta_{mn}$ and C_{mn}. Finally, the derived equations can be written into a conventional text file and compiled using any programming language (we used a C-compiler). Afterwards, those equations can be evaluated allowing to find the roots of the polynomials in (9.10) and to solve the system of linear equations in (9.12).

In summary, instead of numerically solving a coupled-system of M differential equations (9.6), the method proposed requires the finding the roots of a $2(M-1)$ order polynomial, for which efficient and accurate algorithms are available, e.g. Bairstow's method [32], and the solution of a system of linear equations (9.12).

9.3.1 Analytical Expressions for the Three-Modes Case

The analytical expressions for $a_{m,r}$ and $(d_z^r A_m)_{z=0}$ as a function of $\Delta\beta_{m,n}$ and $C_{m,n}$ for $M=3$ are given by Eqs. (9.13) to (9.18), obtained executing the MATLAB code presented in 0. Replacing the $\Delta\beta_{m,n}$ and $C_{m,n}$ numeric values in the $a_{m,r}$ analytical equations, the $s_{m,r}$ values can be calculated using algorithms for the calculation of polynomial roots. Moreover, $(d_z^r A_m)_{z=0}$ values are obtained replacing the $\Delta\beta_{m,n}$ and $C_{m,n}$ numeric values in the analytical formulas. Finally, the system of linear equations (9.12) can be numerically solved.

9.3.2 Analytical Expressions for More Than Three-Modes

The analytical expression for $a_{m,r}$ and $(d_z^r A_m)_{z=0}$ as a function of $\Delta\beta_{m,n}$ and $C_{m,n}$ for $M>3$ can be obtained executing Matlab code similar to that made available in [6]. However, the equations become too long to be printed here in full. See Section 9.13, for the analytical equations for 6-modes and the derivation MATLAB scripts.

$$
\begin{bmatrix} a_{1,0} \\ a_{1,1} \\ a_{1,2} \\ a_{1,3} \\ a_{1,4} \end{bmatrix} =
\begin{bmatrix}
1 \\
-2j\Delta\beta_{12} - j\,\Delta\beta_{13} \\
C_{12}^2 - \Delta\beta_{12}^2 - 2\Delta\beta_{12}\Delta\beta_{13} + C_{13}^2 + C_{23}^2 \\
j\begin{pmatrix} \Delta\beta_{12}^2\Delta\beta_{13} - \Delta\beta_{12}C_{12}^2 - 2\,\Delta\beta_{12}C_{13}^2 \\ -\,\Delta\beta_{13}C_{12}^2 - \Delta\beta_{12}C_{23}^2 - 2\,C_{12}C_{13}C_{23} \end{pmatrix} \\
-\Delta\beta_{12}(\Delta\beta_{12}C_{13}^2 + \Delta\beta_{13}C_{12}^2 + 2C_{12}C_{13}C_{23})
\end{bmatrix}
\tag{9.13}
$$

$$
\begin{bmatrix} a_{2,0} \\ a_{2,1} \\ a_{2,2} \\ a_{2,3} \\ a_{2,4} \end{bmatrix} = \begin{bmatrix} -1 \\ -3\mathrm{j}\Delta\beta_{12}+\mathrm{j}\Delta\beta_{13} \\ 3\Delta\beta_{12}^2 - 2\Delta\beta_{12}\Delta\beta_{13} - C_{12}^2 - C_{13}^2 - C_{23}^2 \\ \mathrm{j}\begin{pmatrix} -\Delta\beta_{12}^2\Delta\beta_{13} - 2\Delta\beta_{12}C_{12}^2 - \Delta\beta_{12}C_{12}^2 + \Delta\beta_{13}C_{12}^2 \\ -2\Delta\beta_{12}C_{23}^2 + \Delta\beta_{12}^3 + 2C_{12}C_{13}C_{23} \end{pmatrix} \\ \Delta\beta_{12}\left(\Delta\beta_{12}C_{12}^2 - \Delta\beta_{13}C_{12}^2 + \Delta\beta_{12}C_{23}^2 - 2C_{12}C_{13}C_{23}\right) \end{bmatrix} \quad (9.14)
$$

$$
\begin{bmatrix} a_{3,0} \\ a_{3,1} \\ a_{3,2} \\ a_{3,3} \\ a_{3,4} \end{bmatrix} = \begin{bmatrix} -1 \\ \mathrm{j}\Delta\beta_{12} - 3\mathrm{j}\Delta\beta_{13} \\ 3\Delta\beta_{13}^2 - 2\Delta\beta_{12}\Delta\beta_{13} - C_{12}^2 - C_{13}^2 - k_{23}^2 \\ \mathrm{j}\begin{pmatrix} -\Delta\beta_{12}\Delta\beta_{13}^2 + \Delta\beta_{12}k_{13}^2 - \Delta\beta_{13}C_{12}^2 \\ -2\Delta\beta_{13}C_{13}^2 - 2\Delta\beta_{13}C_{23}^2 + \Delta\beta_{13}^3 + 2C_{12}C_{13}C_{23} \end{pmatrix} \\ -\Delta\beta_{13}\left(C_{13}^2\left(\Delta\beta_{12} - \Delta\beta_{13}\right) - 2\Delta\beta_{13}C_{23} + 2C_{12}C_{13}C_{23}\right) \end{bmatrix}
$$

$$(9.15)$$

$$
\begin{bmatrix} d_z^0 A_1 \\ d_z^1 A_1 \\ d_z^2 A_1 \\ d_z^3 A_1 \end{bmatrix}_{z=0} = \begin{bmatrix} A_1 \\ -\mathrm{j}C_{12}A_2 - \mathrm{j}\,C_{13}A_3 \\ \begin{Bmatrix} -C_{12}(C_{12}A_1 + C_{23}A_3) - C_{13}(C_{13}A_1 + C_{23}A_2) \\ +dB_{12}C_{12}A_2 + dB_{13}C_{13}A_3 \end{Bmatrix} \\ \mathrm{j}\left\{ \begin{matrix} -C_{12}\begin{pmatrix} -C_{12}(C_{12}A_2 + C_{13}A_3) \\ -C_{23}(C_{13}A_1 + C_{23}A_2) \\ -dB_{12}C_{12}A_1 + dB_{23}C_{23}A_3 \end{pmatrix} \\ -C_{13}\begin{pmatrix} -C_{13}(C_{12}A_2 + C_{13}A_3) \\ -C_{23}(C_{12}A_1 + C_{23}A_3) \\ -dB_{13}C_{13}A_1 - dB_{23}C_{23}A_2 \end{pmatrix} \\ -2dB_{12}C_{12}(C_{12}A_1 + C_{23}A_3) \\ -2dB_{13}C_{13}(C_{13}A_1 + C_{23}A_2) \\ +dB_{12}{}^2C_{12}A_2 + dB_{13}{}^2C_{13}A_3 \end{matrix} \right\} \end{bmatrix} \quad (9.16)
$$

$$\begin{bmatrix} d_z^0 A_2 \\ d_z^1 A_2 \\ d_z^2 A_2 \\ d_z^3 A_2 \end{bmatrix}_{z=0} = \begin{bmatrix} A_2 \\ -jC_{12}A_1 - jC_{23}A_3 \\ \left\{ \begin{array}{c} -C_{12}(C_{12}A_2 + C_{13}A_3) - C_{23}(C_{13}A_1 + C_{23}A_2) \\ -\Delta\beta_{12}C_{12}A_1 + \Delta\beta_{23}C_{23}A_3 \end{array} \right\} \\ j \left\{ \begin{array}{c} -C_{12}\begin{pmatrix} -C_{12}(C_{12}A_1 + C_{23}A_3) \\ -C_{13}(C_{13}A_1 + C_{23}A_2) \\ +\Delta\beta_{12}C_{12}A_2 + \Delta\beta_{13}C_{13}A_3 \end{pmatrix} \\ -C_{23}\begin{pmatrix} -C_{13}(C_{12}A_2 + C_{13}A_3) \\ -C_{23}(C_{12}A_1 + C_{23}A_3) \\ -\Delta\beta_{13}C_{13}A_1 - \Delta\beta_{23}C_{23}A_2 \end{pmatrix} \\ + 2\Delta\beta_{12}C_{12}(C_{12}A_2 + C_{13}A_3) \\ - 2\Delta\beta_{23}C_{23}(C_{13}A_1 + C_{23}A_2) \\ + \Delta\beta_{12}{}^2 C_{12}A_1 + j\Delta\beta_{23}{}^2 C_{23}A_3 \end{array} \right\} \end{bmatrix} \quad (9.17)$$

$$\begin{bmatrix} d_z^0 A_3 \\ d_z^1 A_3 \\ d_z^2 A_3 \\ d_z^3 A_3 \end{bmatrix}_{z=0} = \begin{bmatrix} A_3 \\ -jC_{13}A_1 - jC_{23}A_2 \\ \left\{ \begin{array}{c} -C_{13}(C_{12}A_2 + C_{13}A_3) - C_{23}(C_{12}A_1 + C_{23}A_3) \\ -\Delta\beta_{13}C_{13}A_1 - \Delta\beta_{23}C_{23}A_2 \end{array} \right\} \\ j \left\{ \begin{array}{c} -C_{13}\begin{pmatrix} -C_{12}(C_{12}A_1 + C_{23}A_3) \\ -C_{13}(C_{13}A_1 + C_{23}A_2) \\ +\Delta\beta_{12}C_{12}A_2 + \Delta\beta_{13}C_{13}A_3 \end{pmatrix} \\ -C_{23}\begin{pmatrix} -C_{12}(C_{12}A_2 + C_{13}A_3) \\ -C_{23}(C_{13}A_1 + C_{23}A_2) \\ -\Delta\beta_{12}C_{12}A_1 + \Delta\beta_{23}C_{23}A_3 \end{pmatrix} \\ + 2\Delta\beta_{13}C_{13}(C_{12}A_2 + C_{13}A_3) \\ + 2\Delta\beta_{23}C_{23}(C_{12}A_1 + C_{23}A_3) \\ + \Delta\beta_{13}{}^2 C_{13}A_1 + j\Delta\beta_{23}{}^2 C_{23}A_2 \end{array} \right\} \end{bmatrix} \quad (9.18)$$

9.3.3 Algorithm Complexity

The RK45 method requires a step-size of a fraction of the beat-length between the two mode-groups most furthest apart, which can easily be of the order of a millimeter or less [24]. In this way, to resolve a 1 mm beat-length, more than 10^4 and 10^6 steps are required for a transmission length of 1 and 100 meters, respectively. Each RK45 step requires six evaluations of a system of M Eq. (9.6), each equation with $2(M - 1)$ multiplications, thus totalizing $12M(M - 1)$ multiplication operations per step.

The semi-analytical method proposed uses Bairstow's method to find the roots of M polynomials of order $2(M - 1)$. This method consists on the progressive division of the original polynomial by quadratic polynomials while adjusting the coefficients of the later. Thus, the method requires $(M - 1)$ polynomial divisions of progressively lower complexity. Assuming the number of multiplications required to be the product of the number of terms of the polynomials involved, the i-th-division requires $[(2(M - 1) + 1) - 2(i - 1)](2 + 1)$, adding up to $3(M^2 - 1)$ multiplication operations. Finally, this figure must be multiplied by the number of iterations for coefficients adjustment, which we cap to be lower than 100, and observed that in general only 20 repetitions were required. Thus, the total complexity is on the order of $60M(M^2 - 1)$.

Finally, the proposed semi-analytical method reduces the number of multiplications required by a factor from 280 to 28,000 when transmitting over 1 to 100 meters, for $M = 6$. These factors agree with the observed simulation times.

9.4 Single-Section Modelling

In this section, the semi-analytical solutions of Section 9.3 are validated using the Runge–Kutta–Fehlberg (RK45) method [33]. The fiber considered for guides six linearly polarized (LP) modes: LP_{01}, LP_{11a}, LP_{11b}, LP_{21a}, LP_{21b}, and LP_{02}. It has a relative index gradient at the core–cladding interface 4.5×10^{-3} and a core radius (w_1) of 12.83 µm, optimization details [34]. Table 9.1 shows the fiber characteristics at 1550 nm (for the sake of clarity, the modes were numbered from one to six). Figure 9.3 depicts the amplitude of $C_{m,n}$ as a function of the fiber displacement vector for a radial displacement from 0 to $0.3 \cdot w_1$. Note that the coupling coefficients were found to be real and symmetric has concluded in [31], therefore only $C_{m,n}$ with $n > m$ are shown. From Fig. 9.3, the pairs of modes with higher coupling strength can be identified, and it can be verified that the coupling between symmetric modes (LP$_{01}$, for example) and anti-symmetric modes (LP$_{11}$, for example) requires a nonsymmetrical perturbation. The surfaces shown in Fig. 9.3 allow the rapid calculation of the coupling coefficients $C_{m,n}$ using interpolation for a random displacement, as required for integration in a modified split-step Fourier method (SSMF).

Table 9.1 Fiber Properties at 1550 nm

		$u = $ LP$_{01}$	$u = $ LP$_{02}$	$u = $ LP$_{11a}$	$u = $ LP$_{11b}$	$u = $ LP$_{21a}$	$u = $ LP$_{21b}$
$\Delta\beta_{1,n}/\kappa(\times 10^{-3})$		0	7.4	3.7	3.7	7.4	7.4
$\beta^{(1)}_{1,n}$(ps/km)		0	−2.6	−0.4	−0.4	2.6	2.6
D_n (ps/km/nm)		22.2	21.5	22.2	22.2	21.8	21.8
S_n (ps/km/nm^2)		66.4	61.5	66.2	66.2	63.7	63.7
	$v = $ LP01	0.72	0.36	0.36	0.36	0.18	0.18
	$v = $ LP02	0.36	0.36	0.18	0.18	0.18	0.18
γ_{uv} (W^{-1}/km)	$v = $ LP11a	0.36	0.18	0.55	0.55	0.27	0.27
	$v = $ LP11b	0.36	0.18	0.55	0.55	0.27	0.27
	$v = $ LP21a	0.18	0.18	0.27	0.27	0.41	0.41
	$v = $ LP21b	0.18	0.18	0.27	0.27	0.41	0.41

Note: κ is the wave number.

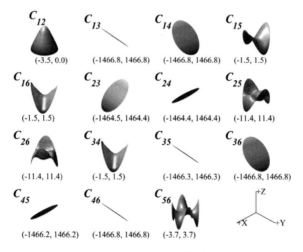

Figure 9.3 $C_{m,n}$ as a function of the fiber displacement vector, minimum and maximum values as (min, max). Modes numbered as: 1→LP$_{01}$, 2→LP$_{02}$, 3→LP$_{11a}$, 4→LP$_{11b}$, 5→LP$_{21a}$, 6→LP$_{21b}$.

In the following, the semi-analytical (SA) solutions are compared to the numerical solutions obtained using the RK45 method. The variable-step width of the RK45 method is specified considering a relative error tolerance of 10^{-6} and an absolute error tolerance of 10^{-9}. Figure 9.4 shows the overlap of the modal powers given by the SA solutions and the modal powers given by the numerical method as a function of z, considering a fiber core displacement of $\rho_d = 0.08 \cdot w_1$ and $\varphi_d = \pi/3$, for an even power distribution between the modes at the input. A very good agreement between the SA and numerical solutions can be noticed in Fig. 9.4 inset which zooms in the mode power evolution around 0.8 m. Similar agreement is obtained for different input conditions.

$$\text{MSE}_m = \frac{1}{N} \sum_{n=1}^{N} \left| A_m^{SA}(z_n) - A_m^{NUM}(z_n) \right|^2, \qquad (9.19)$$

where A_m^{SA} is the SA mode amplitude solution, A_m^{NUM} is the numerical mode amplitude solution, and z_n are the discrete points considered in a specific fiber length. MSE_m has been calculated considering 10^5 discrete points equally spaced along a fiber with 1 m, considering ρ_d varying 0 and $0.08 \cdot w_1$ (1000 points equally spaced), and φ_d varying from $-\pi$ to π (1000 points equally spaced). In all the cases tested the MSE_m was always of the order of magnitude of the RK45 absolute tolerance, as verified by repeating the error calculation for different tolerance values. Therefore, it can be concluded that the semi-analytical method proposed provides an accurate estimative of the linear mode coupling taking place along a FMF. More importantly, using the semi-analytical method the computation time required to calculate the linear coupling along a fiber with a few meters is reduced by three orders of magnitude compared to the RK45 method which required tens of seconds executing on a standard personal computer operating at 2.8 GHz.

In conclusion, the semi-analytical solutions obtained enable a time efficient and accurate computation of the linear coupling occurring along the fiber length. They are therefore a valuable alternative to the numerical solution, which would not be practical due to computation time constraints.

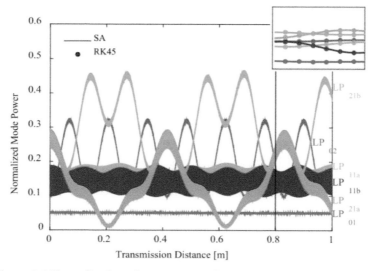

Figure 9.4 Normalized mode power as a function of the fiber length, for uneven power distribution at the fiber input. The subscripts SA and RK45 were used for semi-analytical and numerical solutions, respectively. The inset shows the excellent agreement between SA and RK45 around 0.8 m.

9.5 Multi-Section Modelling

We propose a multi-section model were the coupling strength is set using a given radial displacement and a uniformly distributed azimuthal displacement for each section. The radial displacement to be used depends not only on the target coupling strength but also on the fiber step length to be used. But first we quantitatively define the mode coupling strength and present its dependence on the radial displacement.

9.5.1 Setting Mode Coupling Strength and Correlation Length

The statistical nature of polarization mode dispersion (PMD) in SMFs is mainly determined by the correlation length, which is defined in terms of fiber mode coupling. In SMFs, the mode coupling is easily defined as there are only two polarizations, and the L_c is defined as the length for which the average power in the orthogonal polarization is within e^{-2} of the power in the launching polarization. In FMFs, the mode coupling strength can be quantified as the ratio between the average power in all the other orthogonal modes and average power remaining in the launching mode, after a certain distance. Thus, there are as many coupling strength values and L_c as the number of modes. Inevitable, the fiber mode m showing higher coupling strength will set an important reference for the study of the mode group-delay statistics. Finally, the mode coupling strength definition for FMFs is:

$$XT_m = \sum_{v \neq m}(P_v/P_m), \qquad (9.20)$$

where P_v is the power of mode v, after a given fiber segment under test, when only the m mode was launched, where m is the mode that shows higher coupling strength. In the FMF case, we generalize L_c for mode m as the length for which $(P_m - \sum_{v \neq m} P_v) = e^{-2}$, this is $XT_m = [e^2 - 1]/[e^2 + 1]$ (-1.18 dB).

In our multi-step model, a given amount of coupling is set by selecting a fixed amount of radial displacement and selecting a random azimuth displacement given by a uniform distribution. In this way, the proposed model introduces a random amount of crosstalk per step that in average approximates the desired level. Figure 9.5 shows the *mode coupling strength* averaged over the azimuth displacement, as a function of the normalized radial displacement, for a 6 LP mode fiber presented in Section 9.4. Note that, coupling strengths are calculated considering degenerate modes such as LP_{11a} and LP_{11b} as one mode, e.g. $XT_{LP11a,b}$ equals to $\sum_{v \neq LP11a,b}\{P_v/(P_{LP11a} + P_{LP11b})\}$. In Fig. 9.5, the mode coupling strength only depends significantly on the mode being considered for displacements higher than 1%. Such higher coupling for LP_{02} and LP_{21} can be explained noting they belong to the same LP mode group. Moreover, $XT_{LP21} \leq XT_{LP02}$ for any displacement in Fig. 9.5 because any power launched in LP_{21a} couples preferentially with LP_{21b} (and vice-versa) and in the second place to LP_{02}. Given the higher values of XT_{LP02}, we define L_c for this mode. Note that XT_m values above 10 dB mean that almost all power launched in mode m has been transferred to other modes.

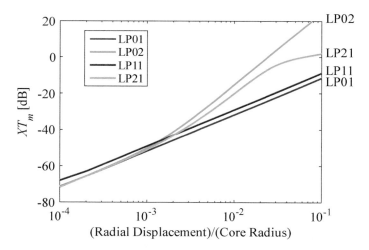

Figure 9.5 XT_m averaged over the azimuth displacement as a function of the radial displacement.

In the literature, the mode coupling values of fabricated FMFs range from –50 dB/100 m to –40 dB/100 m for fibers with step-index or graded-index profiles [20, 21], going up to –28 dB/100 m for coupled multi-core fibers [22] and –7 dB/100 m for fibers with ring-index profiles [23].

9.5.2 Mode Coupling Accumulation over Transmission Length

In a multi-section model, the mode coupling accumulates section after section in such a way that in average should follow the same continuous growing function that was first derived to describe the accumulation of polarization coupling in polarization-maintaining fibers [35]:

$$XT = \tanh(hz), \tag{9.21}$$

where h is the mode coupling parameter (measured in m$^-$ units) and z is the fiber length.

To validate our multi-section model, we have run 10,000 transmission simulations considering the 6-mode fiber presented in Section 9.4. Figure 9.6 shows the average XT_{LP02} as a function of the fiber length (L) from 10 m to 1000 km, considering a fiber section of 10 m and different values of coupling strength. Note that the dashed lines in Fig. 9.6 represent the evolution predicted for (9.21) using the respective h coefficient. A very good agreement between the proposed multi-section model and (9.21) is noticeable. Furthermore, similar matches were obtained for other section sizes and respective radial displacements.

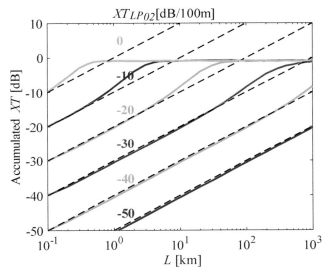

Figure 9.6 Accumulated XT as a function of the fiber length, for different coupling strength, averaged over 10,000 runs.

9.5.3 Polarization Mode Coupling

After a few meters, the two nearly degenerate polarization modes *of each spatial mode strongly couple to each other and the FMF enters the* polarization coupling state [4, 31]. Such propagation mode coupling can be described by a block diagonal matrix with a sequence of $M/2$ submatrices along the diagonal. Each of these 2 × 2 random unitary submatrices is a PMD transfer matrix [4].

   The full coupling matrix for the i-th-section is modelled as the product of two matrices: one block diagonal matrix describing the polarization mode coupling, and one matrix describing finite inter- and intra-mode group coupling (as described in

Sections 9.3, 9.4 and 9.5). This approach follows a similar reasoning to that in [4, 24] to deal with coupling processes having different correlation lengths.

9.6 GD Statistics in Non-Delay-Managed Links

In previous work [36], it has been shown that the approach of considering principal states of polarization (PSPs) with well-defined GDs in SMFs, can be extended to FMFs. In FMFs, the coupled modes having well defined GDs are called principal modes (PMs). In both cases, the statistics of the GDs are dependent on the linear coupling strength, thus the correlation length L_c. The coupling regimes may then be broadly defined as strong coupling when $L \gg L_c$, weak coupling when $L \ll L_c$, and intermediate coupling otherwise. In the FMF case, the statistical properties of the GDs are well known for the two extreme regimes [4, 11–14]. In the weak coupling regime, the GD spread grows linearly with distance and in the strong coupling regime grows with the square root of the distance. In the intermediate coupling regime, we have shown in [27] through simulation that the GDs statistics in SMFs can be extended to FMFs, at least for fibers guiding 3 LP modes. At the same time, the complete analytical derivation of such extension was presented in [37].

The temporal spread of propagating pulse is determined by the modal dispersion (MD) vector τ, as defined for a generalized $(M^2 - 1)$-dimensional Stokes space in [13] (M modes). Knowledge of the MD vector allows the extraction of the PMs and respective GDs as explained in [13]. Moreover, the square modulus of the MD vector is proportional to the sum of the GDs τ_i (with $\sum \tau_i = 0$) [13]:

$$\|\tau\|^2 = M \sum_{i=1}^{M} \tau_i^2 \tag{9.22}$$

In this way, it can be noted that $\|\tau\|/M$ is the standard deviation of the GD vector $[\tau_1, \tau_2, \dots \tau_M]$, σ_{gd}. The MD vector has been used to explicitly determine the delay spread T in two limiting cases: one in which the PMs change rapidly across the signal bandwidth, and one in which the andwidth of the PMs is much larger than the signal bandwidth. In the first case, T is a deterministic quantity and determined by σ_{gd}, $T^2 = E\{||\tau||^2\}/M^2 = E\{\sigma_{gd}{}^2\}$ 0, where $E\{\cdot\}$ denotes expectation. In the latter case, T is a random quantity given by $\max_i\{\tau_i\} - \min_i\{\tau_i\}$ [11, 13], determined by the GD PDF.

In the following, we review the known MD statistics and use them to validate the multi-section model proposed in Section 9.5 for a fiber guiding 6 LP modes despite the different coupling strengths between different pairs of modes belonging to different mode groups. The FMF presented in Section 9.4 is considered again, the modal and chromatic dispersion values are given in Table 9.1. The fiber presents a DMD of 5.19 ps/km and we assumed zero DMD between degenerate LP modes and between orthogonal polarizations. As explained in Section 9.5.2, the polarization mode coupling is considered in each section using a block diagonal matrix. Regarding the coupling matrix describing finite inter- and intra-mode group coupling, the XT_{LP02} value was varied from -50 to 0 dB/100 m by using a given radial displacement and a uniformly distributed azimuthal displacement for each section (see Fig. 9.5), assuming a section length of 10 m. This range fully covers the range of coupling values presented in the literature [20–23]. Finally, the GDs of the PMs are the eigenvalues of the semi-analytically simulated transmission matrix. Note that the simulated transmission matrix must be compensated for chromatic dispersion as introduced by (9.3).

9.6.1 GD Standard Deviation and Intensity Impulse Response

Knowledge of the modulus of the MD vector $||\tau||$ allows to determine the standard deviation of the GD vector $[\tau_1, \tau_2, ... \tau_M]$ σ_{gd}, since $\sigma_{gd} = ||\tau||/M$. $E\{||\tau(z)||^2\}$ can be found by integration of two deterministic differential equations (z dependence is omitted) [5, 37]:

$$\partial_z E\{||\tau||^2\} = E\{2\partial_\omega\beta\tau\} = 2\partial_\omega\beta\, E\{\tau\} \tag{9.23a}$$

$$\partial_z E\{\tau\} = \partial_\omega\beta - 1/L_c\, E\{\tau\}, \tag{9.23b}$$

where $\partial_\omega\beta$ term represents the uncoupled GDs per unit length and L_c is the correlation length characteristic of the fiber, considering the same L_c for all groups of modes.

For non-DMD-managed spans (that is, $\partial_\omega\beta$ constant), $E\{||\tau(z)||^2\}$ can be found through analytical integration of (9.23), and is given by [5, 37]

$$E\{||\tau||^2\} = 2||\partial_\omega\beta||^2 L_c^2 (e^{-z/L_c} + z/L_c - 1). \tag{9.24}$$

Equation (9.24) was proposed and validated by simulation in [27] for fibers guiding 3 LP modes, and at the same time its analytical derivation being presented in [37].

Figure 9.7 shows the standard deviation of the GD vector ($[\tau_1, \tau_2, \ldots \tau_{12}]$) as a function of distance up to 1000 km, obtained by averaging over 6000 different realizations of lateral offsets giving rise to a given XT_{LP02} value. These results were obtained using the fiber presented in Table 9.1, treating the polarization mode coupling as described in Section 9.5.2. Figure 9.7 shows a good agreement between simulation and (9.24), for any coupling value studied and for any distance up to 1000 km (even 10,000 km has further results shown). Similar agreement between (9.24) and simulation results has been presented in Fig. 9.3 of [5]. This provides mutual validation of (9.24) and the proposed multi-section model proposed. In Fig. 9.7, for coupling values ranging from −50 to −40 dB/100 m, σ_{gd} scales approximately linearly with distance. But, at −40 dB/100 m the deviation from linear growth is already noticeable around 1000 km, thus even with such a low coupling, the FMF is operating in intermediate coupling regime. Increasing XT_{LP02}, σ_{gd} gradually converges to the strong coupling regime. However, even for a XT_{LP02} equal to −7.01 dB/100 m (the highest value found in literature [23]), the fiber is still not well modelled by random unitary matrices every 100 m, it would underestimate σ_{gd} by a factor of 2.76.

For FMFs where the PMs change rapidly across the signal bandwidth, MD can be conveniently characterized by exciting each spatial channel (one at a time) with a short optical pulse and measuring the received intensities in each of the output spatial channels. Such process leads to $M \times M$ intensity waveforms, whose sum $I(t)$ has been used to assess the signal delay spread caused by MD [5, 14]. For strong mode-coupling and typical MD values, it has been shown that $I(t) = r(t) * I_0(t)$ [14], where $*$ represents convolution, $I_0(t)$ is the launching signal intensity waveform, and $r(t)$ is FMF's intensity impulse response (IIR). Also in [14], it was shown theoretically and experimentally that $r(t)$ is a Gaussian function with variance equal to $T^2 = E\{||\tau||^2\}/M^2 = E\{\sigma_{gd}^2\}$, thus,

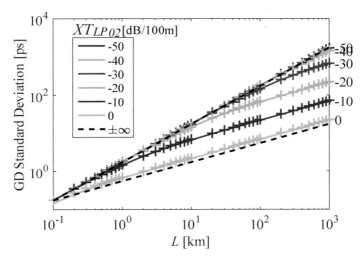

Figure 9.7 Standard deviation of the GDs of the PMs as a function of transmission distance showing simulation results (markers) and analytical results (solid lines).

$$r(t) = \frac{1}{\sqrt{2\pi T^2}} \exp\left(\frac{-t^2}{2T^2}\right).$$

(9.25)

Equation (9.25) is valid as long as the correlation bandwidth ($B_{MD} = 1/2\Pi T$) of the fiber transfer matrix is much smaller than the channel bandwidth ($B \sim$ tens of GHz). Figure 9.8 shows the mode-averaged intensity waveform for $M = 12$ modes after transmission of a Nyquist signal $I_0(t) = \sqrt{B}\ \sin(\Pi B t)/(\pi B t)$, with $B = 20$ GHz, over a 1000 km link with coupling values ranging from −30 to 0 dB/100 m. Simulation results, $r(t) * I_0(t)$ waveform, and $r(t)$ IIR are plotted using colored full lines, black dashed lines, and red dashed lines, respectively. Figure 9.8 displays simulation results for 100 different fiber realizations for each

XT_{LP02} value. All the waveforms were normalized so that their peak value is one. Figure 9.8 shows an excellent agreement between simulations and theory (experimentally validated) as obtained in [14]. Note that the deviations from theory reduce as the coupling strength increases and the PMs bandwidth decreases. Finally, further results show that the deviation of $T(z)$ from theory is in agreement with the theory in [14].

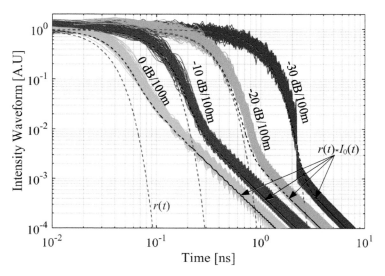

Figure 9.8 Mode-averaged intensity waveform for different coupling values after transmission of a Nyquist signal. Simulation results, $r(t) * I_0(t)$ and $r(t)$ plotted using colored full lines, black dashed lines and red dashed lines, respectively.

9.6.2 GD Probability Density Function and Maximum GD Spread

The probability density function (PDF) of the GDs has been derived analytically for strong coupling [11] where the coupling matrix can be described as a Gaussian unitary ensemble. The ordered joint PDF of the eigenvalues (τ_i) of a $M \times M$ Gaussian unitary ensemble with zero trace $(\sum \tau_i = 0)$ is:

$$p_M(\tau_1, ..., \tau_{M-1}) = \rho_M \prod_{M \ge i > j > 0} (\tau_i - \tau_j)^2 e^{\left(-\sum_{i=1}^{M} \tau_i^2\right)} \tag{9.26}$$

with order constrain $\tau_1 \leq \tau_2 \leq \ldots \leq \tau_M$ and where the constant ρ_M is defined by requiring (9.26) to integrate to unity. The unordered joint PDF is just $1/M!$ of (9.26) but without the order constraint. In this way, the marginal PDF of τ is can be obtained by integrating over $\tau_2, \ldots, \tau_{M-1}$:

$$p_{M\tau}(\tau) = \frac{1}{M!} \int_{-\infty}^{+\infty} \ldots \int_{-\infty}^{+\infty} p_M(\tau, \tau_2, \ldots, \tau_{M-1}) d\tau_2 \ldots d\tau_{M-1} \tag{9.27}$$

Analytical solutions of (9.27) can be found in [4, 11] for any M.

Figure 9.9 shows the PDF of the ordered GDs (τ_m, $\tau_1 \leq \tau_2 \leq \ldots \leq \tau_6$) obtained for 6000 different fiber matrix realizations, normalized by the σ_{gd}, after 1000 km for two different coupling values, overlapped with the analytical joint PDF (thin black line) derived for the strong coupling regime (9.27). Note that the normalization factor (σ_{gd}) depends on the XT_{LP02} (L_c) value, see (9.24). Exceptionally, these results were obtained for single-polarization to facilitate the visualization of the individual GDs evolution in Fig. 9.9, but similar matching between simulation and theory was obtained when considering dual-polarization. Figure 9.9a shows that for −30 dB/100 m the GDs of the PMs vaguely resemble the GDs of the LP modes given the impulse-like PDF of τ_2 ("LP$_{11a}$") and τ_3 ("LP$_{11b}$"). Further results for lower coupling values shown that all GDs present impulse-like PDFs. In Fig. 9.9b, for −20 dB/100 m, the match between the simulated PDFs and the analytical PDF for strong coupling is good, even though the GDs have been normalized by different factors (9.24). Further increase of the coupling strength leads to improved matching between the simulated PDFs and the analytical PDF, as observed in additional results.

In a MDM system for which the bandwidth of the PMs is much larger than the signal bandwidth, the digital equalizer must span a temporal memory at least as long as the the difference between the maximum and the minimum group delay $(\tau_M - \tau_1)_{total}$. As shown in [12], the probability of having a GD spread lower than x, $P(\tau_M - \tau_1 \leq x)$—the cumulative distribution function, can be computed as a function of the joint probability of having all eigenvalues falling within an arbitrary interval $[x, y]$, $P(\tau_M \leq x, \tau_1 \geq y)$, this is:

$$P(\tau_M - \tau_1 \le x) = -\int_{-\infty}^{+\infty} \partial_y P(\tau_M \le y + x, \tau_1 \ge y) dy \qquad (9.28)$$

$$P(\tau_M \le x, \tau_1 \ge y) = \int_y^x \dots \int_y^x p_M(\tau) \, d\tau_1 \dots d\tau_M \qquad (9.29)$$

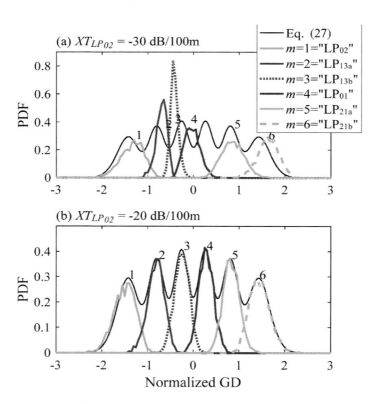

Figure 9.9 Probability density function of the ordered normalized GDs (τ_m / σ_{gd}), obtained through simulation after 1000 km, with different XT_{LP02} values.

According to [12], (9.27)–(9.29) can be evaluated using at least three methods: Fredholm determinant, Andréief identity or one approximation based on Tracy–Widom distribution. Finally, from (9.27), we can obtain the equalizer memory length x required to accommodate the GD spread with a given outage probability p, this is: $P(\tau_M - \tau_1 > x) = p = 1 - P(\tau_M - \tau_1 \le x)$.

Figure 9.10 shows the complementary cumulative distribution function (CCDF) of the normalized GD spread, $P[(\tau_6 - \tau_1)/\sigma_{gd} > p]$, obtained through simulation after 1000 km for different coupling values (averaging over 6000 different realizations). These results were obtained for single-polarization to be consistent with the PDFs in Fig. 9.9. Figure. 9.10 shows that for $XT_{LP02} \geq -30$ dB/100 m the CCDFs are very similar to the analytical approximation obtained for strong coupling (9.28) (dashed line). Conversely, for XT_{LP02} lower than -30 dB/100 m the normalized GD spread is significantly smaller than the normalized GD spread for strong coupling. Finally, we can conclude that the required temporal equalizer memory length (in time units) to span a channel with an outage probability smaller than 10^{-4} is equal to 4.5 σ_{gd}, for any coupling strength, where σ_{gd} depends on the mode coupling strength, see (9.24).

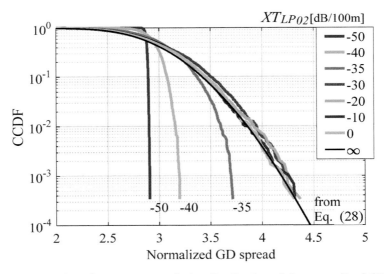

Figure 9.10 Complementary cumulative distribution of the normalized GD spread, obtained through simulation after 1000 km, with different XT_{LP02}.

9.7 GD Statistics in Delay-Managed Links

In DMD managed spans, GD spread is reduced by cascading fibers with opposite sign DMD. In the absence of mode coupling, the GD spread at the end of the span would be zero. However, in the presence of coupling, the DMD compensation is no longer fully effective. In order to minimize the impact of coupling, the length of the segments over which DMD sign is inverted has to be made much smaller than the correlation length set by the coupling, L_c.

To compensate for linear mode coupling and group delay spread, MIMO-DSP can be used, but DSP complexity increases with the number of modes and the total GD spread. In order to minimize complexity, the total GD spread should typically be reduced to less than 10 ns [38].

For DMD-managed spans, where uncoupled GDs (per unit length) $\partial_\omega \beta$ are a piecewise constant function of z a general analytical solution of (9.23) for $E\{||\tau(z)||^2\}$, rapidly becomes too complex as the number of fiber segments increases. Therefore, numerical integration should be performed as in [37].

In order to verify the deterministic numerical integration of (9.23), we made use of the multi-section model presented in Section 9.5. The simulations considered that each span of length L comprised S segments, where each segment was itself composed by two fibers of length $L/S/2$ with the same characteristics but opposite sign GD. The first fiber is the same presented in Section 9.5. The second fiber is not obtained through optimization but just by negating the GD vector, keeping the remaining characteristics of the first fiber. Finally, to make analysis straightforward DMD value is sweep by scaling the GD vector in Table 9.1 as required after normalization by the highest GD value in the vector.

Figure 9.11 shows the evolution of the standard deviation of the GD vector ($[\tau_1, \tau_2, \ldots \tau_{12}]$) with propagation distance, assuming compensation length of 20 km (10 km with the positive GD vector followed by 10km with the negative GD vector), for different values of coupling strength. In Fig. 9.11, there are two sets of results, one obtained for transmission using the proposed multi-section model (dot markers) and one given by the deterministic numerical integration of (9.23) (full lines). A section length of 10 m was used as smaller section lengths generated similar results. In Fig. 9.11, we can observe a very good match between the deterministic numerical integration and the proposed multi-section model. It can be inferred from these results that the semi-analytical solutions in Section 9.4 multi-section model in Section 9.5 are accurate under any coupling regime for DMD managed links.

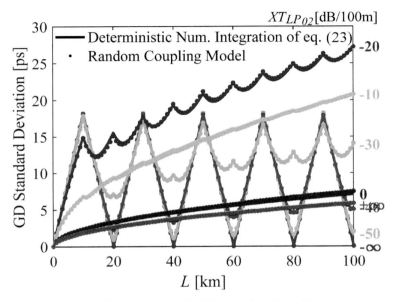

Figure 9.11 Standard deviation of the GDs as a function of the propagation distance, for fibers with a compensation length of 20 km and different values of coupling strength.

In order to study a broader range of DMD scenarios, the GD vector was normalized by the highest GD value in the vector. Figure 9.12 shows the combinations of (DMD, XT) that allow for a GD spread lower than 800 ps after 100 km with a probability higher than 95%. In Fig. 9.12, for a given span configuration, GD spread is lower than 800 ps for (DMD, XT) pairs below the respective curve. For non-DMD-managed spans, the maximum tolerable DMD increases with the coupling strength, being very low for weak coupling. For DMD-managed spans, as the number of segments increases, increasingly high DMD values are tolerable for weak coupling. For higher levels of coupling (above –20 dB/100 m), the tolerable DMD converges to that of the non-DMD-managed spans. Importantly, the tolerable DMD for the DMD-managed spans is always greater than or equal to the non-DMD-managed spans.

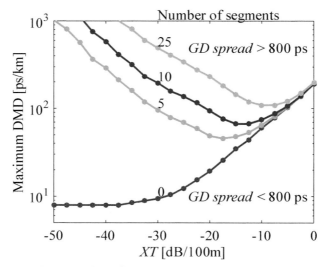

Figure 9.12 Contour plot of the pairs (*DMD*, *XT*) that allow for a GD spread lower than 800 ps after 100 km with a probability higher than 95%.

9.8 Nonlinear Propagation Modelling

The generalized nonlinear Schrödinger equation (GNLSE) for FMFs can be written as [7, 8, 17]

$$
\partial_z A_{ui} + \overbrace{\left(\beta_{ui}^{(1)} \partial_z - \frac{\mathbf{j}\beta_{ui}^{(2)}}{2} d_t^2 + \cdots + \frac{\alpha_{ui}}{2} \right)}^{=\hat{D}} A_{ui} = \overbrace{-\mathbf{j}[\gamma_{uuii} \mid A_{ui} \mid^2}^{=\hat{N}} \cdots
$$

$$
\cdots \left[+2\gamma_{uvii} \sum_{v \neq u} \mid A_{vi} \mid^2 + \frac{2}{3}\gamma_{uvij} \sum_v \mid A_{vj} \mid^2 \right] A_{ui} - \mathbf{j}\sum_{vk} C_{uvik} A_{vk} e^{j(\beta_{ui}^{(0)} - \beta_{vk}^{(0)})z}, \quad (9.30)
$$

where *i* and *j* are the orthogonal polarizations of mode *u*. $A_{ui}(z,t)$, $\beta_{ui}^{(1)}$, $\beta_{ui}^{(2)}$ and α_{ui} are the slowly varying field envelope, GD, GD dispersion and attenuation, respectively. γ_{uvij} is the nonlinear coefficient between *ui* and *vj*, which depends on the intermodal effective area as shown in [17]. In (9.30), $\hat{D}$ is the differential operator that accounts for dispersion and attenuation, and $\hat{N}$ is the nonlinear operator that accounts for all the intramodal and intermodal nonlinear effects [17]. The last term on the right-hand side accounts for the linear mode coupling arising from fiber structure imperfections, where C_{uvij} are the coupling coefficients as derived in [6].

9.8.1 Modified Split-Step Fourier Method

To numerically solve (9.30), we use a modified version of the split-step Fourier method (SSFM) developed for SMFs. In the SMF case, an approximate solution of the Schrödinger equation is obtained by assuming that over a small distance h the dispersive and nonlinear effects act independently. For FMFs, we extend such an approach by assuming that the mode coupling also acts independently. Such approximation requires h to be much shorter than: the dispersion length $T_0^2/|\beta_u^{(2)}|$, the walk-off length $T_0/|\beta_u^{(1)} - \beta_v^{(1)}|$ (T_0 is the pulse width), and the correlation length L_c defined in [6] such that $XT(L_c) = [e^2 - 1]/[e^2 + 1]$.

To include the linear mode coupling, the SSFM is now modified to include an additional step. Figure 9.13 presents a schematic illustration of the symmetric SSFM considered for numerical simulations in this paper. By using a symmetric SSFM, the effect of nonlinearity is included in the middle of the segment rather than at the segment boundary providing higher accuracy [39]. Finally, the step-size was selected by bounding the local error [39], more computationally efficient at high accuracy than the other methods, e.g. nonlinear phase rotation.

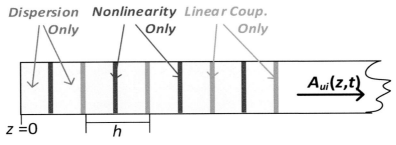

Figure 9.13 Schematic illustration of the symmetric SSFM used for numerical simulations.

9.8.2 Extreme Coupling Strength Regimes

In the presence of extreme mode coupling strength (weak or strong), it has been shown that the nonlinear distortion can be modelled using averaged coefficients instead of explicitly considering and solving for random coupling matrices. In [7, 8], new Manakov equations were derived for FMFs.

In the weak-coupling (WC) regime [8], it has been found that only the averaging over birefringence fluctuations must be considered, reducing the intramodal degeneracy factor to 8/9 and the intermodal degeneracy factor to 4/3.

$$\hat{N} = -j \left[\frac{8}{9} \sum_{k=\{i,j\}} \gamma_{uuik} \left| A_{uk} \right|^2 + \frac{4}{3} \sum_{\substack{v \neq u \\ k=\{i,j\}}} \gamma_{uvik} \left| A_{vk} \right|^2 \right], \tag{9.31}$$

In the strong coupling (SC) regime [7, 8], the averaging must include all propagation modes. For *N*-modes, the nonlinear operators for WC and SC are, respectively [7, 8],

$$\hat{N} = -j \sum_{\substack{v \\ k=\{i,j\}}} \kappa \left| A_{vk} \right|^2, \ \kappa = \frac{4}{3} \frac{2N}{2N+1} \left(\frac{1}{N^2} \sum_{\substack{u,v \\ k,l=\{i,j\}}} \gamma_{uvkl} \right). \tag{9.32}$$

9.8.3 Intermediate Coupling Strength Regime

In the intermediate coupling regime, (9.30) must be solved explicitly applying every step new random matrices characteristic of a given coupling strength. In [6], the authors proposed a semi-analytical solution method for the coupled linear differential equations that describe the linear modal coupling in FMFs, this is a solution of (9.30) assuming the linear mode coupling acting independently as explain in Section 9.8.1, the linear mode coupling step in Fig. 9.13. The semi-analytical solution method [6] has been proved accurate in the linear power regime. It accurately matched the analytical predictions for the statistics of GDs in FMF links for different transmission lengths 10 m to

10,000 km, in any coupling regime −50 dB/100 m to 0 dB/100 m, without and with group delay management. For convenience, the mode coupling strength (XT) is quantified taking the LP01 mode as reference, this is: $XT = \Sigma_{n \neq \text{LP01}} \, P_n/P_{\text{LP01}}$, where P_n is the power of mode n, after a given segment under test, when only mode LP01 was launched.

Here, we use the semi-analytical model [6] to implement the linear mode coupling step in Fig. 9.13. Using this method, the accuracy of full stochastic solutions of (9.30) will be compared with different analytical expectations, regarding the total nonlinear noise and the nonlinear transmission performance of quadrature amplitude modulated signals.

9.8.4   Total Nonlinear Noise: Analytical Integration

The total nonlinear noise generated can be analytically calculated using a generalization of SMFs four-wave mixing (FWM) theory to FMFs [40]. That is, when considering three waves denoted p, q, r propagating in modes denoted a, b, c respectively, the nonlinear signal A_{ds} generated at angular frequency $\omega_s = \omega_p + \omega_q - \omega_r$ in mode d is:

$$A_{ds} = \xi_{abcd} A_{ap} A_{bq} A_{bq} A_{cr}^* \frac{1 - e^{-\alpha L} e^{-j\Delta\beta_{abcd} L}}{j\Delta\beta_{abcd,pqrs} L + a} \cdot e^{-\alpha/2L} e^{-j\beta_{ds} L}, \qquad (9.33)$$

where a is the attenuation and L is the span length. ξ_{abcd} is the total nonlinear coefficient between modes a, b, c, d given by the product of γ_{abcd} and the degeneracy factor dependent on the coupling strength (9.30)–(9.32). $\Delta\beta_{abcd,pqrs}$ is the phase mismatch between waves p, q, r, s. The phase mismatch is given by $\Delta\beta_{abcd,pqrs} = \beta_{ap} + \beta_{bq} - \beta_{cr} - \beta_{ds}$ where β_{ap} is the propagation constant of mode a at angular frequency ω_p.

Finally, assuming an optical super-channel with a total bandwidth B the total nonlinear noise between a given set of modes can be calculated by integrating the product of (9.33) with the signal power spectral density (PSD) in each mode. A closed form solution for this integral was obtained (and experimentally validated) for the case of a signal with a rectangular spectrum

(OFDM or Nyquist WDM super channel) in each interacting mode, and the overall efficiency parameter η_{abcd} was shown to be [40]:

$$\eta_{abcd} = \frac{\gamma_{abcd}^2}{\pi\alpha\,|\,\beta_\square^{(2)}\,|}\left[\ln\left(\frac{B^2 + 2B\Delta f_{abcd}}{2f_w}\right) + s\ln\left(s\frac{B^2 - 2B\Delta f_{abcd}}{2f_w}\right)\right], \qquad (9.34)$$

where $\Delta f_{abcd} = (\beta_a^{(1)} + \beta_b^{(1)} - \beta_c^{(1)} - \beta_d^{(1)})/2p + \beta_\square^{(1)}$, $s = \text{sign}(B - 2\,\Delta f_{abcd})$, $f_w = |\,\alpha/4\pi^2\,|\,\beta_\square^{(2)}\,|$, $\beta_a^{(1)}$, is the group delay of mode α, $\beta_\square^{(2)}$ is the second-order dispersion coefficient, and Δf_{abcd} is the velocity-matched frequency offset. In the derivation of (9.34) it is assumed that [40, 41]: the second-order dispersion coefficient $\beta_\square^{(2)}$ is mode independent; mode group velocities $\beta_a^{(1)}$ are frequency independent; given large $\beta_a^{(0)}$ differences, strong inter-mode phase-matching is only possible for interactions of pairs of modes ($a = d$, $b = c$ or $b = d$, $a = c$). According to (9.34), the FWM efficiency is maximized for frequency offsets Δf_{abcd} at which the walk-off induced by chromatic dispersion and the walk-off induced by mode delay cancel out exactly. Finally, the total nonlinear power generated in mode d is given by $(\Sigma_{a,b,c}\eta_{abcd})P_aP_bP_c$, where P_a is the signal power spectral density in mode a.

Figure 9.14 shows the nonlinear noise power generated at the center of the WDM signal as a function of the overall bandwidth (B) for a particular six linearly polarized (LP) mode fiber with no linear mode coupling. In addition to the logarithmically increasing background expected for a SMF [42], a number of discontinuities are apparent whenever B becomes sufficiently large to allow strong phase matching among an additional pair of modes. In Fig. 9.14, the dashed vertical lines identify these discontinuities ($B/2 = |\Delta f_{abcd}|$). To enhance the visualization of all possible phase matchings, the results in Fig. 9.14 were obtained with an arbitrary GD vector: (0, 8, 13, 14, 17, 18) ps/km for (LP$_{01}$, LP$_{02}$, LP$_{11a}$, LP$_{11b}$, LP$_{21a}$, LP$_{21b}$), respectively. All the other fiber characteristics follow Table 9.1.

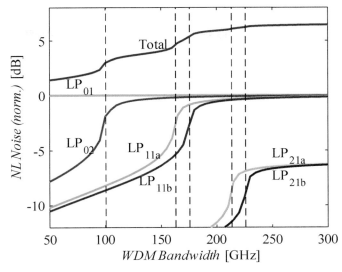

Figure 9.14 Contribution to the nonlinear noise power in the LP_{01} mode for signals propagating in higher order modes normalized by the LP_{01} intra-modal nonlinear noise power, as a function of WDM bandwidth.

9.9 Linear Coupling Impact Nonlinear Noise for Delay Uncompensated Spans

In this section, the nonlinear noise power is found by solving (9.30) for a range of different linear coupling strengths ranging from the weak to the strong coupling regimes using the modified SSFM presented in Section 9.8, and by using (9.34). The simulations assume an optical super-channel with: a rectangular power spectral density (e.g. OFDM), a total WDM bandwidth B, and a subcarrier spacing of 50 MHz (smaller spacing generated similar results). The XT value was varied from −70 to 0 dB/100 m covering all coupling values presented in the literature. To the best of our knowledge, the lowest XT values reported are around −50 dB/100 m [20] and the highest XT value reported is −7 dB/100 m [23]. Finally, simulations considered the same fiber characteristics as used in Fig. 9.14.

Figure 9.15 shows the nonlinear noise power at the center frequency of the WDM band carried by the LP_{01} mode versus the total WDM bandwidth. The modified SSFM step size was selected by bounding the local error to be lower than 10^{-5} (smaller local errors generated similar results). The simulation results in Fig. 9.3 lay between two analytical lines obtained with (9.34) using: the ordinary fiber nonlinear coefficients (for weak mode coupling) [8], dotted line, and using the average nonlinear coefficients derived in [7] for strong mode coupling, dashed line. It can be noted in Fig. 9.15 that the rate of decrease of the nonlinear noise with XT increasing is higher for larger bandwidths than for smaller bandwidths, which shows that the averaging of the nonlinear coefficients among the higher-order modes occurs more rapidly. For small values of *XT*, –70 and –60 dB/100 m the steps associated with the inter-mode interactions of LP_{01} with LP_{02} and LP_{21a}/LP_{21b}, become smooth, but the step associated with LP_{11a}/LP_{11b} remains unchanged. This is in line with the asymmetries on the coupling strength between pairs of modes from the same mode groups (stronger) and from different mode groups (weaker) (modes LP_{02} and LP_{21a}/LP_{21b} belong to the same mode group). Increasing *XT* up to –40 dB/100 m, smooths the step associated with inter-mode interactions of LP_{01} with LP_{02}. Furthermore, increasing *XT* above –20 dB/100 m reduces nonlinear noise power below the LP_{01} intra-mode nonlinear noise power in the absence of linear coupling which was used to normalize the results. In the limit, strongly coupling all modes, using unitary matrices every 10 m (and shorter steps, as verified), the nonlinear noise power matches the analytical results (dashed line) obtained with (9.34) and the average nonlinear coefficients in [7].

In conclusion, for the crosstalk values shown by the majority of FMFs (from –50 to –20 dB/100 m), the nonlinear noise is not accurately estimated by either the weak linear coupling regime or the strong coupling regime. However, the overall conclusion that the stronger coupling reduces nonlinear noise power remains valid. Finally, the reduction of nonlinear noise below that of uncoupled single-mode propagation for linear coupling requires XT values above –20 dB/100 m.

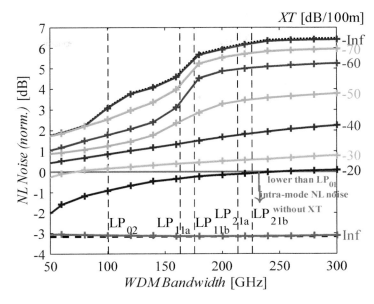

Figure 9.15 Total nonlinear noise power in the LP_{01} mode as a function of WDM bandwidth showing analytical predictions from strong (dashed) and weak (dotted) regimes along with numerical simulations (solid) for different mode coupling strengths (colors).

9.10 Linear Coupling Impact on Nonlinear Noise for Delay Compensated Spans

In this section, we revisit the GD managed spans studied in Section 9.7, to analyze their nonlinear performance. As mode delay compensation is used and the GD spread is reduced, the total nonlinear noise is expected to increase as phase matching becomes possible for smaller WDM bandwidths.

Figure 9.16, shows the impact of mode delay compensation on the total nonlinear noise, in the absence of linear mode coupling, simulations considered the same fiber characteristics as used in Fig. 9.14 with different GD vector scaling for each DMD value. We can see that when the WDM bandwidth is small enough such that not all phase matching conditions are meet, the introduction of mode delay compensation increases the total nonlinear noise comparing with the noncompensated case. But if the WDM bandwidth is such that already satisfies all possible phase matchings for noncompensated links, then the introduction of mode delay compensation does not significantly increases the total noise. From Fig. 9.16 it can be concluded that there is a trade-off between DSP complexity given the GD spread and an increase of the nonlinear noise.

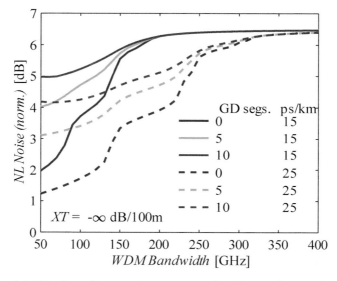

Figure 9.16 Total nonlinear noise power in the LP_{01} mode as a function of WDM bandwidth, for [0 5 10] compensation segment for DMD [15 25] ps/km, and no XT.

To design the optimum FMF link, we will compare the total nonlinear noise falling at the center of the LP_{01} mode, considering the transmission of an optical super-channel with a bandwidth of 1 THz over a given FMF link to that a SMF link, for a broad range of local DMD values and XT values. The nonlinear signal field generated after 100 km was found by following the numerical method presented in Section 9.8.

Figure 9.17 shows a contour plot of the normalized nonlinear noise power (in dB) generated at the center of the WDM band in the LP_{01} mode, as a function of mode coupling strength and DMD. Simulations considered the fiber characteristics in Table 9.1,

DMD value sweeping was obtained by scaling the GD vector in Table 9.1. The nonlinear noise was normalized to the LP_{01} intra-modal nonlinear noise power obtained in the absence of coupling. In Fig. 9.17, two DMD-managed scenarios are shown: (a) 1 segment and (b) 25 segments. Note that, the contour line highlighting the regions from Fig. 9.12 in Section 9.7 where the GD spread was higher than 800 ps has been overlapped. In Fig. 9.17, it can be seen that the nonlinear noise decreases by increasing either the DMD value or the XT value. Moreover, it can be noted that the nonlinear noise increases with the number of segments, as the contour lines move to higher DMD and XT values analogously to the enhancement observed for resonant chromatic dispersion managed systems. However, such increase is generally lower than 0.5 dB for the same (DMD, XT) value as found in [41]. For long period GD maps (Fig. 9.17a), the optimum design appears to be to maximize the mode coupling, and operate at the highest possible DMD. However, for shorter period GD-managed maps (Fig. 9.17b), for XT ranging from –40 to –30 dB/100 m since the DMD tolerance increases faster with the number of segments than the nonlinear noise, system performance can be improved by increasing the number of segments and allowing for higher DMD values. Importantly, the optimum solution for each GD map (at the highest tolerable DMD for the highest XT considered) shows negligible difference in the predicted nonlinear noise.

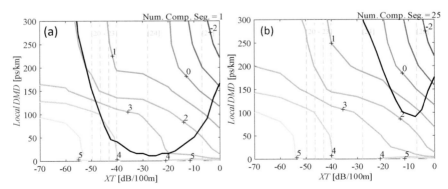

Figure 9.17 Contour plots of the nonlinear noise power in the LP_{01} mode, as a function of mode coupling strength and DMD, normalized by the LP_{01} intra-modal nonlinear noise power (normalized noise values are in dB). Four DMD maps are considered: (1) 1 segment, and (b) 25 segments.

Figure 9.18 shows the nonlinear performance for the highest tolerable DMD (such that GD spread < 800 ps as in Fig. 9.17) for a broad range of DMD maps. It can be seen that for the XT values given suppression of NL below that of uncoupled propagation (XT > –10 dB/100 m), the usage of DMD compensation plays no role. Thus, it can be concluded that the usage of high XT fibers is preferable given that the deployment complexity associated with GD compensation is removed and fibers with relatively with DMD (up to 150 ps/km according to Fig. 9.17) can still be used.

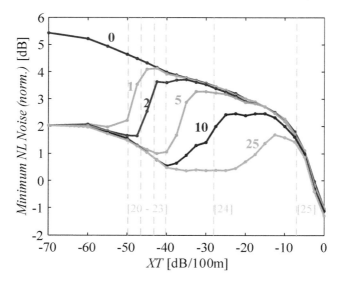

Figure 9.18 Minimum nonlinear noise power (normalized) as a function of the mode coupling strength, for a broad range of GD maps ranging from unmanaged to short period.

9.11 Manakov Approximation vs. Fully Stochastic Propagation

In this section, the link conditions under which the Manakov approximation are accurately established, in terms of uncoupled DMD and linear coupling strength. The validation results in the following consider only mode delay uncompensated links as further results for DMD compensated links generated similar results.

The simulations setup and the linear DSP blocks are summarized in the following.

The simulation setup is shown in Fig. 9.19. Over each polarization mode was transmitted an optical super-channel consisting of 3 channels spaced of 14.1 GHz carrying 14 Gbaud 16-QAM signals, giving a total bit rate of 2 Tb/s (672 Gb/s per wavelength). Together with the information data, a preamble was transmitted consisting of constant amplitude zero autocorrelation (CAZAC) sequences, used for time synchronization and channel estimation. Root raised cosine filters with a roll-off factor of 0.001 were used for pulse shaping. Simulations considered 2^{16} symbols per polarization mode, from which 2^{11} were CAZAC symbols. After homodyne detection, the baseband electrical signals were sampled at 56 GS/s, yielding 12 digital signals at 2 samples/symbol. Afterwards, the coherently received signals were compensated for chromatic dispersion in the frequency domain using the values in Table 9.1. In all cases, mode coupling and (residual) DMD were subsequently compensated for using data-aided channel estimation and equalization, as shown in Fig. 9.19. Coarse time synchronization was performed using the Schmidl & Cox autocorrelation metric. Subsequently, fine-time synchronization and channel impulse response (CIR) estimation were performed by cross-correlating with the training CAZAC sequences. The 12 × 12 CIR estimations were converted into the frequency domain. The MIMO frequency domain equalizer was calculated by inverting the channel matrix, and, finally, the Q-factor for each received signal was calculated using the mean and standard deviation of the received symbols. In the following, the Q-factor was averaged over the 12 polarization modes considering only the center channels.

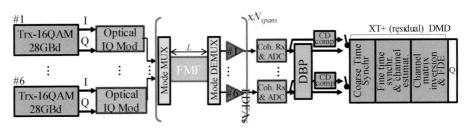

Figure 9.19 Block diagram for system simulations using 6 LP modes.

The fiber attenuation is fully compensated using an array of 6 erbium doped fiber amplifiers [43], considering a noise figure of 3dB and negligible mode dependent gain since the aim of this paper is to assess the isolated impact of mode coupling and mode delay on the Manakov approximation. Moreover, the mode multiplexer (MUX) and de-multiplexer (DEMUX) are assumed ideal for the same reasons.

System performance simulations considered transmission over only 3 spans of 50 km such that at moderate launch signal powers (−5 to 0 dBm) performance was limited by nonlinear noise rather than by spontaneous emission noise, thus enhancing the limitations of the different approximated nonlinear modelling models. The DMD value was varied by scaling the mode group delay values in Table 9.1 to allow for an objective assessment of the Manakov approximation as other fiber characteristics are kept. The XT value was varied from −70 to 0 dB/100 m (following Section 9.8.3) covering the range of coupling values presented in the literature [20, 22, 23]. The step size was selected by bounding the local error to be lower than 10−5, lower error bounds generated negligible results change.

Simulations included four different methods for the solution of (9.30), namely: the WC-Manakov approximation (9.2) [8]; the SC-Manakov (9.3) [7, 8] approximation; the distributed mode coupling model using the approach presented in Section 9.8.3; a lumped mode coupling model according which random unitary matrices are introduce every L_{lumped} (like in [8]), such that $XT_{\text{[dB/100 m]}} + 10\log_{10}(L_{\text{lumped}}[\text{m}]/100[\text{m}]) = 0$ dB. To improve the accuracy of the SC-Manakov model, the uncoupled GD vector was scaled by the ratio of the standard deviation of the coupled GD vector obtained for $XT = -\infty$ dB/100 m and the XT under consideration, using equation (9.23) in [6] derived through an analytical statistical analysis.

Figure 9.20 shows the Q-factor as a function of launching power in the absence of DMD for two different XT values: −70 dB/100 m (WC-regime), and −30 dB/100 m (intermediate coupling regime). The figure shows that all models seem to agree for the WC-regime (−70dB/100 m), but not so much for the intermediate regime (−30 dB/100 m). The SC-Manakov and lumped XT models differ by more than 0.5 dB from the distributed XT model in the nonlinear regime. Further insight can be obtained by varying the XT and DMD while maintaining a given launching signal power in the nonlinear regime.

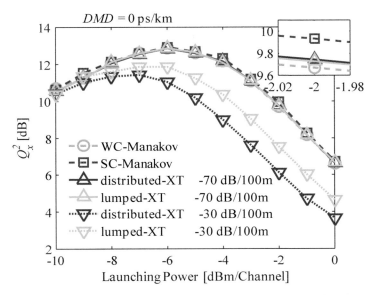

Figure 9.20 Q-factor as a function of launching signal power in the absence of DMD two different *XT* values: –70 dB/100 m (WC-regime), and –30 dB/ 100 m (intermediate coupling regime).

Figure 9.21 shows the Q-factor as a function of XT for different models, with 2 dBm/ch, in: (a) the absence of DMD, and (b) the presence of a low DMD value, 8 ps/km. First, in all cases Fig. 9.21a,b, WC-Manakov and lumped XT models are in agreement with the distributed XT model for XT < –50 dB/100 m, conversely, SC-Manakov and lumped XT models are in agreement with the distributed XT model for XT > –10 dB/100 m. However, in the intermediate coupling regime and for all three DMD cases, the WC- and SC-Manakov models as well as the lumped XT models differ by more than 0.5 dB from the distributed XT model. More importantly, it can be seen that system performance in the nonlinear regime can in fact degrade with increasing XT for low-to-intermediate values (–50 to –30 dB/100 m) before it eventually approaches the SC-regime and performance improves above that of the WC-regime, as in Fig. 9.21a. Such behavior can be explained considering that for a certain range of intermediate XT values, additional pathways to FWM phase matching are created without introducing sufficiently fast random rotations of the polarization state of the field along the fiber length which would reduce the efficiency of the overall nonlinear process.

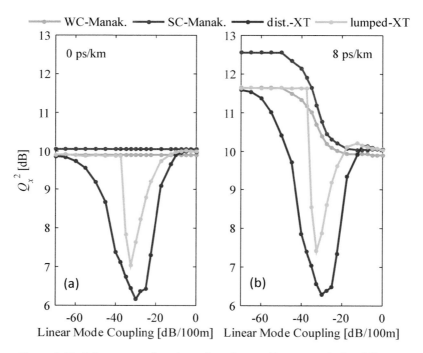

Figure 9.21 Q-factor as a function of mode coupling strength for different channel models/approximations, for –2 dBm/ch and for uncoupled DMD equal to: (a) 0 ps/km, and (b) 8 ps/km. Shadow accounts for 3 times the standard deviation given 20 repetitions.

Figure 9.21b shows yet another scenario, within the SC-regime, performance degrades with increasing XT. In this case, the performance degradation is due to the severe reduction of the overall GD which allows for phase matching between pairs of modes which were not possible for XT = –∞ dB/100 m given the relative narrow bandwidth (42 GHz) of the super-channel considered. Note that the increased penalty is relatively small given that XT reduces GD spread as well as the overall nonlinear coefficients, as explained in Section 9.8.2. Finally, this explanation is in agreement with the behavior of the SC-Manakov model given the GD vector scaling discussed earlier.

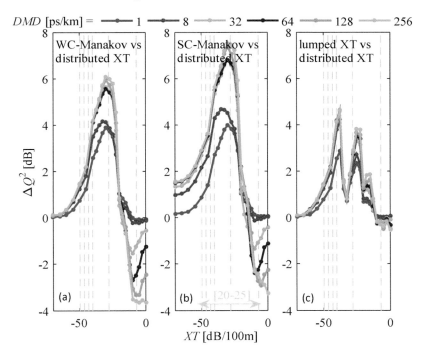

Figure 9.22 Q-factor error as a function of mode coupling strength, with the distributed XT model as reference, for different models: (a) WC-Manakov model, (b) SC-Manakov model, and (c) lumped XT model. Data points have been averaged over 20 repetitions.

In overall, Fig. 9.21 shows that to some extent lumped XT model captures the behavior of the distributed model, even though overestimating performance by slightly more than 0.5 dB. Finally, Fig. 9.22 shows Q-factor *error* as a function of mode coupling strength, with the distributed XT model as reference. For extremely small XT values (<-60 dB/100 m) WC-Manakov generates accurate results; however, practical fibers have XT ≥ -50 dB/100 m. For high XT values (>-10 dB/100 m), SC-Manakov is only accurate if DMD $<$ 10 ps/km; however, practical fibers have higher DMD in particular for more than 3-modes, besides the usage of moderate–to-high DMD fibers in GD-managed links. The lumped XT model is able to accurately model FMF propagation for XT > –10 dB/100 m even for DMD several times higher than 100 ps/km, a practical scenario, thus a useful model. Finally, in the intermediate coupling regime (all other XT values) only a distributed XT model capable of introducing controllable amounts of XT with a small step-size (1-to-100 m) can accurately model transmission.

9.12   Conclusions

This chapter reviewed the modelling of the linear and nonlinear impairments of few-mode fibers, namely mode group-delay spread, linear mode coupling, and intermodal nonlinear effects. Propagation over few-mode fibers is modelled deriving a few-mode split-step Fourier method composed of three steps: dispersion step, nonlinear step, and a linear mode coupling step.

The linear mode coupling step is implemented using semi-analytical solutions capable of introducing arbitrary strength coupling in a distributed manner and allowing a time efficient computation after any real-world fiber length. The model proved to be accurate against analytical predictions for the statistics of group-delays in few-mode fiber links, namely: standard deviation, probability density function, and cumulative distribution function. It proved accurate for different transmission lengths 10 m-to-10,000 km, in any coupling regime –50 dB/100 m to 0 dB/100 m, without and with group-delay management.

The derived few-mode split-step Fourier method, using the linear mode coupling semi-analytical solutions, proved accurate against the analytical integration of the total nonlinear noise for optical super-channel with rectangular power spectral densities. Using the proposed model, the optimum link configurations minimizing the nonlinear penalty at practical levels of equalization complexity were obtained, namely: the coupling strength required to give suppression of nonlinear distortion below the isolated propagation without mode coupling, for different mode delay maps. Furthermore, the proposed model was used to validate the application requirements of models based on Manakov or lumped XT approximations. The Manakov approximations are proved to be accurate only for the extreme regimes not likely in practice ($<$ −50 dB/100 m, or $>$ −10 dB/100 m with DMD $<$ 10 ps/km), and the lumped XT model was found to overestimate the system performance by 0.5-to-4 dB in the intermediate coupling regime.

Finally, the reviewed modelling methods are essential tools for the modelling and development of future high-capacity multimode fiber systems, in particular for the intermediate coupling regime.

Acknowledgements

This work has been partially supported by the EU (654809-HSPACE and 659950-INVENTION), and by EPSRC (EP/L000091/1-PEACE).

Research Data

The Matlab scripts, source C-code, mex compiled C-code, and figure data points are available at https://doi.org/10.17036/researchdata.aston.ac.uk.00000338.

References

1. Richardson, D. J., J. M. Fini, and L. E. Nelson, Space-division multiplexing in optical fibres. *Nat. Photonics*, 2013. **7**: p. 354.

2. Li, G., et al., Space-division multiplexing: The next frontier in optical communication. *Adv. Opt. Photonics*, 2014. **6**(4): pp. 413–487.

3. Marcuse, D., A. Telephone, and T. Company, *Theory of Dielectric Optical Waveguides*, 1991: Academic Press.

4. Ho, K.-P., and J. M. Kahn, Linear propagation effects in mode-division multiplexing systems. *J. Lightwave Technol.*, 2014. **32**(4): pp. 614–628.

5. Antonelli, C., A. Mecozzi, and M. Shtaif, The delay spread in fibers for SDM transmission: Dependence on fiber parameters and perturbations. *Opt. Express*, 2015. **23**(3): pp. 2196–202.

6. Ferreira, F. M., et al., Semi-analytical modelling of linear mode coupling in few-mode fibers. *J. Lightwave Technol.*, 2017. **35**(18): pp. 4011–4022.

7. Mecozzi, A., C. Antonelli, and M. Shtaif, Nonlinear propagation in multi-mode fibers in the strong coupling regime. *Opt. Express*, 2012. **20**(11): pp. 11673–11678.

8. Mumtaz, S., R.-J. Essiambre, and G. P. Agrawal, Nonlinear propagation in multimode and multicore fibers: Generalization of the Manakov Equations. *J. Lightwave Technol.*, 2013. **31**(3): pp. 398–406.

9. Rademacher, G., and K. Petermann, Nonlinear Gaussian Noise model for multimode fibers with space-division multiplexing. *J. Lightwave Technol.*, 2016. **34**(9): pp. 2280–2287.

10. Ferreira, F. M., et al., Advantages of strong mode coupling for suppression of nonlinear distortion in few-mode fibers. *2016 Optical Fiber Communications Conference and Exhibition (OFC)*, 2016.

11. Ho, K.-P., and J. M. Kahn, Statistics of group delays in multimode fiber with strong mode coupling. *J. Lightwave Technol.*, 2011. **29**(21): pp. 3119–3128.

12. Keang-Po, H., and J. M. Kahn, Delay-spread distribution for multimode fiber with strong mode coupling. *IEEE Photonics Technol. Lett.*, 2012. **24**(21): pp. 1906–1909.

13. Antonelli, C., et al., Stokes-space analysis of modal dispersion in fibers with multiple mode transmission. *Opt. Express*, 2012. **20**(11): pp. 11718–11733.

14. Mecozzi, A., C. Antonelli, and M. Shtaif, Intensity impulse response of SDM links. *Opt. Express*, 2015. **23**(5): pp. 5738–5743.

15. Ye, F., S. Warm, and K. Petermann. Differential mode delay management in spliced multimode fiber transmission systems. *Optical Fiber Communication Conference/National Fiber Optic Engineers Conference 2013*. 2013. Anaheim, California: Optical Society of America.

16. Juarez, A. A., et al., Modeling of mode coupling in multimode fibers with respect to bandwidth and loss. *J. Lightwave Technol.*, 2014. **32**(8): pp. 1549–1558.

17. Ferreira, F., et al., Nonlinear semi-analytical model for simulation of few-mode fiber transmission. *IEEE Photonics Technol. Lett.*, 2012. **24**(4): pp. 240–242.

18. Ferreira, F., P. Monteiro, and H. Silva. Semi-analytical model for linear modal coupling in few-mode fiber transmission. *2012 14th International Conference on Transparent Optical Networks (ICTON)*. 2012. IEEE.

19. Ferreira, F., et al., Reach improvement of mode division multiplexed systems using fiber splices. *IEEE Photonics Technol. Lett.*, 2013. **25**(12): pp. 1091–1094.

20. Gruner-Nielsen, L., et al., Few mode transmission fiber with low DGD, low mode coupling, and low loss. *J. Lightwave Technol.*, 2012. **30**(23): pp. 3693–3698.

21. Mori, T., et al., Low DMD four LP mode transmission fiber for wide-band WDM-MIMO system. *2013 Optical Fiber Communication Conference and Exposition and the National Fiber Optic Engineers Conference (OFC/NFOEC)*, 2013.

22. Ryf, R., et al., Space-division multiplexed transmission over 4200 km 3-core microstructured fiber. *National Fiber Optic Engineers Conference*, 2012. Los Angeles, California: Optical Society of America.

23. Fontaine, N. K., et al., Experimental investigation of crosstalk accumulation in a ring-core fiber. *2013 IEEE Photonics Society Summer Topical Meeting Series*, 2013.

24. Antonelli, C., et al., Random coupling between groups of degenerate fiber modes in mode multiplexed transmission. *Opt. Express*, 2013. **21**(8): pp. 9484–9490.

25. Agrawal, G., Chapter 2 - Pulse propagation in fibers, in *Nonlinear Fiber Optics (Fifth Edition)*, 2013, Academic Press: Boston. p. 27–56.

26. Ferreira, F., S. Sygletos, and A. Ellis. Impact of linear mode coupling on the group delay spread in few-mode fibers. *2015 Optical Fiber Communications Conference and Exhibition (OFC)*, 2015.

27. Ferreira, F. M., et al., Few-mode fibre group-delays with intermediate coupling. *2015 European Conference on Optical Communication (ECOC)*, 2015.

28. Suibhne, N. M., et al., Experimental verification of four wave mixing efficiency characteristics in a few mode fibre. *39th European Conference and Exhibition on Optical Communication (ECOC 2013)*, 2013.

29. Antonelli, C., A. Mecozzi, and M. Shtaif. Scaling of inter-channel nonlinear interference noise and capacity with the number of strongly coupled modes in SDM systems. *2016 Optical Fiber Communications Conference and Exhibition (OFC)*, 2016.

30. Ferreira, F., et al., Nonlinear transmission performance in delay-managed few-mode fiber links with intermediate coupling. *Optical Fiber Communication Conference*, 2017. Los Angeles, California: Optical Society of America.

31. Palmieri, L., and A. Galtarossa, Coupling effects among degenerate modes in multimode optical fibers. *IEEE Photonics J.*, 2014. **6**(6): pp. 1–8.

32. H., W., et al., *Numerical Recipes 3rd Edition: The Art of Scientific Computing*, ed. N.Y.C. University, 2007.

33. Mathews, J., and K. Fink, *Numerical Methods Using Matlab*, ed., P. Hall, 1999.

34. Ferreira, F. M., D. Fonseca, and H. J. A. da Silva, Design of few-mode fibers with M-modes and low differential mode delay. *J. Lightwave Technol.*, 2014. **32**(3): pp. 353–360.

35. Noda, J., K. Okamoto, and Y. Sasaki, Polarization-maintaining fibers and their applications. *J. Lightwave Technol.*, 1986. **4**(8): pp. 1071–1089.

36. Fan, S., and J. M. Kahn, Principal modes in multimode waveguides. *Opt. Lett.*, 2005. **30**(2): pp. 135–137.

37. Arik, S. O., K.-P. Ho, and J. M. Kahn, Delay spread reduction in mode-division multiplexing: mode coupling versus delay compensation. *J. Lightwave Technol.*, 2015. **33**(21): pp. 4504–4512.

38. Randel, S., et al., Mode-multiplexed 6×20-GBd QPSK transmission over 1200-km DGD-compensated few-mode fiber. *Optical Fiber Communication Conference*, 2012. Los Angeles, California: Optical Society of America.

39. Sinkin, O. V., et al., Optimization of the split-step fourier method in modeling optical-fiber communications systems. *J. Lightwave Technol.*, 2003. **21**(1): pp. 61–68.

40. Ellis, A. D., et al., Expressions for the nonlinear transmission performance of multi-mode optical fiber. *Opt. Express*, 2013. **21**(19): pp. 22834–2246.

41. Rademacher, G., S. Warm, and K. Petermann, Analytical description of cross-modal nonlinear interaction in mode multiplexed multimode fibers. *IEEE Photonics Technol. Lett.*, 2012. **24**(21): pp. 1929–1932.

42. Sinkin, O. V., et al., Impact of broadband four-wave mixing on system characterization. *2013 Optical Fiber Communication Conference and Exposition and the National Fiber Optic Engineers Conference (OFC/NFOEC).* 2013.

43. Rademacher, G., et al., Long-haul transmission over few-mode fibers with space-division multiplexing. *J. Lightwave Technol.*, 2017. **PP**(99): pp. 1–1.

Index